W9-BXM-235

MATHEMATICS FOR ELEMENTARY TEACHERS

PRELIMINARY EDITION

VOLUME 2

GEOMETRY AND OTHER TOPICS

Sybilla Beckmann

University of Georgia

Boston San Francisco New York
London Toronto Sydney Tokyo Singapore Madrid
Mexico City Munich Paris Cape Town Hong Kong Montreal

Reproduced by Addison-Wesley from electronic files supplied by the author.

Copyright © 2003 Pearson Education, Inc.

All rights reserved. No part of this publication may be reproduced, stored in a retrieval system, or transmitted, in any form or by any means, electronic, mechanical, photocopying, recording, or otherwise, without the prior written permission of the publisher. Printed in the United States of America.

ISBN 0-321-14914-9

1 2 3 4 5 6 7 8 9 10 VHG 04 03 02 01

This book is dedicated to Will, Joey, and Arianna

Contents

Features

Text: The text introduces and discusses the concepts and principles.

Exercises and Answers to Exercises: Almost every section has exercises that are followed by detailed answers to these exercises. Students should try to solve the exercises on their own first, consulting the text for the concepts and principles as needed. After solving, or attempting to solve, the exercises, students should consult the answers to the exercises. The answers serve as examples of good explanations.

Problems: Almost every section has problems for which answers are not provided.

Companion Book of Class Activities: There is a companion book of class activities which are designed to be used in class. In the text, you will find references to the class activities. The reference to an activity is placed at a point in the text where it makes sense for students to work on the activity.

Preface

I wrote this book to help elementary teachers develop a deep understanding of the mathematics that they will teach. It is easy to think that the mathematics of elementary school is simple, and that it shouldn't require college-level study in order to teach it well. But to teach mathematics well, teachers must know more than just *how* to carry out basic mathematical procedures, they must be able to explain *why* mathematics works the way it does. Knowing *why* requires a much deeper understanding than knowing *how*. For example, it is easy to multiply fractions—multiply the tops and multiply the bottoms—but *why* do we multiply fractions that way? After all, when we *add* fractions we can't just add the tops and add the bottoms. The reasons why are not obvious: they require study. By learning to explain why mathematics works the way it does, teachers will learn to make sense of mathematics. I hope they will carry this sense-making into their own future classrooms.

This book focuses on "explaining why". Prospective elementary teachers will learn to explain why the standard procedures and formulas of elementary mathematics are valid, why non-standard methods can also be valid, and why other seemingly plausible ways of reasoning are not correct. The book emphasizes key concepts and principles, and it guides prospective teachers to give explanations that draw on these key concepts and principles. I hope that this will help teachers organize their knowledge around the key concepts and principles of mathematics, so that they will be able to help their students do likewise.

A number of problems and activities examine common misconceptions. I hope that by having studied and analyzed these misconceptions, teachers will be able to explain to their students why an erroneous method is wrong instead of just saying "you can't do it that way".

In addition to knowing how to explain mathematics, prospective teachers should also know how mathematics is used. Therefore I have included various examples of how we use mathematics, such as using proportions to compare prices across years with the Consumer Price Index, using visualization skills to explain the phases of the Moon, using angles to explain why spoons reflect upside down, and using spheres to explain how the Global Positioning System works.

I believe that this book is an excellent fit for the recommendations of the Conference Board of the Mathematical Sciences (CBMS) on the mathematical preparation of teachers, and that it will help prepare teachers to teach to the Principles and Standards of the National Council of Teachers of Mathematics (NCTM).

In writing this book I have benefited from much help and advice. I would like to thank Malcolm Adams, Ed Azoff, David Benson, Cal Burgoyne, Denise Mewborn, Ted Shifrin, Shubhangi Stalder, Gale Watson, and Paul Wenston for many helpful comments on an earlier drafts. I also owe special thanks to Kevin Clancey, John Hollingsworth, and Tom Cooney for their advice and encouragement. I thank my family Will, Joey and Arianna Kazez for their patience, support, and encouragement. My children long ago gave up asking, "Mom, are you still working on that book?", and are now convinced that books take forever to write.

Sybilla Beckmann

Athens, Georgia
May, 2002

Chapter 7

Geometry

This chapter is about foundational concepts of geometry. Geometry is the study of space and shapes in space. The word **geometry** comes from the Greek and means *measurement of the Earth* (geo = Earth, metry = measurment). Mathematicians of ancient Greece developed fundamental concepts of geometry in order to answer basic questions about the Earth and its relationship to the Sun, the Moon, and the planets. For example: how big is the Earth? how far away is the Moon? how far away is the Sun? The geometers of ancient Greece had to be able to visualize the Earth and the heavenly bodies in space, they had to extract the relevant relationships from their mental pictures, and they had to analyze these pictures mathematically. The methods the ancient Greeks developed are still useful today for solving modern problems such as in construction, road building, and even medicine.

In this chapter we will study both two- and three-dimensional shapes. We will focus on visualizing shapes, on analyzing features and properties of shapes, on constructing shapes, and on relating shapes.

Traditionally, very little geometry has been taught in elementary school in the United States. Therefore you may feel that you won't need to know much geometry in order to teach mathematics in elementary school. However, the recommendations of the National Council of Teachers of Mathematics [3, p. 41] include a strong component in geometry.

Geometry Standard

Instructional programs from prekindergarten through grade 12 should enable all students to—

- analyze characteristics and properties of two- and three- dimensional geometric shapes and develop mathematical arguments about geometric relationships;
- specify locations and describe spatial relationships using coordinate geometry and other representational systems;
- apply transformations and use symmetry to analyze mathematical situations;
- use visualization, spatial reasoning, and geometric modeling to solve problems.

7.1 Visualization

One important reason to study geometry is that it promotes the ability to visualize and mentally manipulate objects in space. This is a necessary skill for a number of professions. For example, a surgeon or dentist must be able to visualize the steps in and outcomes of an operation, a carpenter must be able to see different designs in his or her mind's eye, an architect must be able to visualize many different possibilities for a building that satisfies certain design criteria, a clothes designer must be able to visualize how pieces of fabric will fit together to make a garment.

According to the report "What Work Requires of Schools" of the U. S. Department of Labor, [4], being able to "see things in the mind's eye" is a foundational skill for solid job performance. In order to be able to foster this in your students, you must first foster it in yourself.

This section aims to help you improve your visualization skills. For some people, visualization comes naturally, for others it is more of a struggle. No matter where you find yourself in this spectrum, you can improve your visualization skills by working at it. If this is not an area of strength for you, be patient and keep trying!

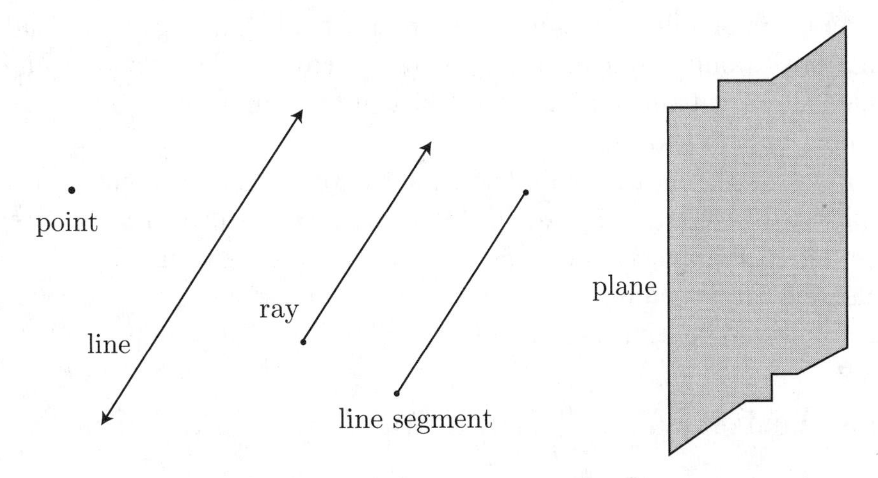

Figure 7.1: Points, Lines, Planes

In the class activities, exercises, and problems of this section you will encounter some basic objects and shapes studied in geometry: points, lines, planes, triangles, squares, rectangles, rhombuses, quadrilaterals, pentagons, hexagons, and circles, as well as prisms, cones, pyramids, and cylinders. Figures 7.1, 7.2, and 7.3 show examples of most of these.

The terms **point**, **line**, and **plane** are usually considered primitive, undefined terms: after all, you have to start somewhere. Even so, we can say how to visualize points, lines, and planes.

- To visualize a **point**, think of a tiny dot, such as the period at the end of a sentence. A point is an idealized version of a dot, having no size or shape.

- To visualize a **line**, think of an infinitely long, stretched string that has no beginning or end. A line is an idealized version of such a string, having no thickness.

- To visualize a **plane**, think of an infinite flat piece of paper that has no beginning or end. A plane is an idealized version of such a piece of paper, having no thickness.

Related to lines are the concepts of *line segment* and *ray*, as pictured in Figure 7.1. A **line segment**, is the part of a line lying between two points on a line. These two points are called the **endpoints** of the line segment. You can think of a line segment as having both a beginning and an end, even though both points are called endpoints. A **ray** is the part of a line lying on one side of a point on the line. So you can think of a ray as having a beginning but no end.

Careful mathematical definitions for the other shapes mentioned above, and pictured in Figures 7.2 and 7.3, will be given later in this chapter. For now, refer to Figures 7.2 and 7.3 so that when you encounter a shape, you can use the proper name for it.

The following activities will help you practice visualizing.

Class Activity 7A: Visualizing Lines and Planes

Class Activity 7B: Parts of a Pyramid

Class Activity 7C: What Shapes Do These Patterns Make?

Class Activity 7D: More Patterns for Shapes

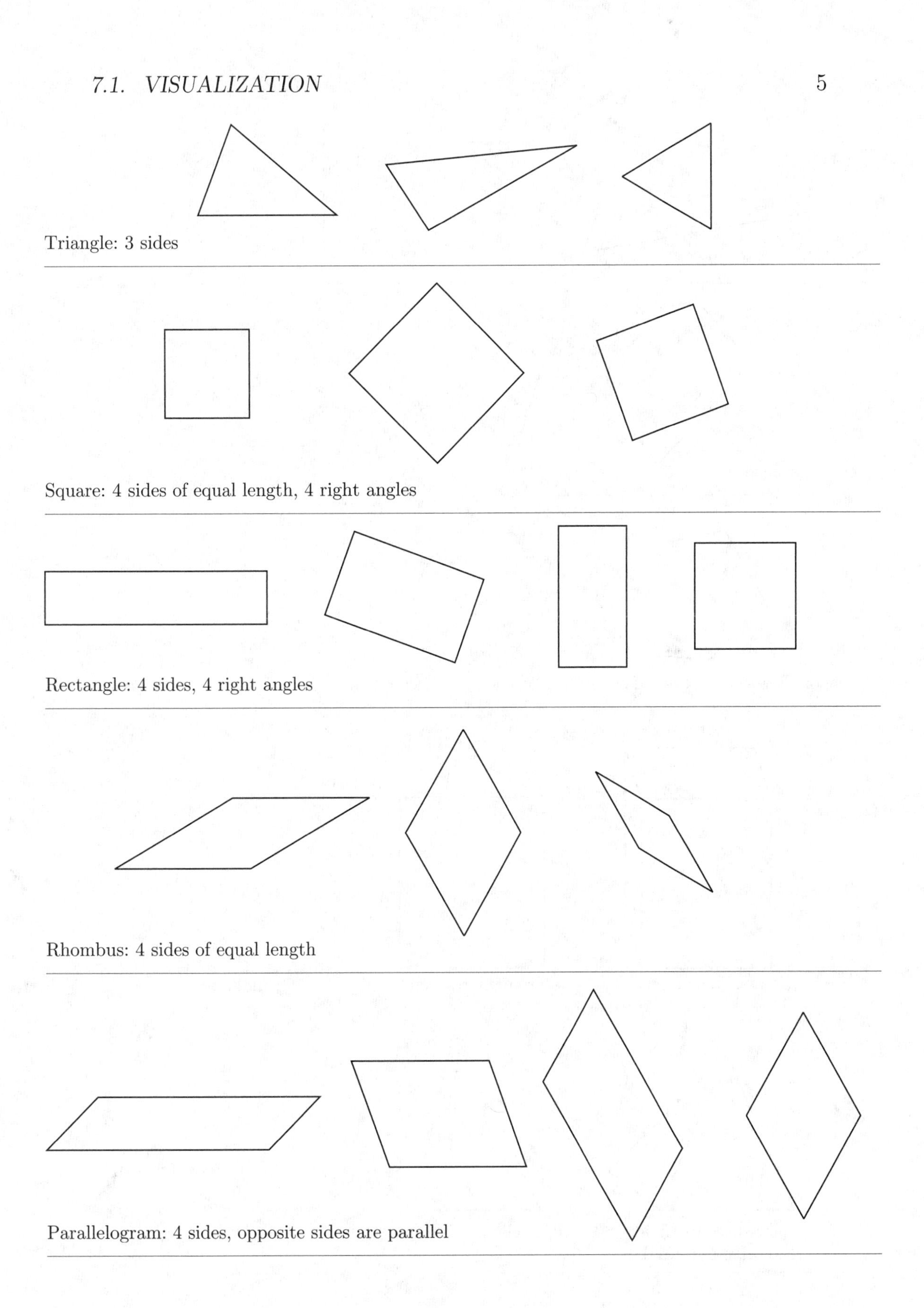

Figure 7.2: Examples of Shapes

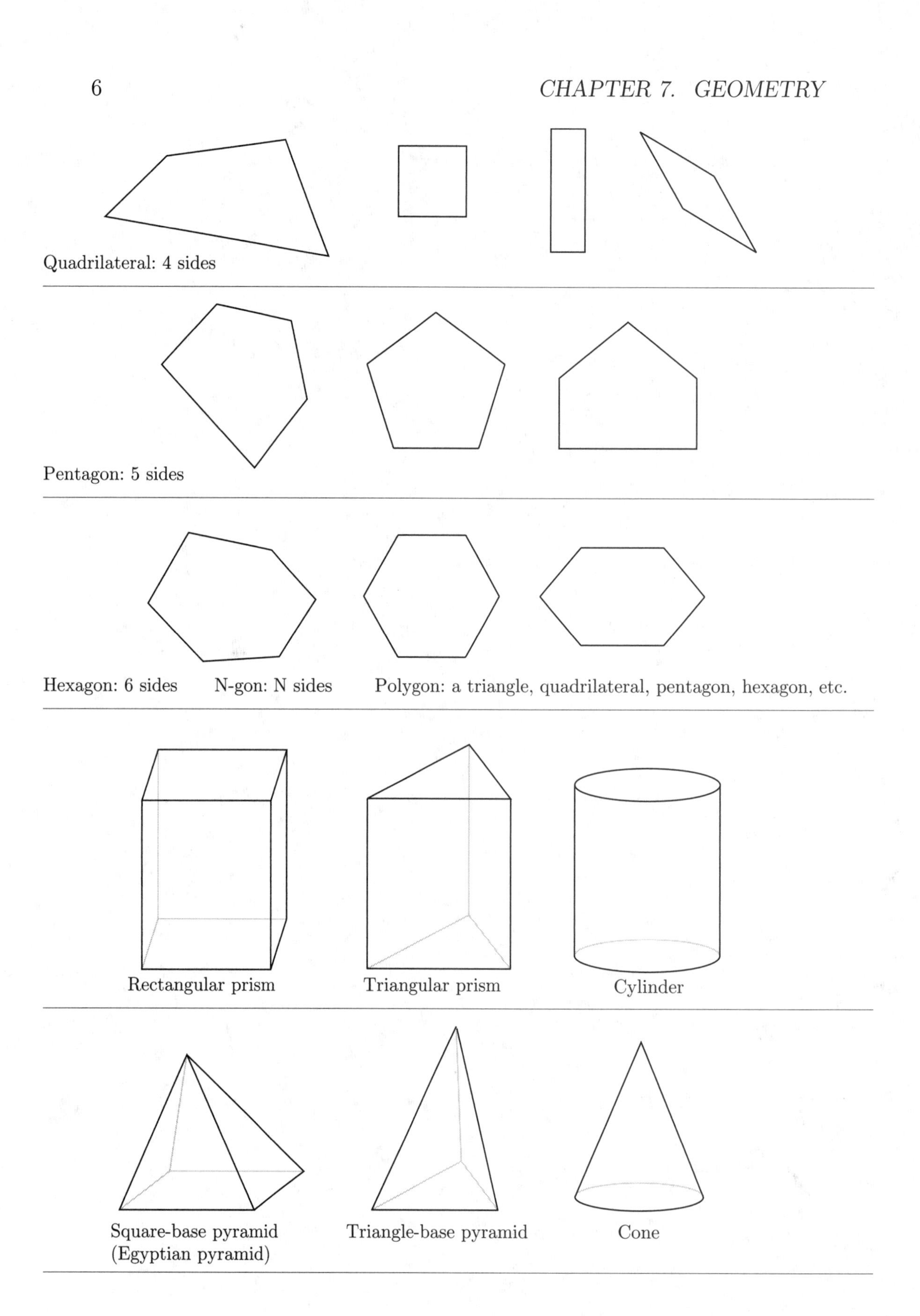

Figure 7.3: More Examples of Shapes

Class Activity 7E: Remember Those Shapes!

Class Activity 7F: Time Zones

Class Activity 7G: Explaining the Phases of the Moon

Exercises for Section 7.1 on Visualization

1. Figure 7.4 shows small versions of several patterns for shapes. Larger versions of these patterns are shown in Figure A.1 on page 365. *Before* you cut, fold, and tape the larger patterns in Figure A.1, *visualize* the folding process in order to help you visualize the final shapes. Then cut out the patterns on page 365 along the solid lines, fold down along the dotted lines, and tape sides with matching labels together.

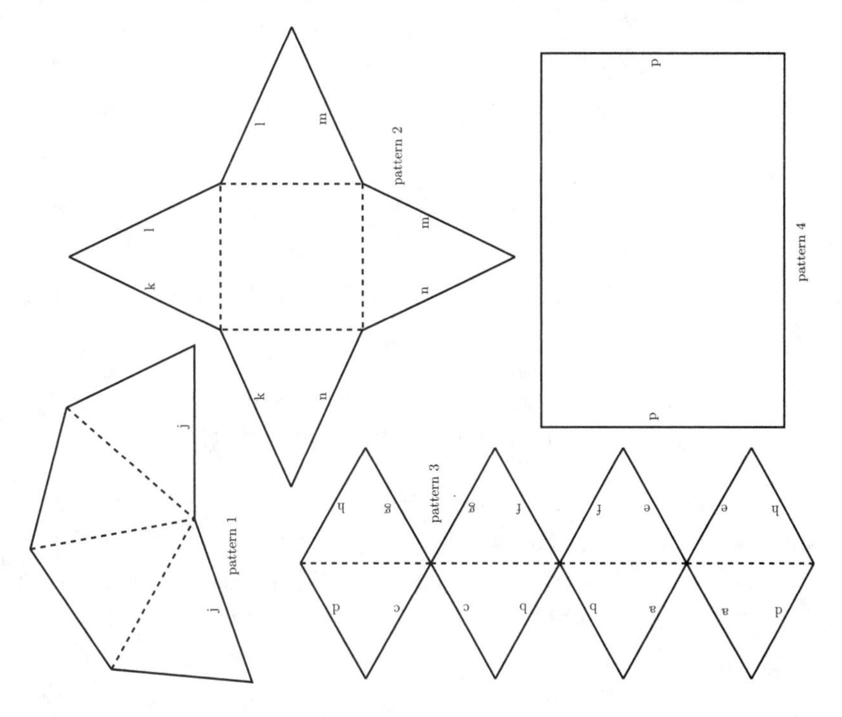

Figure 7.4: Small Version of Patterns for Shapes

2. Figure 7.5 shows small versions of six patterns for shapes. Larger versions of these patterns are shown in Figure A.2 on page 367. *Before* you cut and tape the larger patterns in Figure A.2, try to *visualize* the

final shapes. Predict how the shapes will be alike and how they will be different. Then cut out the patterns on page 367 along the solid lines and tape sides with matching labels together (don't do any folding!).

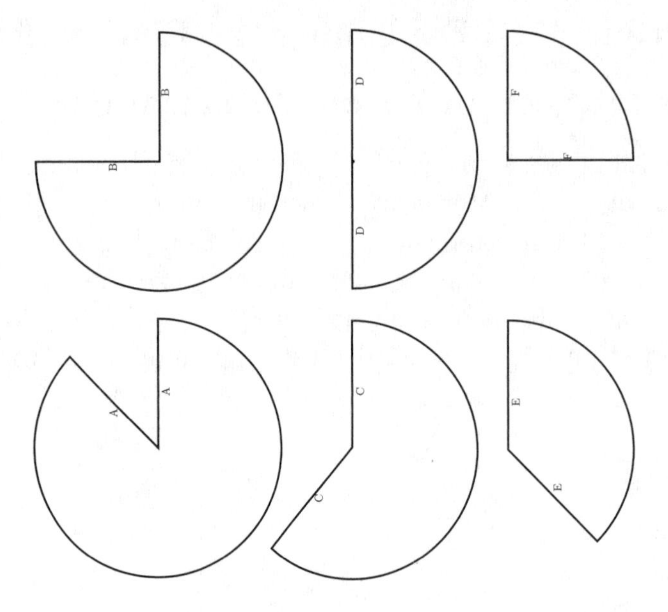

Figure 7.5: Small Version of Patterns for Shapes

3. What familiar parts of clothing are made from the patterns in Figure 7.6 when sides with matching labels are sewn together? Visualize!

4. Make a paper model of a cone (like an ice cream cone). Now visualize a plane slicing through the cone. The places where the plane meets the cone form a figure in the plane. Describe all possible figures in the plane that can be made this way, by slicing the cone. Use your model to help you, but also visualize each case *without* the use of your model.

5. Make a paper model of a cube. Now visualize a plane slicing through the cube. The places where the plane meets the cube form a figure in the plane. Describe how to choose a plane so that the slice is:

— a square

— a rectangle that is not a square

— a triangle

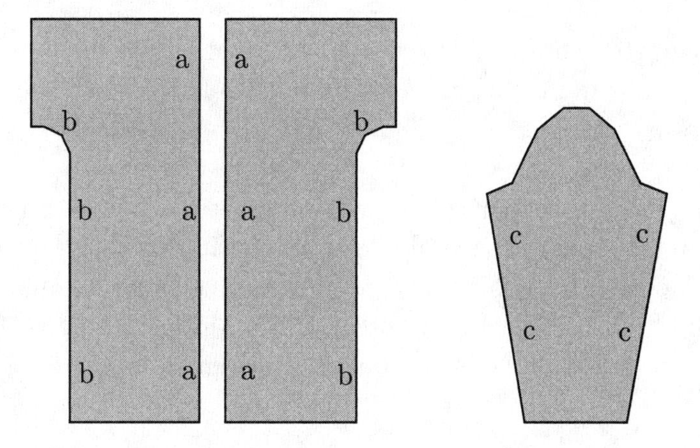

Figure 7.6: What Parts of Clothes Do These Patterns Make?

— a rhombus that is not a square

— a hexagon.

6. Figure 7.7 shows a picture of the Earth as seen from outer space looking down on the North Pole, which is marked N. How does the Sun appear to people at locations A, B, C, and D? Use your answers to explain what time of day it is at A, B, C, and D.

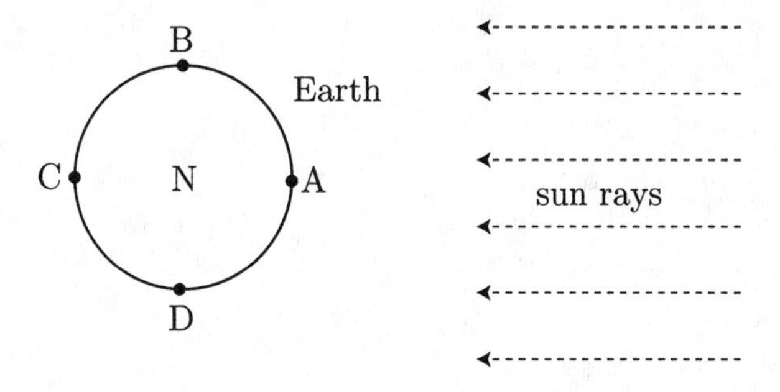

Figure 7.7: What Time Is It?

7. Imagine floating in outer space above the North Pole. Looking down on the Earth, which way is it rotating, clockwise, or counterclockwise? Use the results of the previous exercise to help you figure this out.

What if you were floating above the South Pole instead of the North Pole, would the rotation look different in that case?

8. Figure 7.8 shows four different configurations of the Earth and Moon as seen from outer space, floating above the North Pole (not to scale!). The Moon rotates around the Earth in the direction indicated. In which of pictures A, B, C, and D is the Moon waxing (getting bigger), in which is it waning (getting smaller)? Explain why!

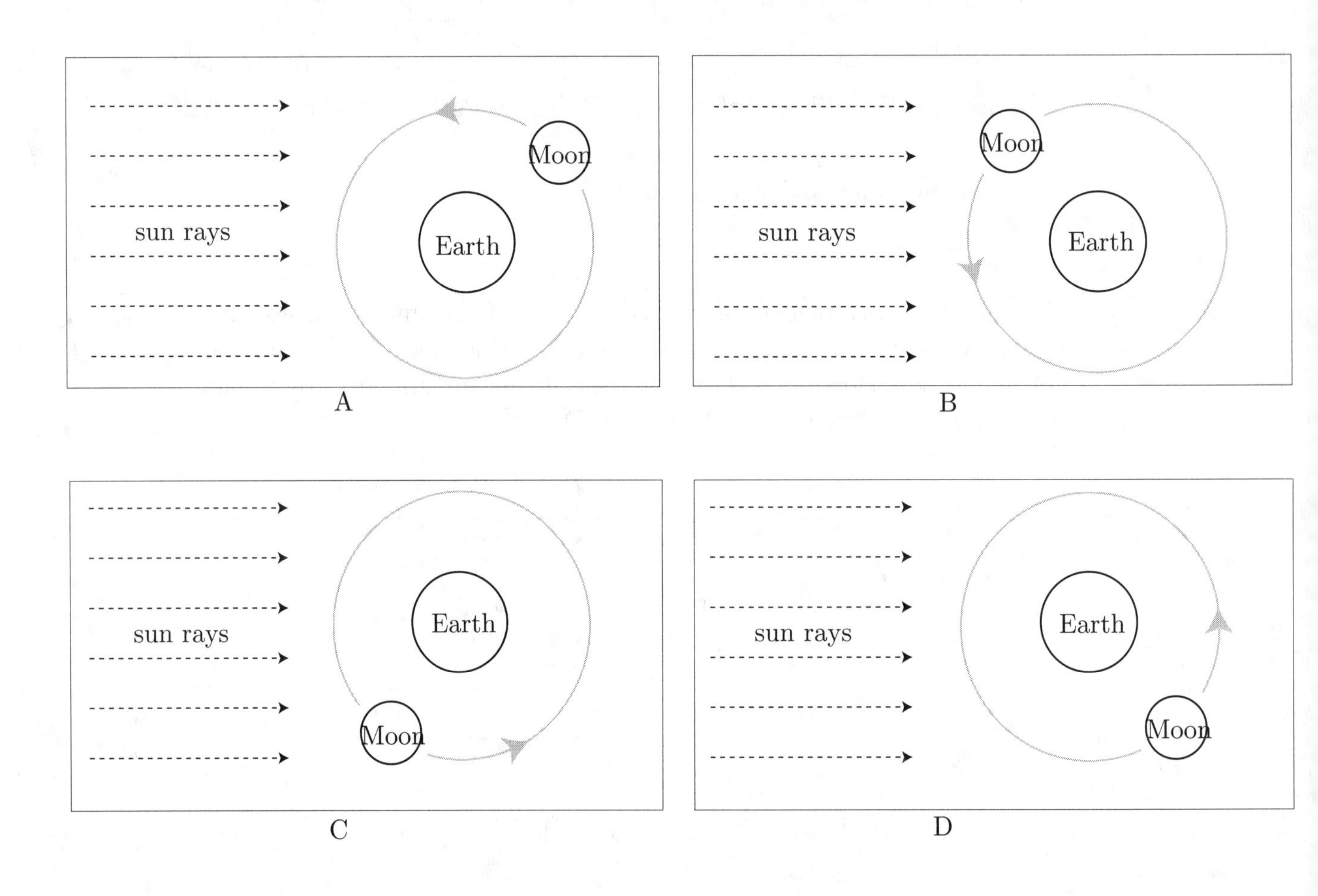

Figure 7.8: Waxing and Waning Moon

Answers to Exercises for Section 7.1

1. Patterns 1 and 2 make pyramids, without and with a base, respectively. Pattern 3 makes a shape that looks like two pyramids stuck together at their bases. This shape is called an **octahedron**. Pattern 4 makes a cylinder.

2. All six patterns make cones. Notice that the patterns at the top of the page make wide, short cones, while the ones at the bottom of the page make narrower, taller cones.

3. The two patterns on the left form a pant leg when they are sewn together. The curved portion of side b makes the crotch. The pattern on the right makes a sleeve. The curved portion at the top (which is not sewed) makes the arm hole. The seam at edge c runs straight down the arm, from the arm pit to the wrist.

4. With the cone positioned as shown in Figure 7.9, horizontal planes slice the cone either in a single point or in a circle (you get a single point if the plane goes through the very bottom, small circles near the bottom, larger circles as you go up). A vertical plane slices the cone in either a V-shape or a curve called a **hyperbola**. A slanted plane slices the cone in either an **ellipse** (oval shape) or in a curve called a **parabola**. This is why circles, ellipses, hyperbolas, and parabolas are collectively referred to as **conic sections**, because they come from slicing a cone.

5. See Figure 7.10. Horizontal planes slice the cube in squares. A vertical plane slices the cube in either a square or a rectangle that is not a square. You can get a triangle, a rhombus, or a hexagon by slicing with angled planes. To see how to get a hexagon, cut out the two patterns in Figure A.3 on page 369 along the solid lines, fold them down along the dotted lines, and tape them so as to create two solid shapes. The two shapes can be put together at the hexagon so as to make a cube. This will show you how a plane can slice a cube so as to create a hexagon at the slice.

6. Picture yourself at point A in Figure 7.7. At this location, the Sun's rays are directly overhead, which means it must be about noon. Location C is not receiving any sunlight and is directly opposite the location marked noon, so it must be about midnight at point C. Now picture

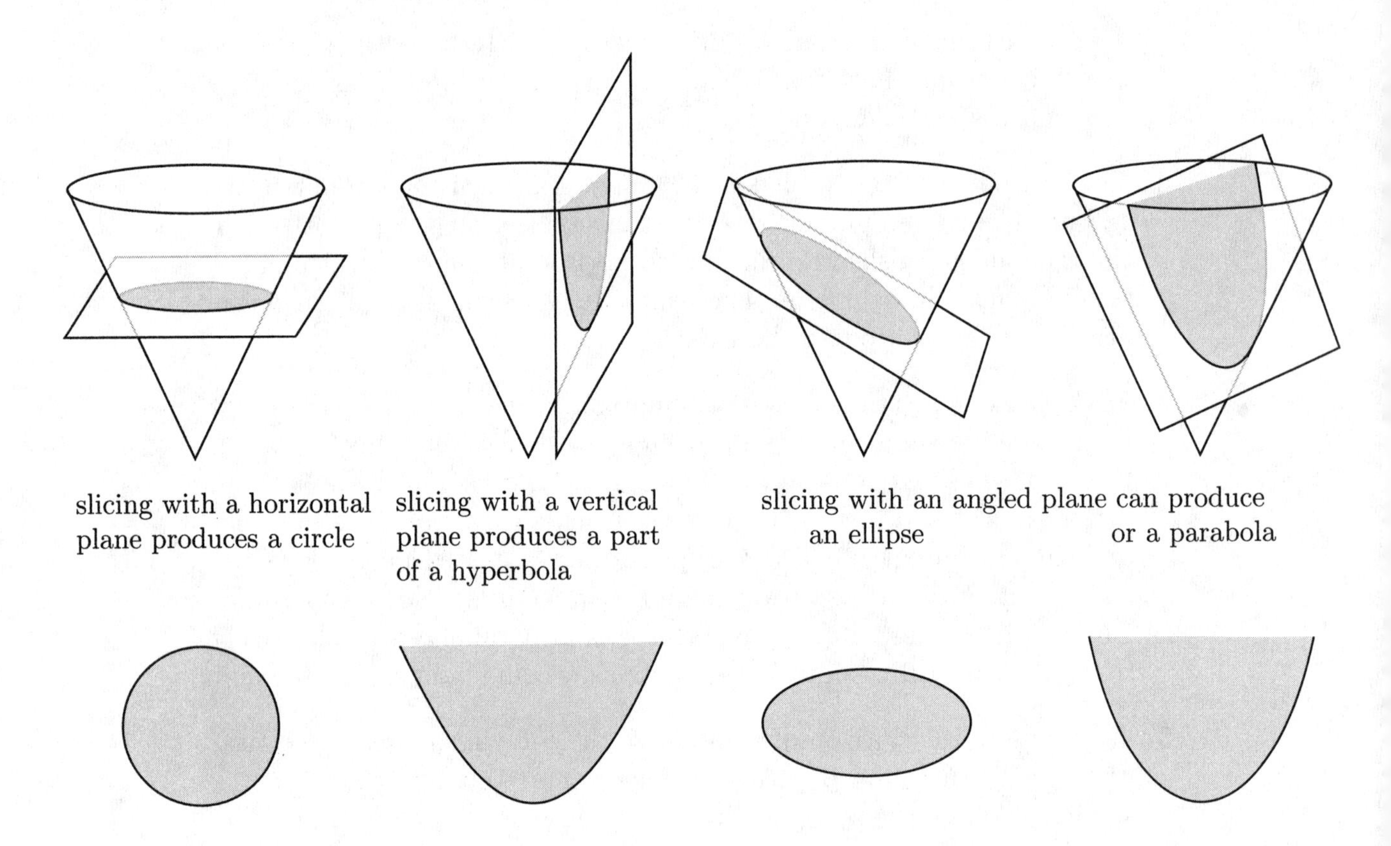

Figure 7.9: Slicing a Cone With a Plane

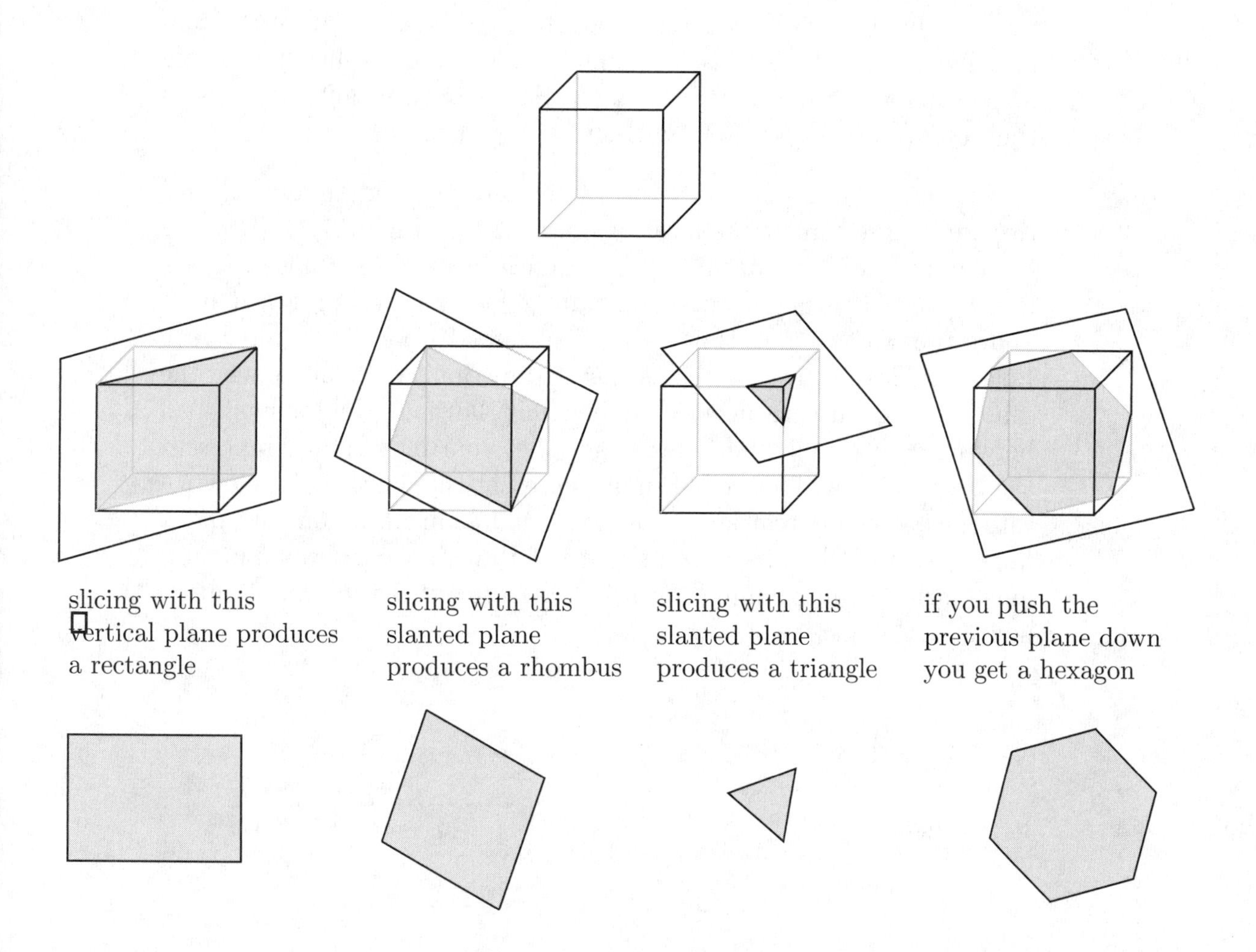

Figure 7.10: Slicing a Cube

yourself at points B and D. At both locations the Sun is on the horizon (because the Sun's rays just graze the surface of the Earth there), so it must be sunrise at one point and sunset at the other. How can we tell which is which? Picture yourself standing at point D and facing north. The Sun is on your right, which is to the east. Since the Sun rises in the east, it is sunrise at point D. Now picture yourself standing at point B and facing north. Now the Sun is on your left, which is to the west. Since the Sun sets in the west, it is sunset at point B.

7. According to the results of the previous exercise, noon, midnight, sunrise, and sunset are at the locations indicated in Figure 7.11. Because the day progresses from midnight to sunrise to noon to sunset, therefore the Earth must be rotating counterclockwise when viewed from above the North Pole. However, when viewed from above the South Pole, the Earth rotates in the *clockwise* direction. You can see why the sense of rotation is reversed by doing the following. Hold a small ball between your thumb and index finger. Let your index finger represent the North Pole and let your thumb represent the South Pole. Now rotate the ball counterclockwise when viewed looking down on your index finger (North Pole). Keep rotating the ball in the same way, but now look down on your thumb (South Pole). Notice that the same rotation now appears clockwise!

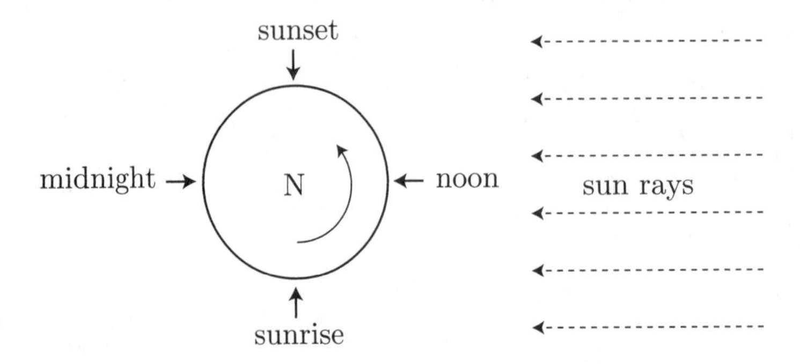

Figure 7.11: The Earth's Rotation as Seen Looking Down on the North Pole

8. See Figure 7.12. The shaded portions in the Moon show what part of the Moon is visible from Earth. The portion of the Moon that is visible from Earth is the portion that is facing the Earth *and also* illuminated

by the Sun. The dashed lines on the Moon help you see the part of the Moon that is illuminated by the Sun, and the part of the Moon that is facing the Earth. In picture A, the Moon is just past being full (the Moon is full when it is opposite the Sun, with the Earth in between). As the Moon moves from the configuration in A to the configuration shown in B and beyond, less of the Moon becomes visible. Therefore the Moon is waning in A and B. After the configuration shown in B, the Moon becomes new (when it is between the Sun and Earth) and then begins to become visible again, as in C. In D, even more of the Moon is visible than in C. So the Moon is waxing in C and D.

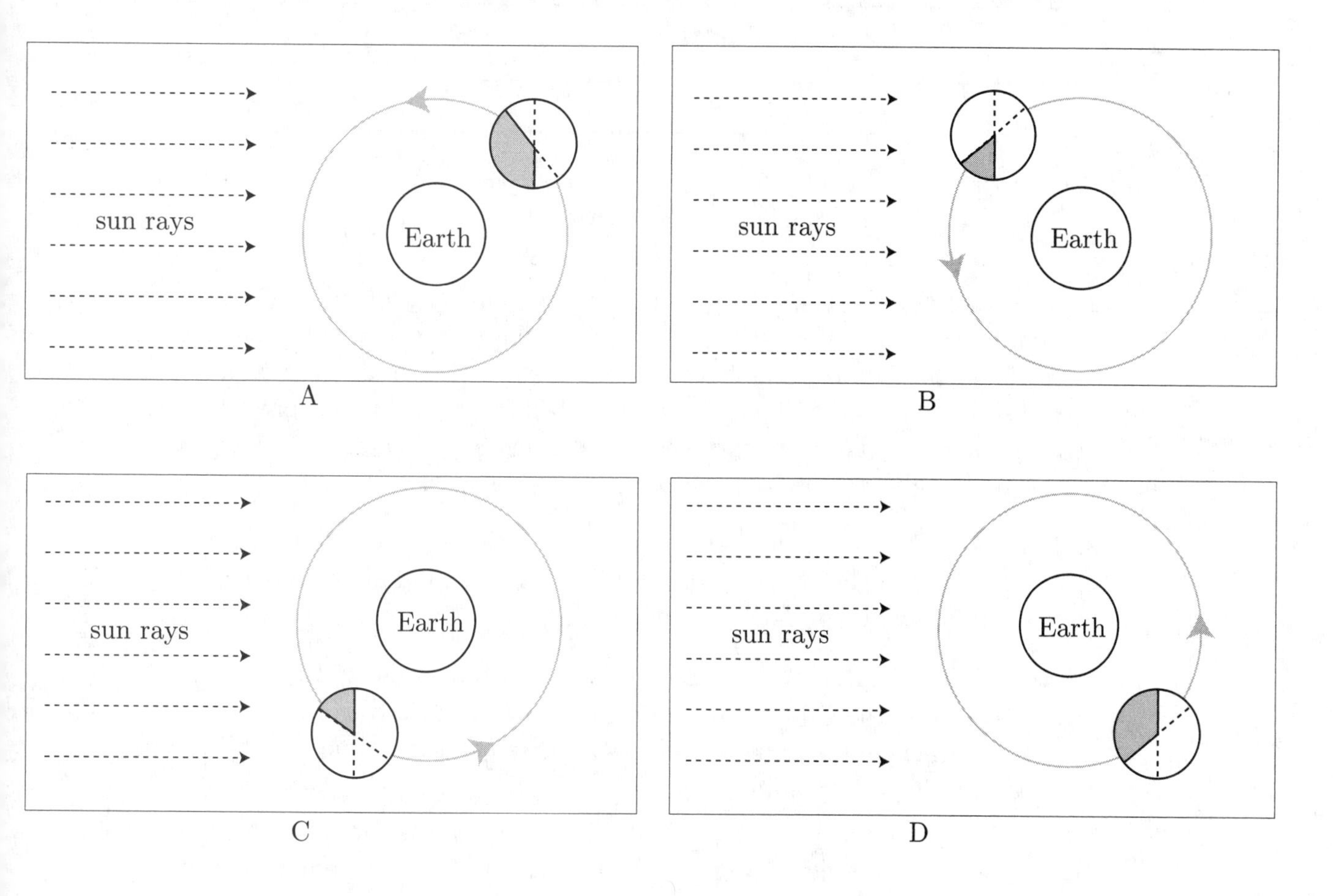

Figure 7.12: Is the Moon Waxing or Waning?

Problems for Section 7.1 on Visualization

1. Suppose you have an ordinary 2 by 4 board, as pictured in Figure 7.13.

   (a) Is it possible to saw the board with a straight cut in such a way that the shape formed by the cut is a square? Explain.

   (b) Is it possible to saw the board with a straight cut in such a way that the shape formed by the cut is *not* a rectangle? (A square *is* a kind of rectangle.) Explain. If the answer is yes, what kind of shape other than a rectangle can you get?

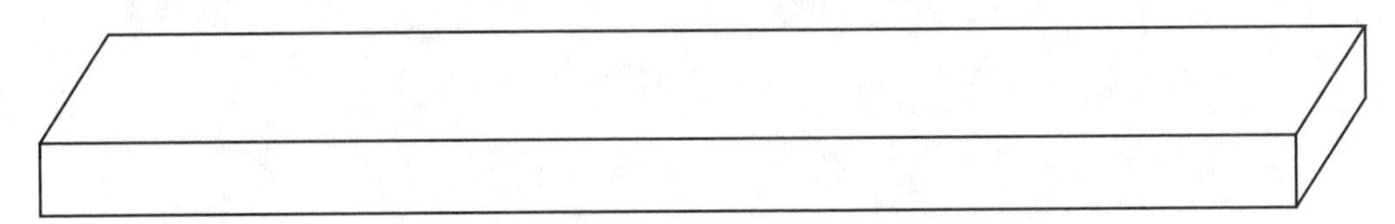

Figure 7.13: A 2 by 4 Board

2. Make three *different* patterns that could be cut out, folded, and taped to make a closed 1 inch by 1 inch by 1 inch cube. In your patterns, each square should be joined to another square along a *whole edge* of the square, not just at a corner. (This kind of pattern is sometimes called a **net**.) For two patterns to be considered different, it should not be possible two match the patterns up when they are cut out.

3. Refer to the previous problem. How many different patterns for a cube as described in the previous problem are there? Find all such patterns and explain why you have found them all.

4. (a) Figure 7.14 shows the Earth as seen from outer space, looking down on the North Pole (labeled N). What does the Sun look like to a person standing at point P? Therefore what time of day is it at point P? Explain.

   (b) If a person at point P in Figure 7.14 can see the Moon, and if the Moon is neither new nor full, then is the Moon waxing or is it waning? Explain.

5. Mike said that he saw a full Moon at 2 pm. Draw a diagram (like one in Figure 7.8) to help you explain why Mike couldn't be right.

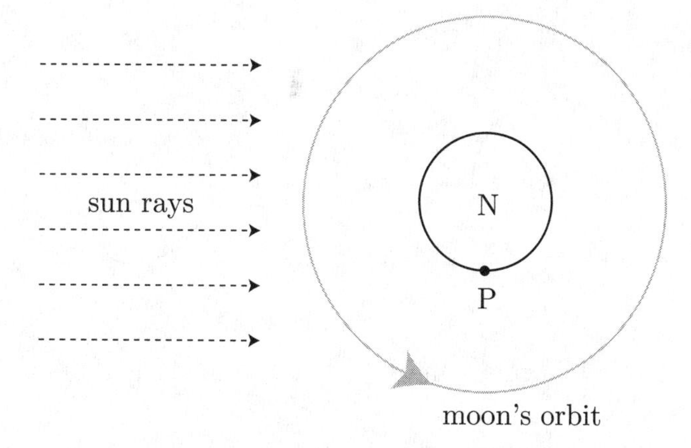

Figure 7.14: The Earth

6. If you go outside shortly after sunrise and you see the Moon up high in the sky, is the Moon waxing or waning? Draw a diagram (like one in Figure 7.8) to help you explain your answer.

7. Write a report on Greenwich Mean Time and the International Date Line. Explain why it is necessary to have a date line somewhere in the world.

7.2 Angles

Angles are used in two ways: to represent an amount of rotation (turning) about a fixed point, and to describe how two rays (or lines, line segments, or even planes) meet at a point. This section discusses basic facts about angles and gives some applications of angles that you might find surprising.

The two ways of thinking about angles, as amounts of rotation and as rays meeting, are closely related. It is equally valid to define angles from either point of view. Since even very young children have experience spinning around, the "rotation" point of view is perhaps more primitive. So our first definition is that an **angle** is an amount of rotation about a fixed point. An angle at a point P is said to be equal[1] to an angle at a point Q if both represent the same amount of rotation—even though this rotation takes place

[1]Some people prefer to say *congruent* rather than *equal* in this situation.

at different points (see Figure 7.15).

Figure 7.15: The Same Amount of Rotation at Different Points

Now to the other point of view about angles. Suppose there are two rays in a plane and these rays have a common endpoint P, as illustrated in Figure 7.16. In this case, the two rays and the region between them is called

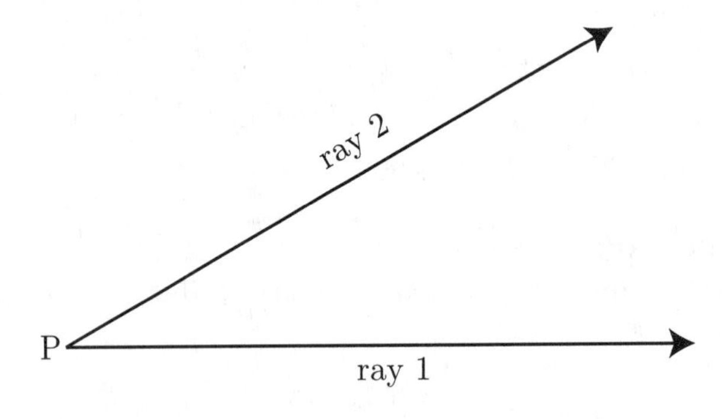

Figure 7.16: Rays Meeting at Point P

the **angle** at P formed by the two rays. This defintion of angle is connected to the "rotation" definition by associating the two rays meeting at a point P to the smallest amount of counterclockwise rotation about P needed to rotate one of the rays to the position of the other ray (see Figure 7.17).

When there are two line segments in a plane and these line segments have a common endpoint P, then we can define the angle between these line segments in the same way as for angles between rays. (Or: simply extend the line segments to rays having P as endpoint and use the definition for angles between rays.)

Angles are commonly measured in **degrees**. Degrees are usually indicated with a small circle: °. For example, 30 degrees is usually written 30°. If you stand at a fixed point and rotate counterclockwise in a full circle so as

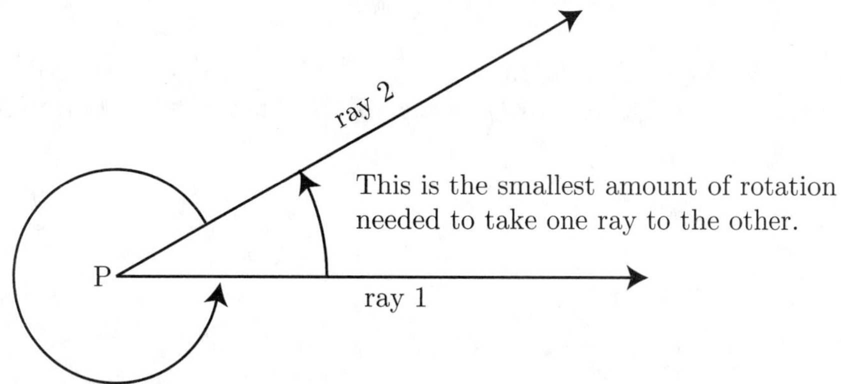

Figure 7.17: Angles Formed by Rays

to return to your starting position, then you will have rotated 360°. If you stand at a point and rotate one half of a full turn, then you will have rotated 180°; a quarter turn is 90°, and so on.

If two rays meet at a common endpoint Q so as to form a straight line, as illustrated in Figure 7.18, then the angle at Q is 180°. So if you were standing at the point Q looking straight down one side of the line and rotated counterclockwise so as to be looking down the other side of the line, you would have rotated 180°.

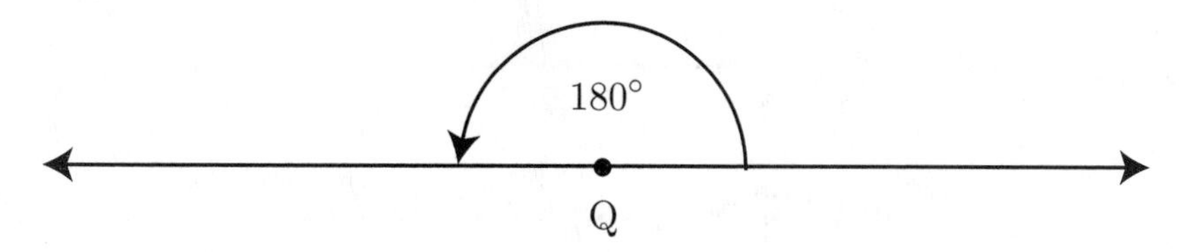

Figure 7.18: Two Rays Forming a Straight Line

A technical point: What about clockwise rotations? Clockwise rotations give rise to negative angles. So for example, if you stand at a point and you rotate a quarter turn clockwise, then you have rotated −90°. You need not be concerned with negative angles since we will not be working with them.

Notice that a small angle formed by rays has a "pointier" look than a large angle formed by rays, as seen in Figure 7.19. Angles less than 90° are called **acute angles**, an angle of 90° is called a **right angle**, while angles greater than 90° are called **obtuse** angles. Right angles are often indicated

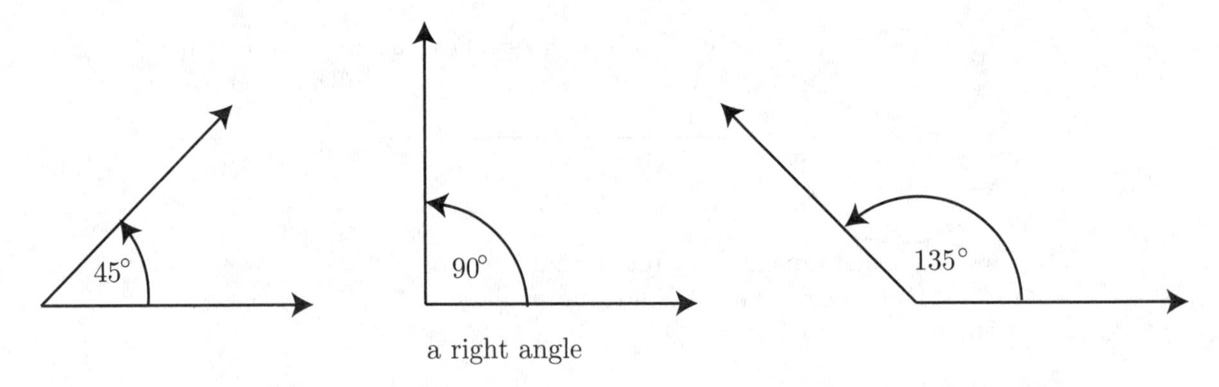

Figure 7.19: Some Angles

by a small square, as shown in Figure 7.20.

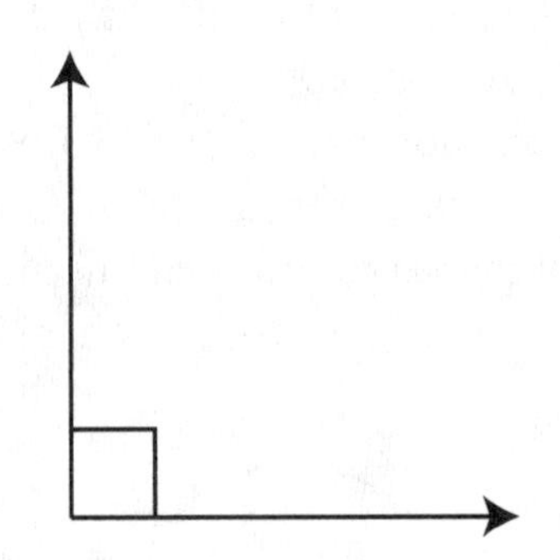

Figure 7.20: Indicating a Right Angle

Angles formed by rays can be measured with a simple device called a **protractor**. Figure 7.21 shows a protractor measuring an angle. Protractors usually have a small hole that should be placed directly over the point where the two rays meet. This hole lies on a horizontal line through the protractor. This horizontal line should be lined up with one of the rays that form the angle to be measured. The protractor in Figure 7.21 shows that the angle it is measuring is 65°.

Some teachers help their students learn about angles by making and using "angle explorers." An angle explorer is made by attaching two strips of

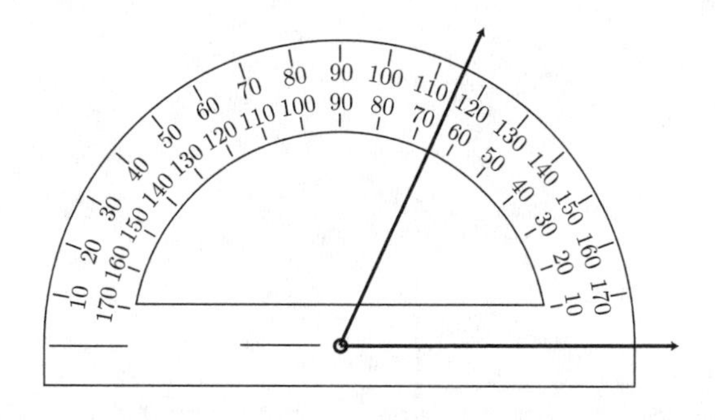

Figure 7.21: A Protractor Measuring an Angle

cardboard with a brass fastener, as shown in Figure 7.22. By rotating the cardboard strips around you get a feel for the relative sizes of angles.

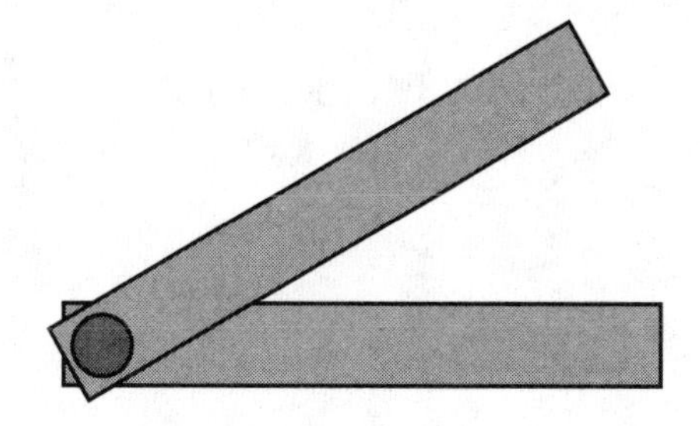

Figure 7.22: An "Angle Explorer"

When two lines in a plane meet, they form four angles. Figure 7.23 shows some examples. When all four of the angles are 90°, we say that the two lines are **perpendicular**.

Two lines in a plane that *never* meet (even somewhere far off the page they are drawn on) are called **parallel**. Figure 7.24 shows examples of lines that are parallel and lines that are not parallel even though you can't see where they meet.

Class Activity 7H: Angles of Sun Rays

Class Activity 7I: Angles in Vision

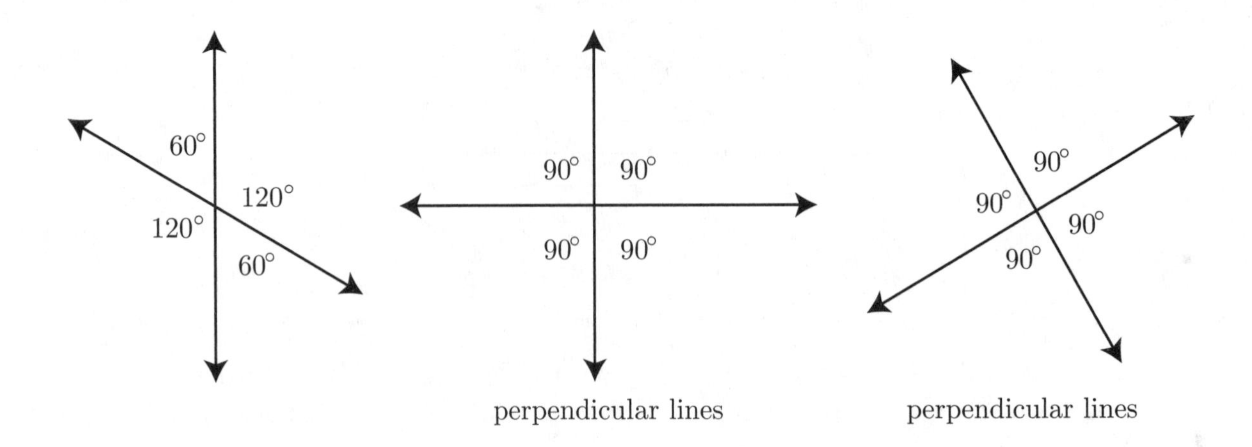

Figure 7.23: Lines Meeting

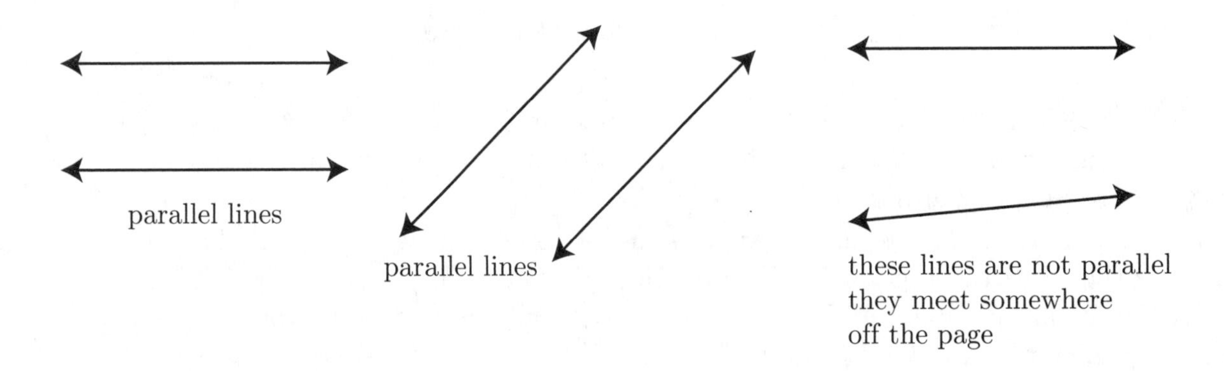

Figure 7.24: Parallel Lines and Lines That Are Not Parallel

Class Activity 7J: Describing Routes Using Distances and Angles

Class Activity 7K: An Angle Misconception

Class Activity 7L: Angles Formed by Two Lines

Angles and Reflected Light

When a light ray hits a smooth reflective surface, such as a mirror, the light ray reflects in a specific way that can be desribed with angles.

In order to describe how reflection of light works, we will need the concept of a **normal line** to a surface. The normal line at a point on a surface is the line that passes through that point and is perpendicular to the surface at that point. In order not to get too technical, you can think of "perpendicular to the surface at that point" as meaning "sticking straight out away from the surface at that point." Figure 7.25 shows a cross-section of a "wiggly" surface and its normal lines at various points.

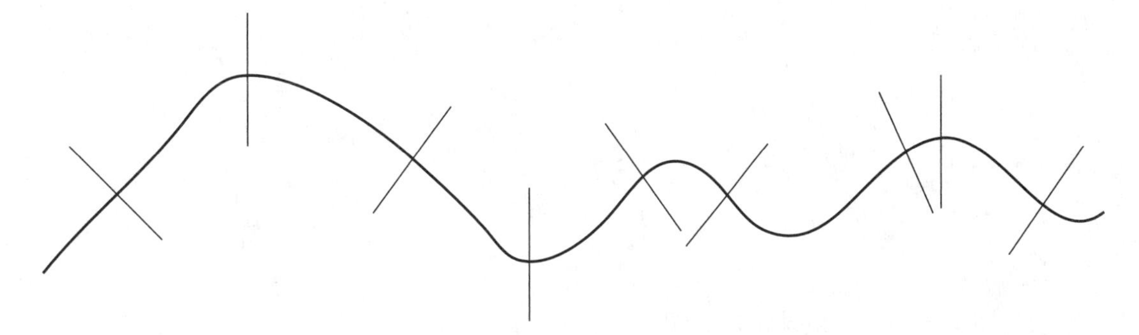

Figure 7.25: Some Normal Lines to a Surface (Cross-Section Shown)

There are two fundamental physical principles that govern the reflection of light from a surface. These principles also apply to the reflection of similar radiation, such as microwaves and radio waves:

1. The incoming and reflected light rays make the *same angle with the normal line* at the point where the incoming light ray hits the surface.
2. The reflected light ray lies in the same plane as the normal line and the incoming light ray. The reflected light ray is not in the same location

as the incoming light ray unless the incoming light ray is in the same location as the normal line.

Figure 7.26 shows examples of how light rays reflect off surfaces. In each case, a cross-section of the surface is shown.

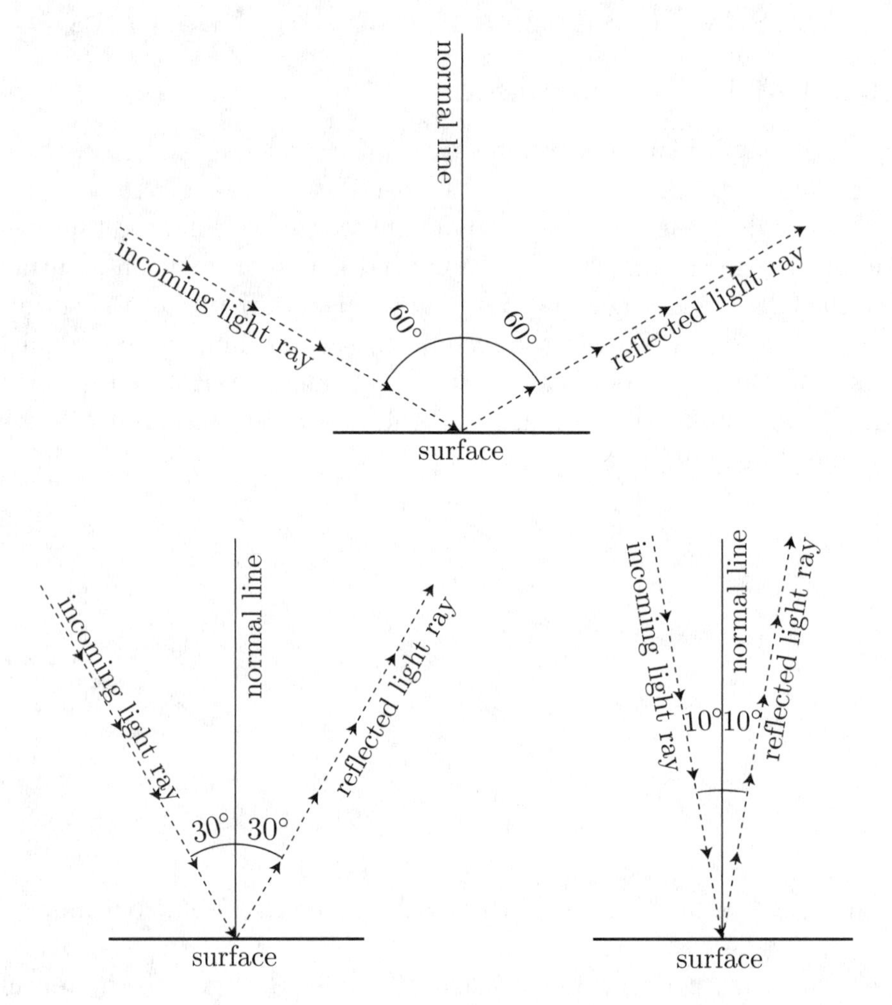

Figure 7.26: Light Rays Reflecting Off Surfaces

You can demonstrate the principles of reflection by putting a mirror on a table, shining a pen-light at the mirror, and seeing where the reflected light hits the wall. (Be careful to keep the light beam away from your and others' eyes.) If you raise and lower the light, while still pointing it at the mirror, the light beam heading toward the mirror will make different angles with the mirror. By observing the reflected light beam on the wall, you can tell that the lightbeams going toward and from the mirror make the same angle with the normal line to the mirror. (If the mirror is on a horizontal table, then the normal lines to the mirror are vertical.)

Class Activity 7M: Looking in a Mirror

Class Activity 7N: Why Do Spoons Reflect Upside Down?

Class Activity 7O: The Special Shape of Satellite Dishes

Exercises for Section 7.2 on Angles

1. My son once told me that some skateboarders can do "ten-eighty's". I said he must mean 180's, not 1080's. My son was right, some skateboarders can do 1080's! What is a 1080 and why is it called that? By contrast, what would a 180 be?

2. Use a protractor to measure the angles formed by the shape in Figure 7.27.

3. Explain why the angle at A in Figure 7.28 is not larger than the angle at B.

4. Draw pictures to show the relationship between the angle that the Sun's rays make with horizontal ground and the length of a shadow of a telephone pole.

5. Suppose that two lines in a plane meet at a point, as in Figure 7.29. Use the fact that the angle formed by a straight line is $180°$ to explain why $a = c$ and $b = d$.

6. Figure 7.30 shows a side view of a flashlight shining on puppet behind a semi-transparent screen. Show why the shadow of the puppet on the screen is bigger than the puppet.

7. Figure 7.31 shows several pictures (from the point of view of a fly looking down from the ceiling) of a person standing in a room, looking into a mirror on the wall. The direction of the person's gaze is indicated. What will the person see in the mirror?

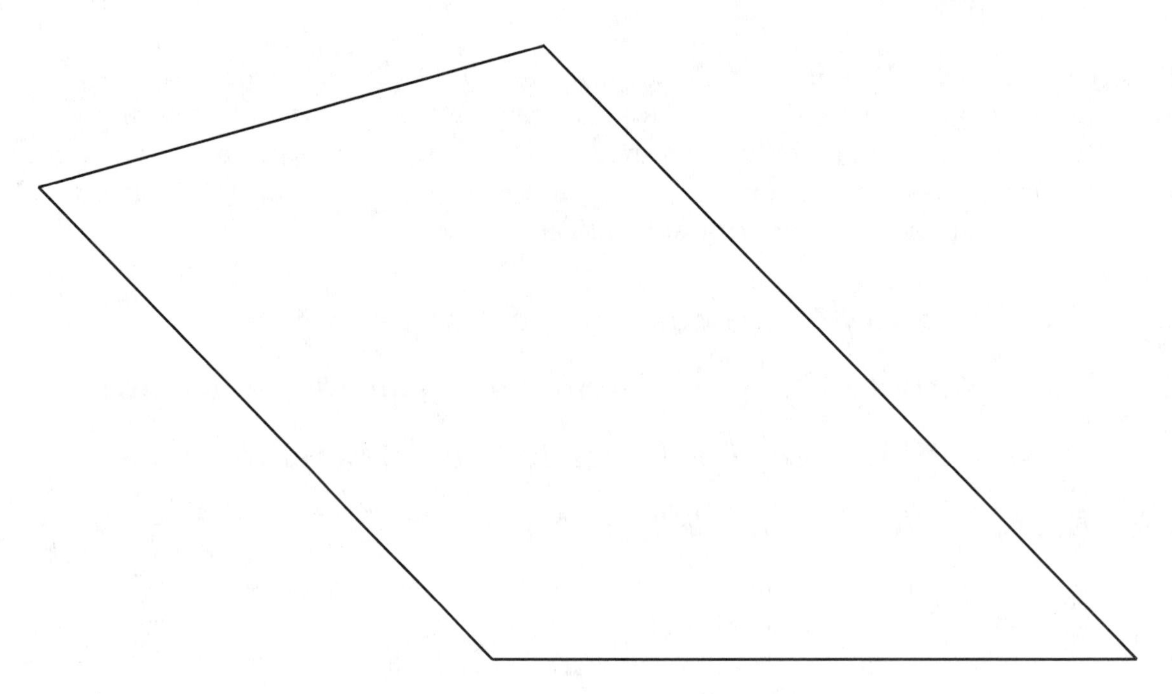

Figure 7.27: Measure the Angles

8. Figure 7.32 shows a mirror seen from the top, and a light ray hitting the mirror. Make a copy of this picture on a clean piece of paper. Use the following paper-folding method to show the location of the reflected light ray:

   - Fold and crease the paper so that the crease goes through the point where the light ray hits the mirror and so that the line labeled "mirror" folds onto itself. (This crease is perpendicular to the line labeled "mirror". You'll be asked to explain why below.)
   - Keep the paper folded and now fold and crease the paper again along the line labeled "light ray".
   - Unfold the paper. The first crease is the normal line to the mirror. The second crease shows the light ray and the reflected light ray.

   (a) Explain why your first crease is perpendicular to the line labeled "mirror".

   (b) Explain why your second crease shows the reflected light ray.

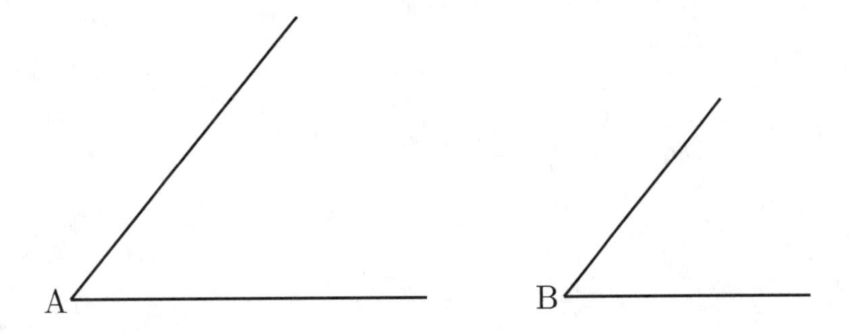

Figure 7.28: Is the Angle on the Left Larger?

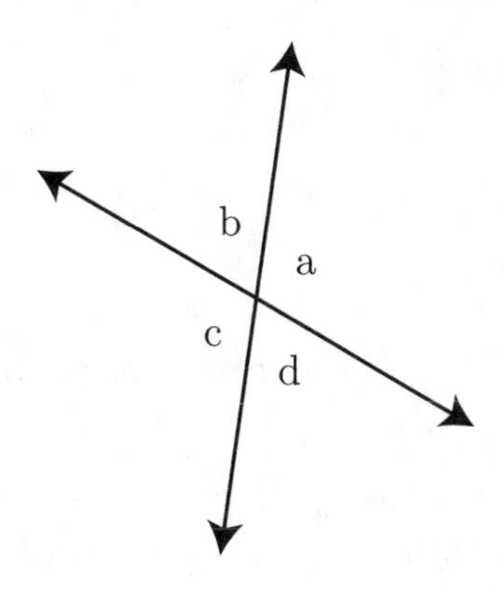

Figure 7.29: Lines Meeting at a Point

Answers to Exercises for Section 7.2

1. A "ten-eighty" is three full rotations. This makes sense because a full rotation is 360° and

$$3 \times 360^\circ = 1080^\circ.$$

A "180" would be half of a full rotation, which is not very impressive by comparison, although I certainly couldn't do it on a skateboard.

2. See Figure 7.33.

3. Even though the *line segments* making up the angle at A are longer, the angle at A is not larger than the angle at B. This is because the lower line segments of both angles would need to be rotated the same

Figure 7.30: How Big is the Shadow of the Puppet?

amount (about points A and B respectively) to get to the location of the upper line segments of the angles.

4. Figure 7.34 shows that when the Sun's rays make a smaller angle with horizontal ground, a telephone pole makes a longer shadow than when the Sun's rays make a larger angle with the ground.

5. Since angles a and b together make up the angle formed by a straight line, therefore

$$a + b = 180°.$$

For the same reason,

$$b + c = 180°.$$

So

$$a = 180° - b$$

and

$$c = 180° - b.$$

Since a and c are both equal to $180° - b$, therefore they are equal to each other, i.e., $a = c$. The same argument (with the letters changed) explains why $b = d$.

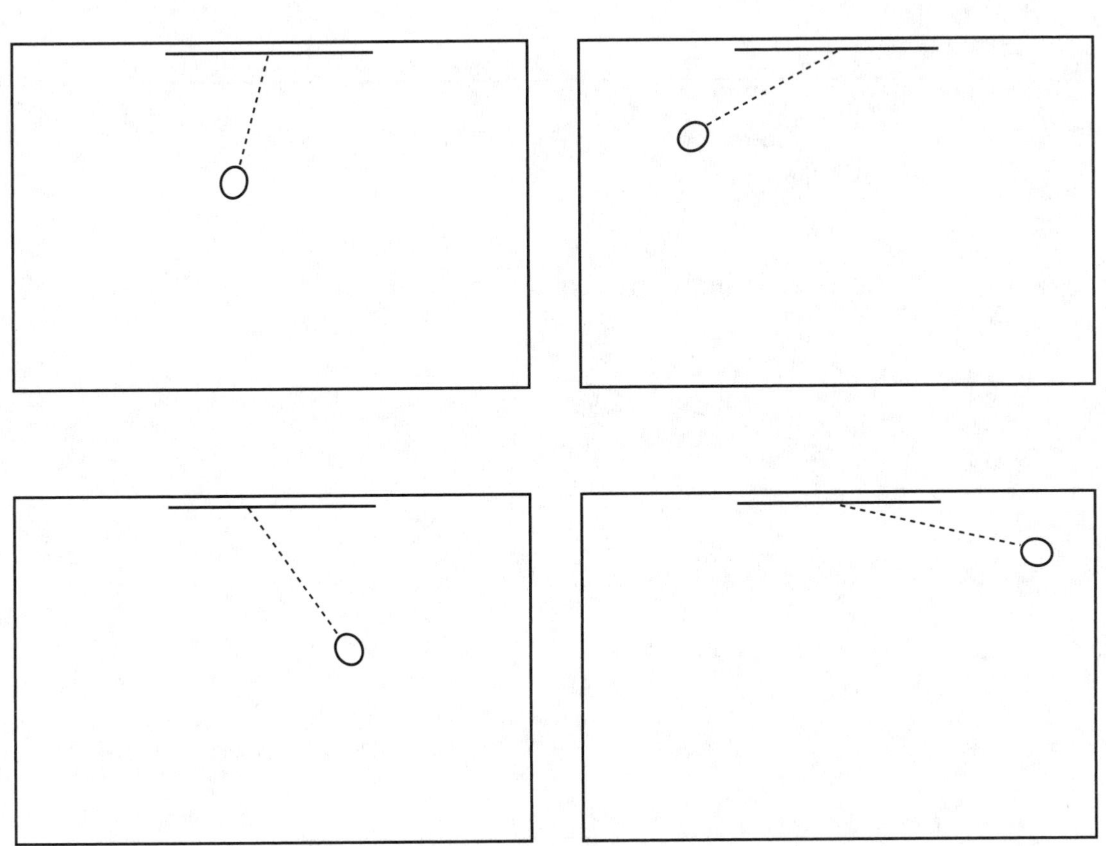

Figure 7.31: Looking Into a Mirror

6. Figure 7.35 shows that the shadow of the puppet is larger than the puppet itself. This is due to the fact that the light rays from the flashlight grazing the top and bottom of the puppet are not horizontal. If the puppet were farther from the screen, or if the flashlight were closer to the puppet, the puppet's shadow would be even larger.

7. Figure 7.36 shows that the person will see points A, B, C, and D, respectively, which are on various walls. Point C is in a corner. As the pictures show, the incoming light rays and their reflections in the mirror make the same angle with the normal line to the mirror at the point of reflection.

8. (a) The first crease is made so that angles a and b shown in Figure 7.37 are folded on top of each other and completely aligned. Therefore

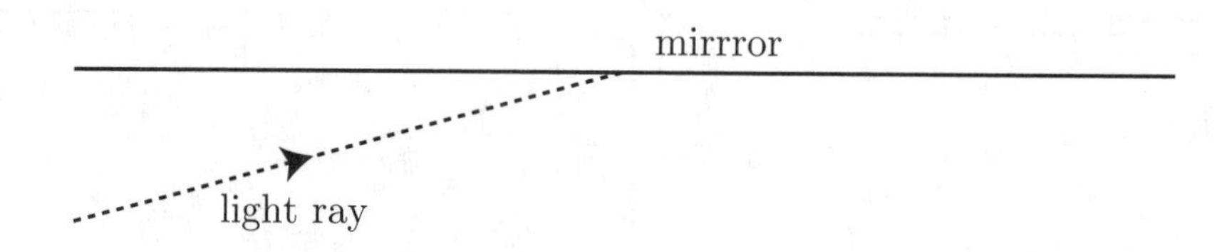

Figure 7.32: Using Paper Folding to Find the Reflected Ray

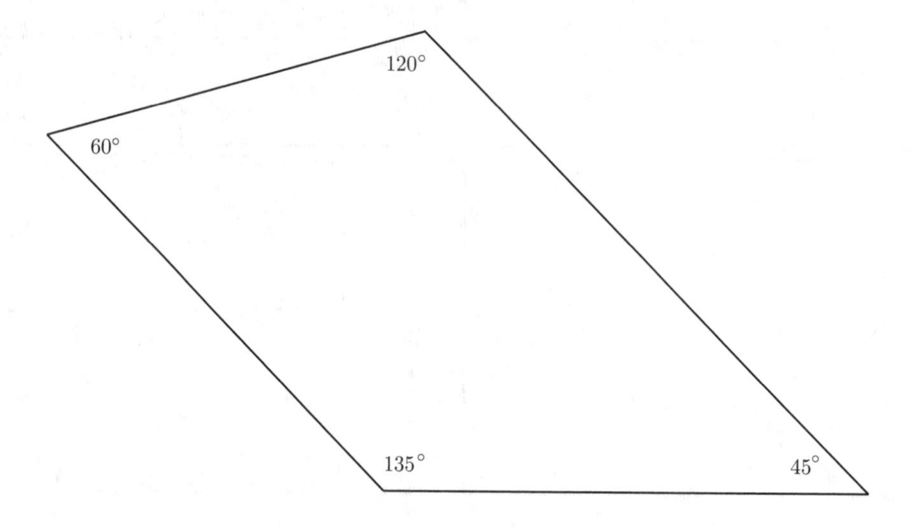

Figure 7.33: Angles Formed by a Shape

angles a and b are equal. But angle a and angle b must add up to 180° because the angle formed by a straight line is 180°. Therefore angle a and angle b must both be half of 180°, which is 90°. Therefore this first crease is the normal line to the mirror at the point where the light ray hits the mirror.

(b) The second crease is made so that angles c and d, shown in Figure 7.38 are folded on top of each other and completely aligned. Therefore angle c is equal to angle d. Since the first crease is a normal line, therefore, by the principles of reflection, the second crease shows the reflected light ray.

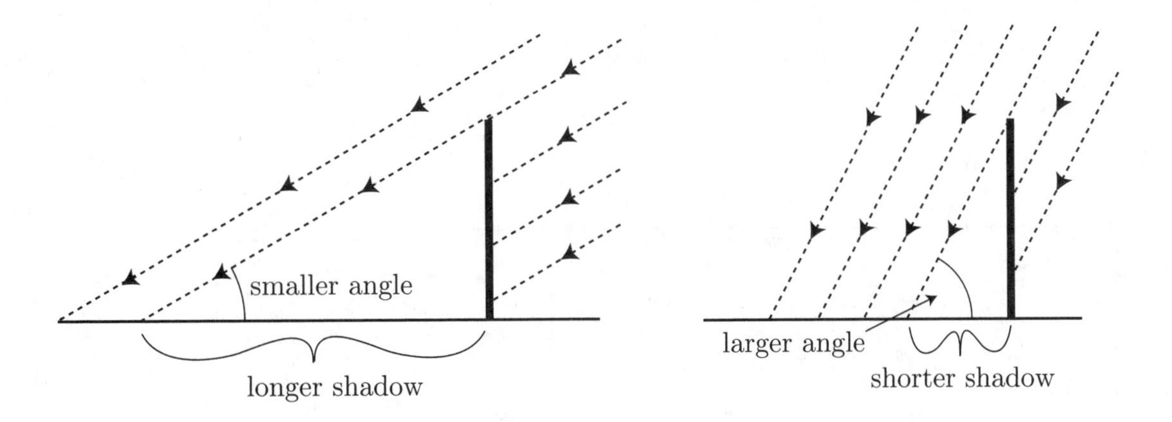

Figure 7.34: Sun Rays Hitting a Telephone Pole

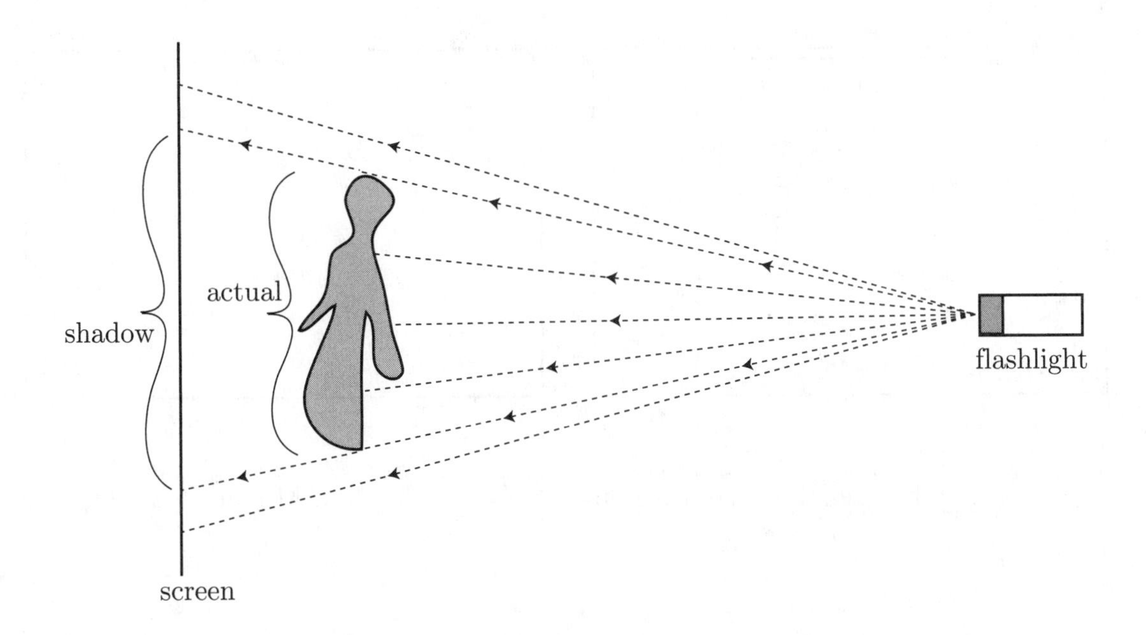

Figure 7.35: The Puppet's Shadow

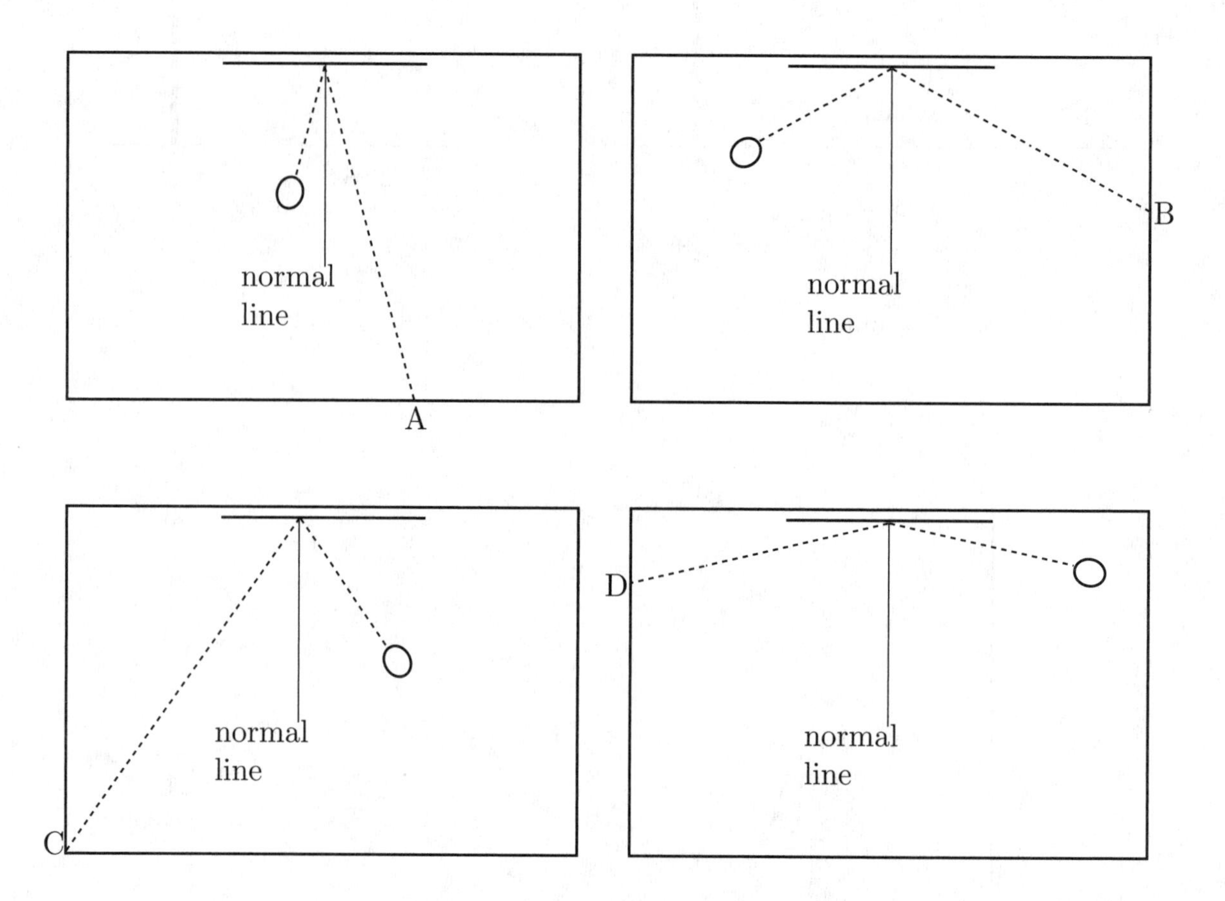

Figure 7.36: What a Person Sees Looking in a Mirror

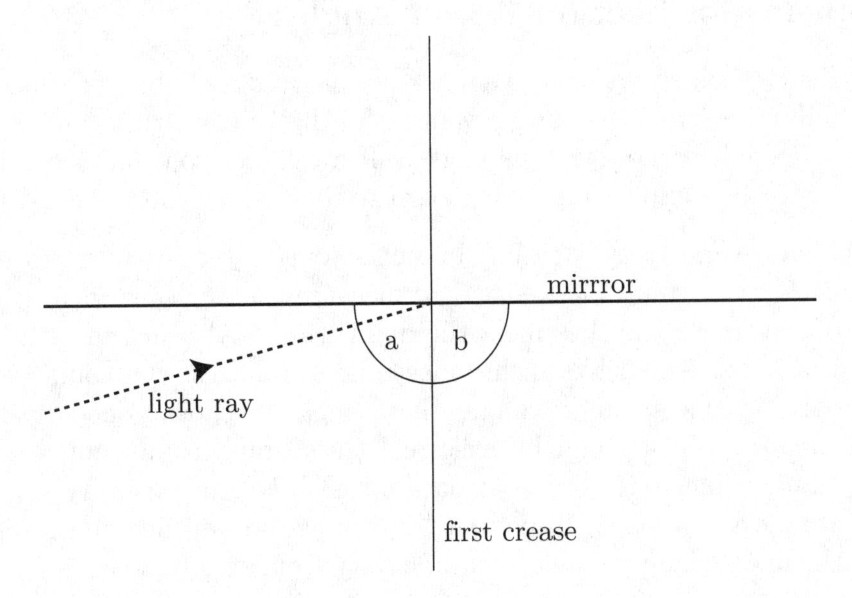

Figure 7.37: The First Crease

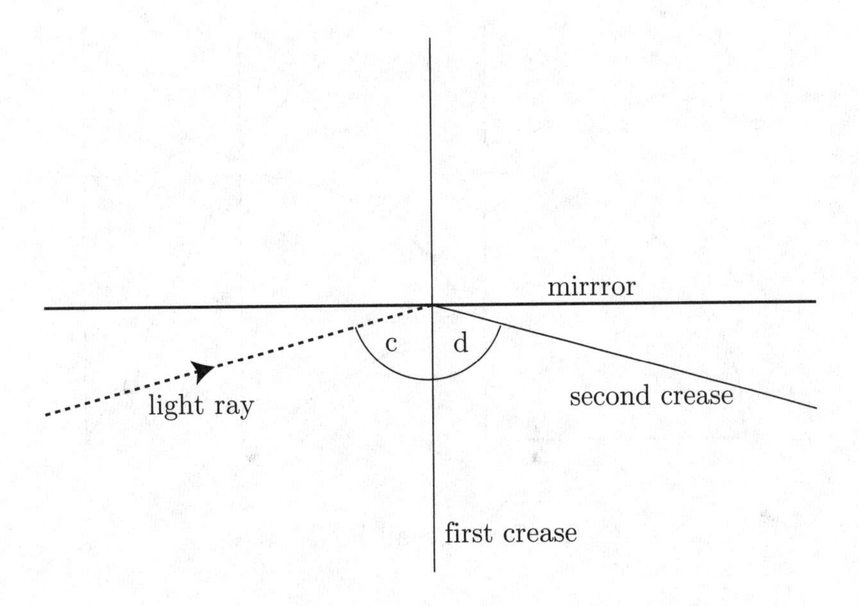

Figure 7.38: The Second Crease

Problems for Section 7.2 on Angles

1. Figure 7.39 shows small versions of "pie wedges." Larger versions are in Figure A.4. These larger versions of the "pie wedges" can be cut out and used to show various angles. If available, you may want to glue these large wedges onto cardboard to make them more sturdy.

   Find several trees with low branches and/or bushes that are not too dense. Use your "pie wedges" to measure the angles with which limbs of the trees or bushes meet the main trunk. Also measure the angles with which smaller branches meet main branches. (Of course you will only be able to approximate these angles because you only have so many pie wedges and because real trees branches are not as neat as straight lines.) Record your data for each tree or bush. Which angles are most common? Which angles, if any, did you not find at all? Do the most common angles you find vary from tree to tree?

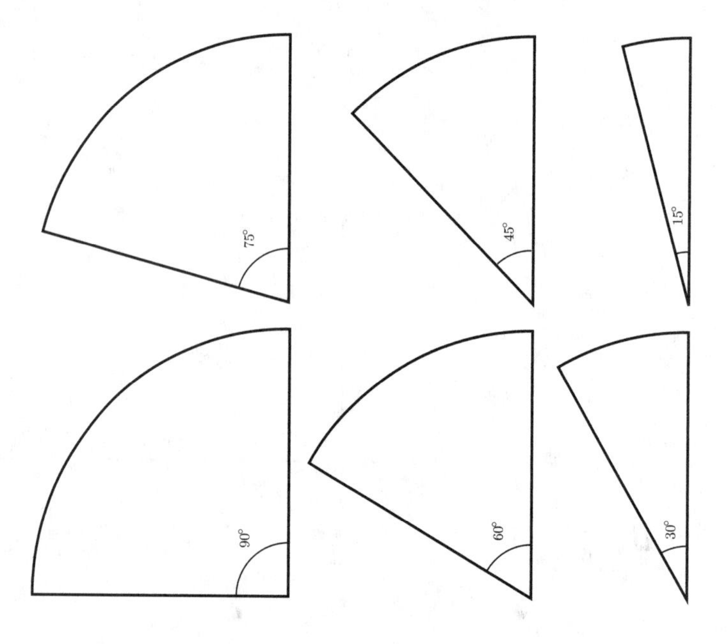

Figure 7.39: Small Pie Wedges

2. Amanda got in her car at point A and drove to point F along the route shown on the map in Figure 7.40.

(a) Trace Amanda's route shown in Figure 7.40; show all of Amanda's angles of turning along her route.

(b) Determine Amanda's total amount of turning along her route by adding the angles you measured in part (a).

(c) Now describe a way to determine Amanda's total amount of turning along her route *without* measuring the individual angles and adding them up. Hint: consider the directions that Amanda faces as she travels along her route.

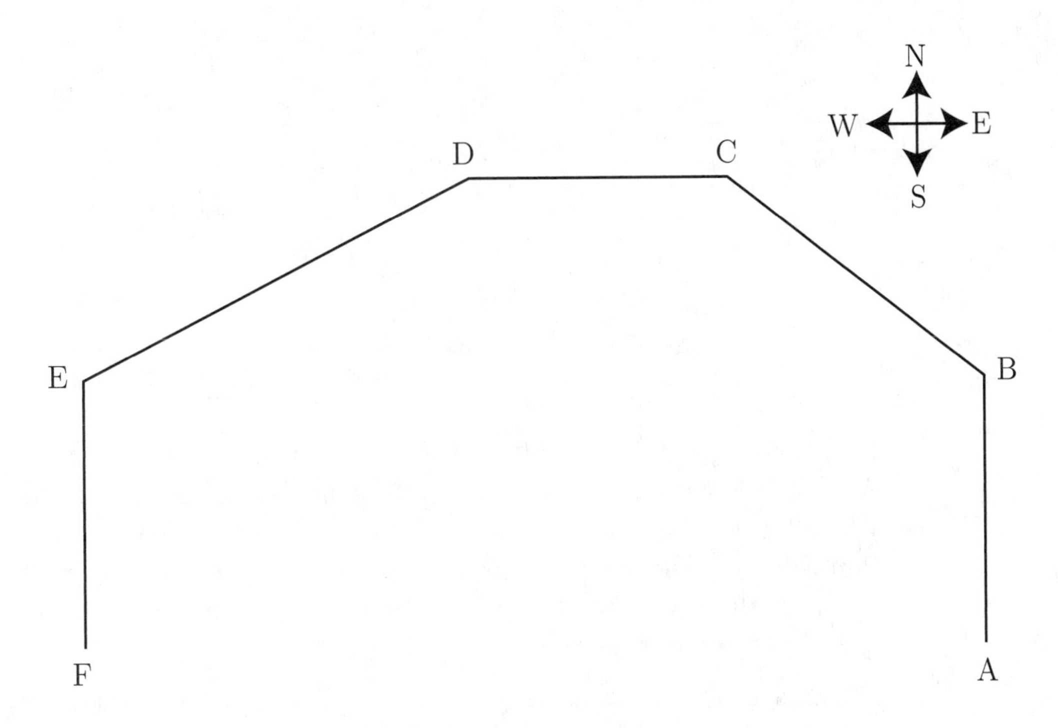

Figure 7.40: Amanda's Route

3. Many people mistakenly believe that the seasons are caused by the Earth's varying proximity to the Sun. In fact, the distance from the Earth to the Sun varies only slightly during the year and the seasons are caused by the tilt of the Earth's axis. As the Earth travels around the Sun during the year, the tilt of the Earth's axis causes the northern hemisphere to vary between being tilted toward the Sun to being tilted away from the Sun.

Figure 7.41 shows the Earth as seen from a point in outer space located in the plane in which the Earth rotates about the Sun. The diagram shows that the Earth's axis is tilted 23.5 degrees from the perpendicular to the plane in which the Earth rotates about the Sun.

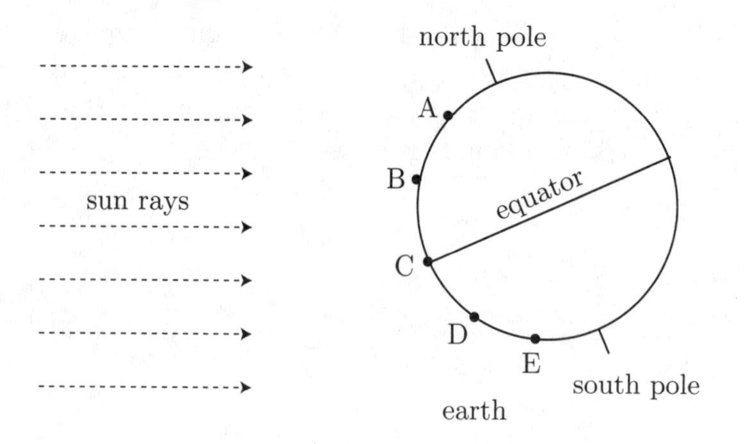

Figure 7.41: The Earth and Sun

(a) Locations A, B, C, D, and E in Figure 7.41 are shown at noon. At which location(s) is the Sun highest in the sky at noon? At which location(s) is the Sun lowest in the sky at noon? When the Sun is high in the sky the Sun rays are more intense than when it is low in the sky. Therefore which location(s) have the most and least intense Sun rays at noon?

(b) During the day, locations A, B, C, D, and E will rotate around the axis through the North and South Poles. How much sunlight will point A receive during the day? How much sunlight will point E receive during the day?

(c) Based on your answers to parts (a) and (b), what season is it in the northern hemisphere and what season is it in the southern hemisphere in Figure 7.41? Explain.

(d) At other times of year, the Earth and Sun are configured as shown in Figure 7.42. At those times, what season is it in the northern and southern hemispheres? Why? (Notice that the second picture in Figure 7.42 still shows the tilt of the Earth's axis.)

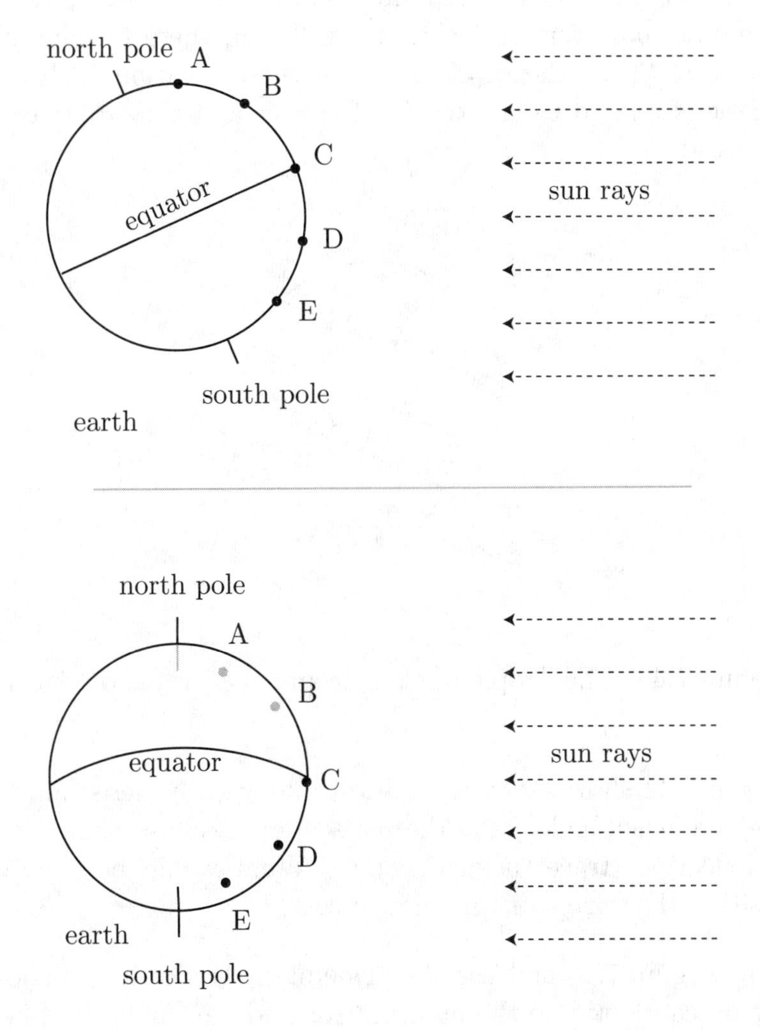

Figure 7.42: The Earth and Sun at Other Times of Year

4. Refer to Figures 7.41 and 7.42 and the results of the previous problem to answer the following. During which season(s) are the Sun's rays most intense at the equator? Look carefully before you answer!

5. Refer to Figures 7.41, 7.42, and 7.43 to help you answer the following. There are only certain locations on Earth where the Sun can ever be seen *directly* overhead. Where are these locations? How are these locations related to the tropic of cancer and the tropic of capricorn? Explain!

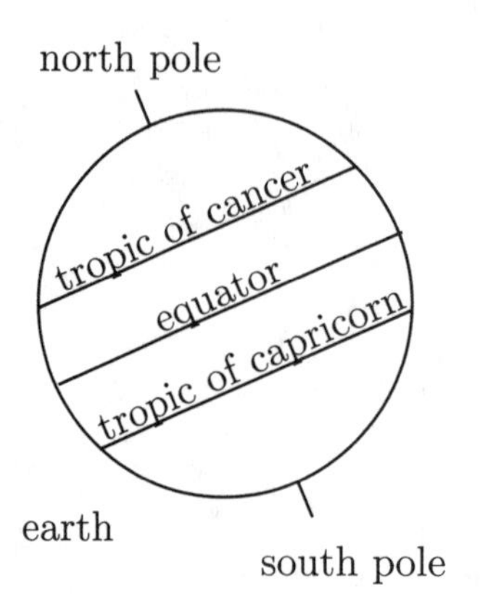

Figure 7.43: The Tropic of Cancer and the Tropic of Capricorn

6. Figure 7.44 shows a cross-section of Joey's toy periscope. What will Joey see when he looks in the periscope? Explain, using the principles of reflection (trace the periscope). What would be a better way to position the mirror in the telescope?

7. Copy Figure 7.45 and use the principles of reflection to show how the person can look into the hand mirror and see the back of her head.

8. Department store dressing rooms often have large mirrors that actually consist of three adjacent mirrors, put together as shown in Figure 7.46 (as seen looking down from the ceiling). Use the principles of reflection to show how you can stand in such a way as to see the reflection of your back. Draw a careful picture, using an enlarged version of Figure 7.46, that shows clearly how light reflected off your back can enter your

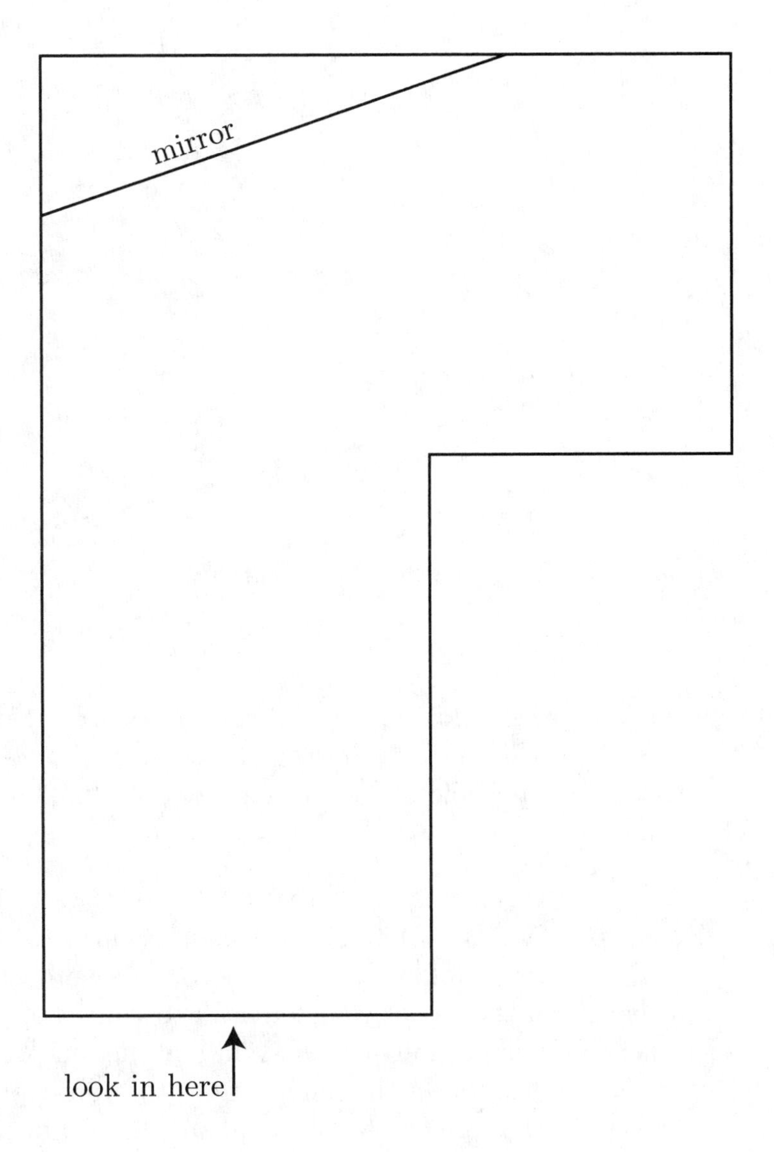

Figure 7.44: Joey's Periscope

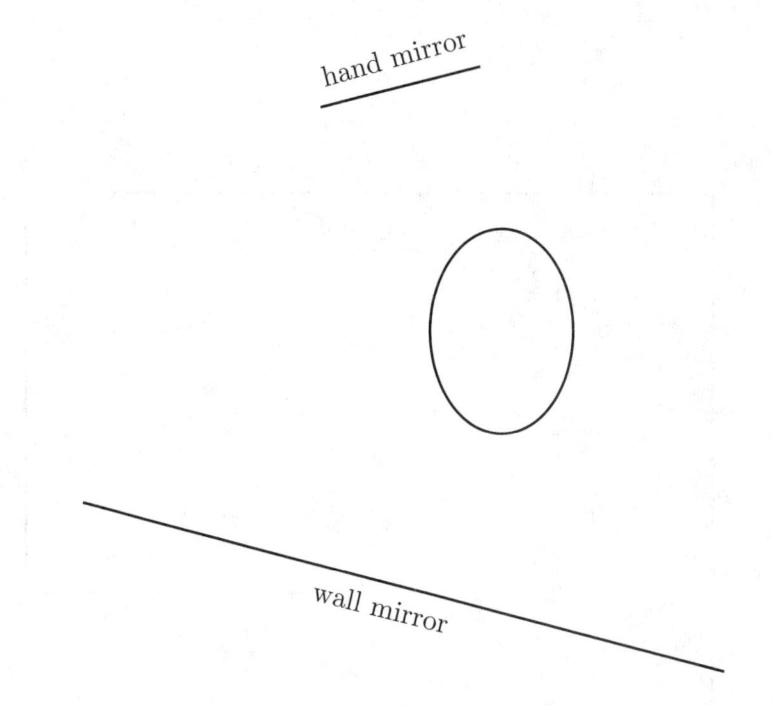

Figure 7.45: Using Mirrors to See the Back of One's Head

eyes. Your picture should show where you are standing, the location of your back and which way your eyes are looking. (You may wish to experiment with paper folding before you attempt a final drawing. See Exercise 8.)

9. (a) How big is the reflection of your face in a mirror? Is it the same size as your face or is it larger or smaller, and if so, how much? Investigate this as follows. Use a ruler to measure the length of your face from the top of your forehead to your chin. Now stand parallel to a mirror hanging on a wall and measure the length of your face's reflection in the mirror, from the top of the forehead to the chin. Measure carefully! Compare the two measurements. Move in closer or farther away and repeat. The *position* of your reflection will probably change, but does the *size* of your reflection change or not?

   (b) Now make a drawing of a person looking into an ordinary (flat) mirror. Draw a sideways view, as in the picture for Class Activ-

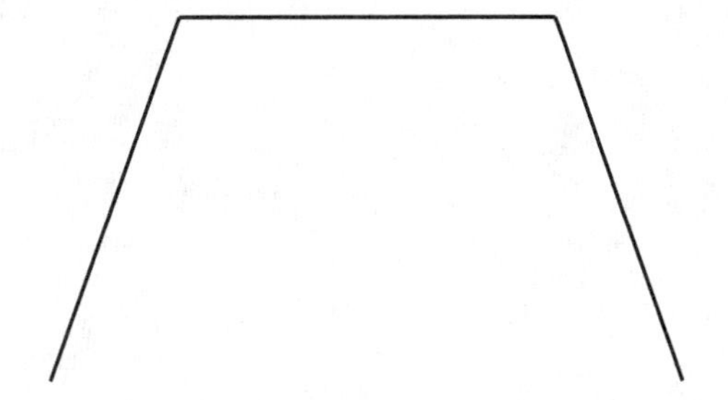

Figure 7.46: A Department Store Mirror

ity 7N (except that the mirror should be flat, not curved, so the mirror should be shown as a straight, vertical line). Use the principles of reflection to show where the person will see the top of his/her forehead and where the person will see his/her chin in the mirror. Measure the length of the reflected face and of the original face and describe how these measurements are related. Does this agree with what you discovered in part (a)?

10. A **concave** mirror is a mirror that curves in, like a bowl, so that the normal lines on the reflective side of the mirror point toward each other. Makeup mirrors are often concave. Figure 7.47 shows an eye looking into a concave makeup mirror. Normal lines to the mirror are shown. Trace this picture and show where a woman applying eye makeup sees her eye in the mirror. Show approximately where she sees the top of her reflected eye and where she sees the bottom of her reflected eye (assume that the woman sees light that enters the center of her eye). Your picture should explain why the reflected eye is seen where it is. Compare how the woman would see her eye in an ordinary flat mirror. Why does a concave mirror make a good mirror for applying makeup? (By the way, although a concave mirror is curved in the same direction as the bowl of a spoon, the smaller amount of curvature in a concave mirror prevents it from reflecting your image upside down, as a spoon does.)

11. A **convex** mirror is a mirror that curves out, so that the normal lines on the reflective side of the mirror point away from each other, as shown in Figure 7.48. Convex mirrors are often used as side-view mirrors on cars and trucks. This problem will help you see why convex mirrors

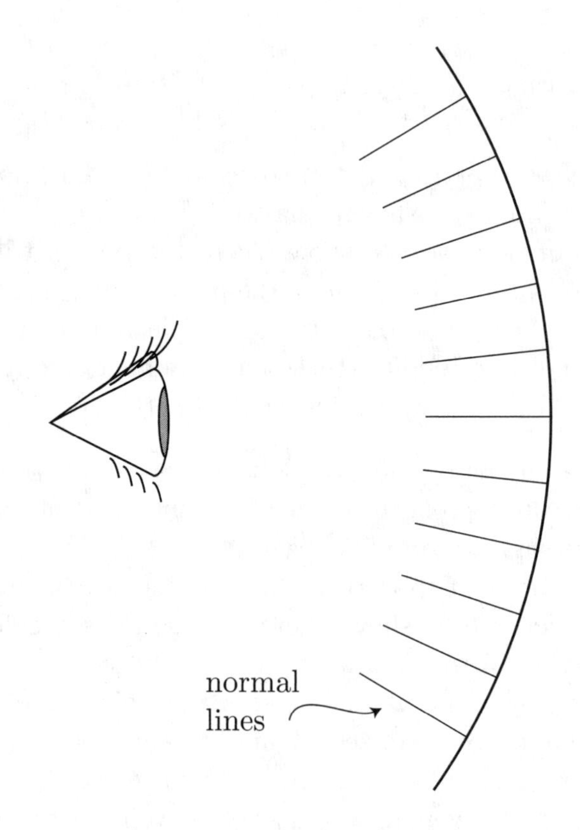

Figure 7.47: An Eye Looking in a Concave Mirror

useful for this purpose.

Figure 7.48 shows a bird's eye view of a cross-section of a convex mirror, a cross-section of a flat mirror, and eyes looking into these mirrors. Trace these mirrors and mark the location of the eyes in your picture. Use the principles of reflection to help you compare how much each eye can see looking into its mirror.

Now explain why convex mirrors are often used as side-view mirrors on cars and trucks.

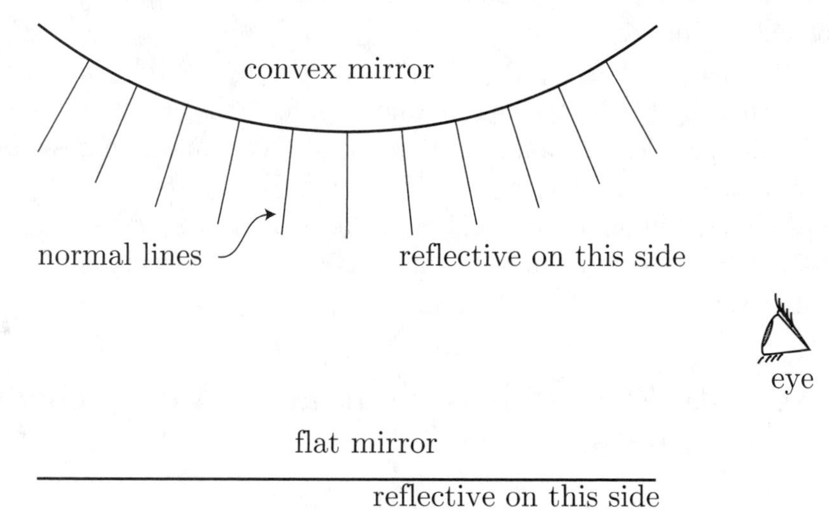

Figure 7.48: A Convex Mirror

12. A **convex** mirror is a mirror that curves out, so that the normal lines on the reflective side of the mirror point away from each other, as shown in Figure 7.48. Convex mirrors are often used as side-view mirrors on cars

and trucks, but these mirrors usually carry the warning sign "objects are closer than they appear". Explain why objects reflected in a convex mirror appear to be farther away than they actually are. Use the fact that they eye interprets a smaller image as being farther away.

7.3 Circles and Spheres

In this section we'll discuss the definitions of circles and spheres and applications of these definitions.

How are circles and spheres defined? To the eye, circles and spheres are distinguished by their perfect roundness. Informally, we might say that a sphere is the surface of a ball. This defines circles and spheres from an informal or artistic point of view, but there is also a mathematical point of view. As we'll see, the mathematical definitions of circles and spheres yield practical applications that cannot be anticipated by considering only the look of these shapes.

Class Activity 7P: Points That are a Fixed Distance From a Given Point

Definitions of *Circle* and *Sphere*

A **circle** is the collection of all the points in a plane that are a certain fixed distance away from a certain fixed point in the plane. This fixed point is called the **center** of the circle, and this distance is called the **radius** of the circle. So the radius is the distance from the center of the circle to any point on the circle. The **diameter** of a circle is two times its radius. Informally, the diameter is the distance "all the way across" the circle.

For example, let's fix a point in a plane and let's call this point P, as in Figure 7.49. All the points in the plane that are 1 inch away from the point P form a circle of radius 1 inch, centered at the point P, as shown in Figure 7.49. The diameter of this circle is 2 inches.

A sphere is defined in almost the same way as a circle, but in space rather than in a plane. A **sphere** is the collection of all the points in space that are a certain fixed distance away from a certain fixed point in space. This fixed point is called the **center** of the sphere, and this distance is called the **radius** of the sphere. So the radius is the distance from the center of the

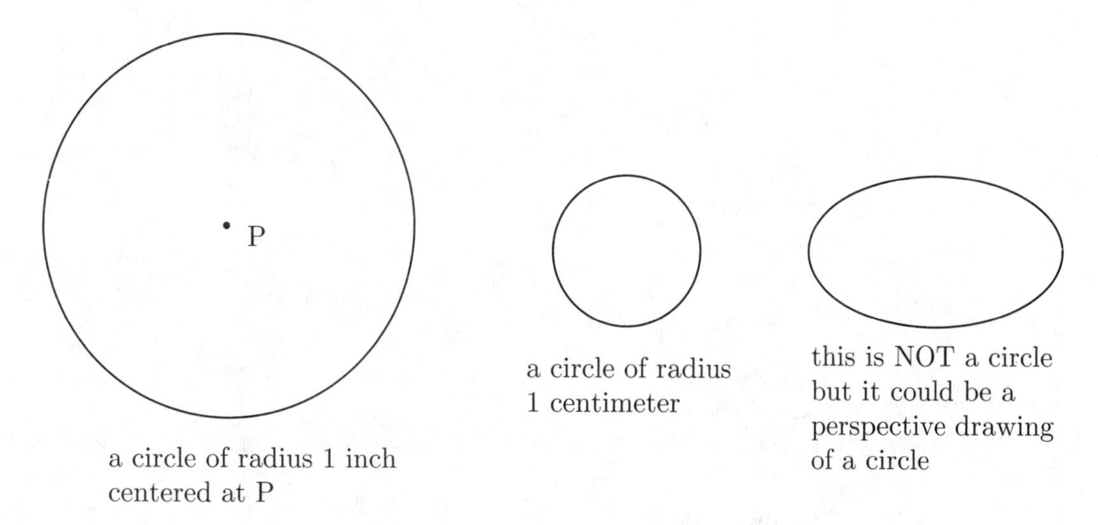

Figure 7.49: Two Circles and a Non-Circle

sphere to any point on the sphere. The **diameter** of a sphere is two times its radius. Informally, the diameter is the distance "all the way across" the sphere.

For example, fix a point P in space. You might want to think of this point P as located in front of you, a few feet away. Then all the points in space that are 1 foot away from the point P form a sphere of radius 1 foot, centered at P. Try to visualize this! What does it look like? It looks like the surface of a very large ball, as illustrated in Figure 7.50. The diameter of this sphere is 2 feet.

It's easy to draw almost perfect circles with the use of a common drawing tool called a **compass**, pictured in Figure 7.51. One side of a compass has a sharp point that you can stick into a point on a piece of paper. The other side of a compass has a pencil attached. To draw a circle centered at a point P and having a given radius, open the compass to the desired radius, stick the point of the compass in the point P, and spin the pencil side of the compass in a full revolution around P, all the while keeping the point of the compass at P.

It is also possible to draw nice circles with simple homemade tools. For example, you can use two pencils and a paperclip to draw a circle, as shown in Figure 7.52. Put a pencil inside one end of a paperclip and keep this pencil fixed at one point on a piece of paper. Put another pencil at the other end

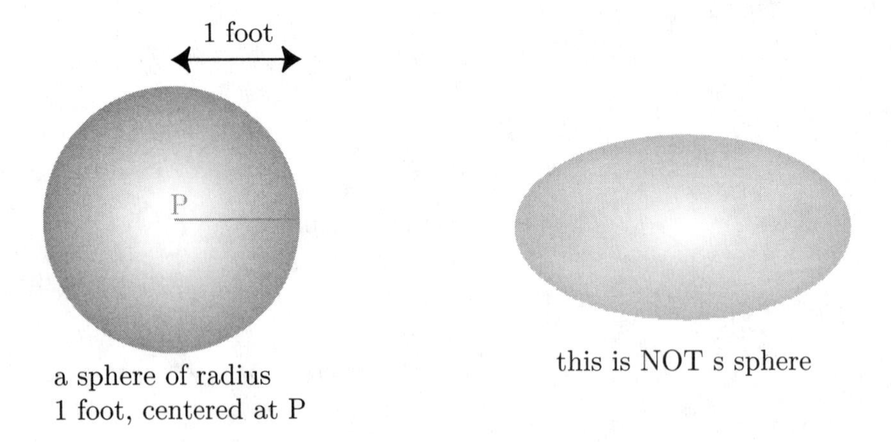

Figure 7.50: A Sphere and a Non-Sphere

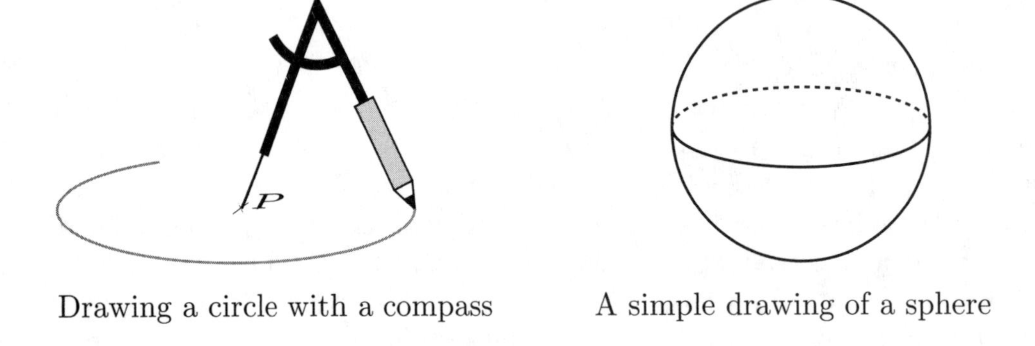

Figure 7.51: Drawing Circles and Spheres

of the paperclip and use that pencil to draw a circle.

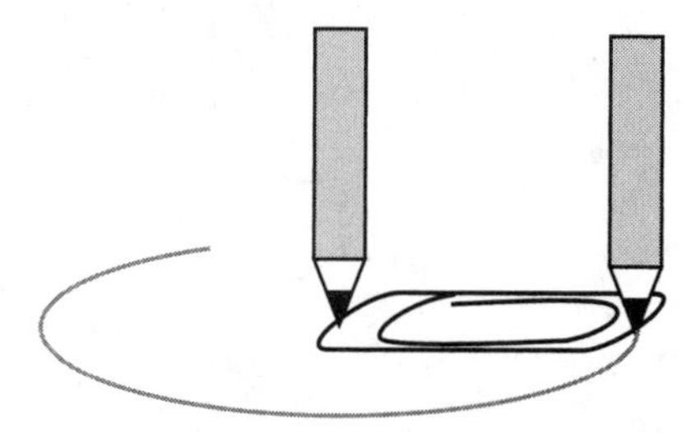

Drawing a circle with a paperclip

Figure 7.52: Drawing a Circle With a Paperclip

Since a sphere is an object in space, and not in a plane, it's harder to draw a picture of a sphere. Unless you are an exceptional artist, something like the simple drawing of a sphere in Figure 7.51 will do.

Class Activity 7Q: Using Circles

When Circles or Spheres Meet

What happens when two circles meet, or two spheres meet? While this may at first seem like a topic of purely theoretical interest, it actually has practical applications.

The last problem of the previous class activity provided a situation where two circles meet. To solve this problem, you could have drawn two circles, one centered at X, with a radius of 30 feet, and one centered at O, with a radius of 50 feet (using the scale shown in the map). The former circle shows all the places that are 30 feet from X, while the latter circle shows all the places that are 50 feet from O. The two locations where these circles meet are all places that are *both* 30 feet from X *and* 50 feet from O.

In general, if you have two circles, how do they meet? Try to visualize what could happen. Draw a few pictures. Returning to last problem of the previous class activity, in that case, the two circles met in two distinct places. Does this always happen with any pair of circles?

As illustrated in Figure 7.53, only three things can happen when you have two circles: either they don't meet at all, or they meet at a single point, or they meet at two distinct points.

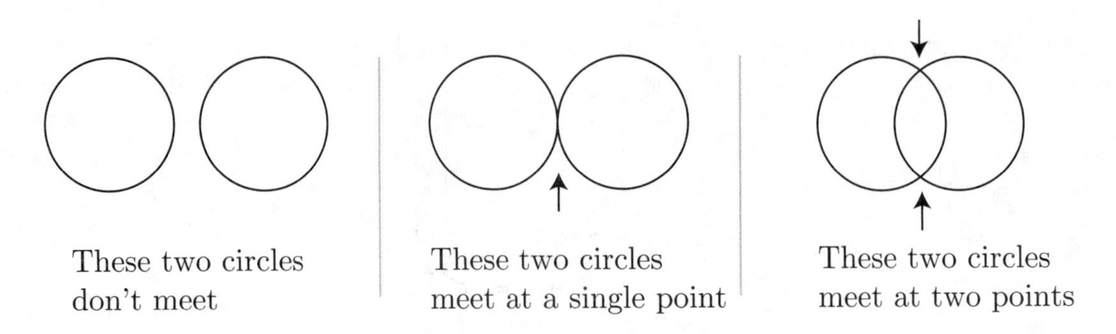

Figure 7.53: The Ways That Two Circles Can Meet

Now what if you have two spheres, how do they meet? This is difficult, but try to visualize two spheres meeting. It might help to think about soap bubbles. When you blow soap bubbles you will occasionally see a "double bubble" which is similar (although not identical) to two spheres meeting. As with circles, two spheres might not meet at all, or they might barely touch, meeting at a single point. The only other possibility is that the two spheres meet along a circle, as shown in Figure 7.54. This is a circle that is common to each of the two spheres.

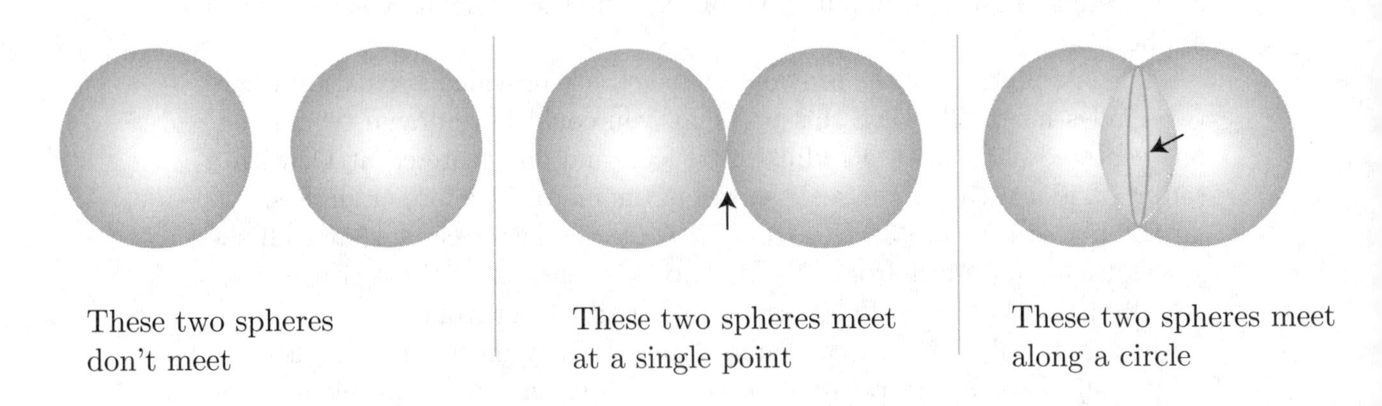

Figure 7.54: The Ways That Two Spheres Can Meet

Class Activity 7R: The Global Positioning System (GPS)

Exercises for Section 7.3 on Circles and Spheres

1. Give the (mathematical) definitions of the terms *circle* and *sphere.*

2. Points P and Q are 4 centimeters apart. Point R is 2 cm from P and 3 cm from Q. Use a ruler and a compass to help you draw a precise picture of how P, Q, and R are located relative to each other.

3. A radio beacon indicates that a certain whale is less than 1 kilometer away from boat A and less than 1 kilometer away from boat B. Boat A and boat B are 1 kilometer apart. Assuming that the whale is swimming near the surface of the water, draw a map showing all the places where the whale could be located.

4. Suppose an airplane is flying along when the pilot realizes that another airplane is 1000 feet away. Describe the shape of the set of all possible locations of the other airplane at that moment.

5. An airplane is in contact with two control towers. The airplane is 20 miles from one control tower and 30 miles from another control tower. Is this information enough to pinpoint the exact location of the airplane? Why or why not? What if the altitude of the airplane is known, then is there enough information to pinpoint the location of the airplane?

Answers to Exercises for Section 7.3

1. See the text.

2. See Figure 7.55. Start by drawing points P and Q on a piece of paper, 4 centimeters apart. Now open a compass to 2 cm, stick its point at P, and draw a circle. Since point R is 2 cm from P it must be located somewhere on that circle. Open a compass to 3 cm, stick its point at Q, and draw a circle. Since point R is 3 cm from Q it must also be located somewhere on this circle. Thus point R must be located at one of the two places where your two circles meet. Plot point R at either one of these two locations.

3. Since the whale is less than 1 km from boat A it must be *inside* a circle of radius 1 km, centered at boat A. Similarly, the whale is *inside* a a

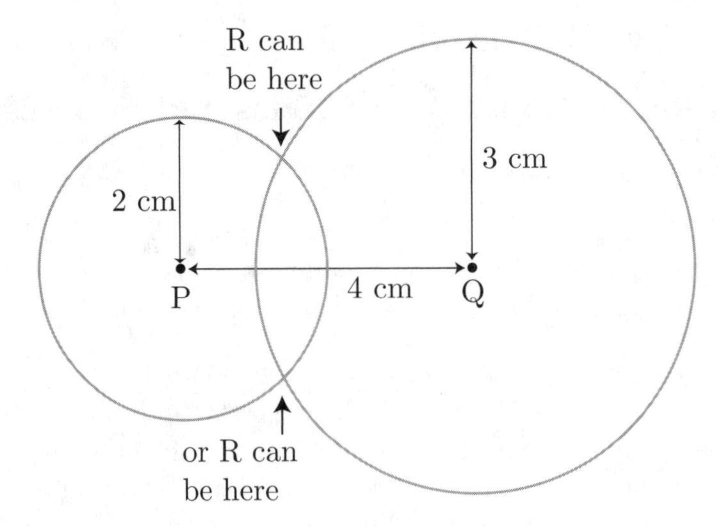

Figure 7.55: Locating Points P, Q, and R

circle of radius 1 km, centered at boat B. The places where the insides of these two circles overlap are all the possible locations of the whale, as shown in Figure 7.56.

4. The other airplane could be at any of the points that are 1000 feet away from the airplane. These points form a sphere of radius 1000 feet.

5. The locations that are 20 miles from the first control tower form a sphere of radius 20 miles, centered at the control tower (some of these locations are under ground and therefore are not plausible locations for the airplane). Similarly, the locations that are 30 miles from the second control tower form a sphere of radius 30 miles. The airplane must be located in a place that lies on *both* spheres, i.e., some place where the two spheres meet. The spheres must meet either at a single point (which is unlikely), or in a circle. So most likely, the airplane could be anywhere on a circle. Therefore we can't pinpoint the location of the airplane without more information.

   Now suppose that the altitude of the airplane is known, let's say it's $20,000$ feet. The locations in the sky that are $20,000$ feet from the ground is like a plane (actually, it's a very large sphere around the whole Earth, but locally, near the control towers, it is approximately

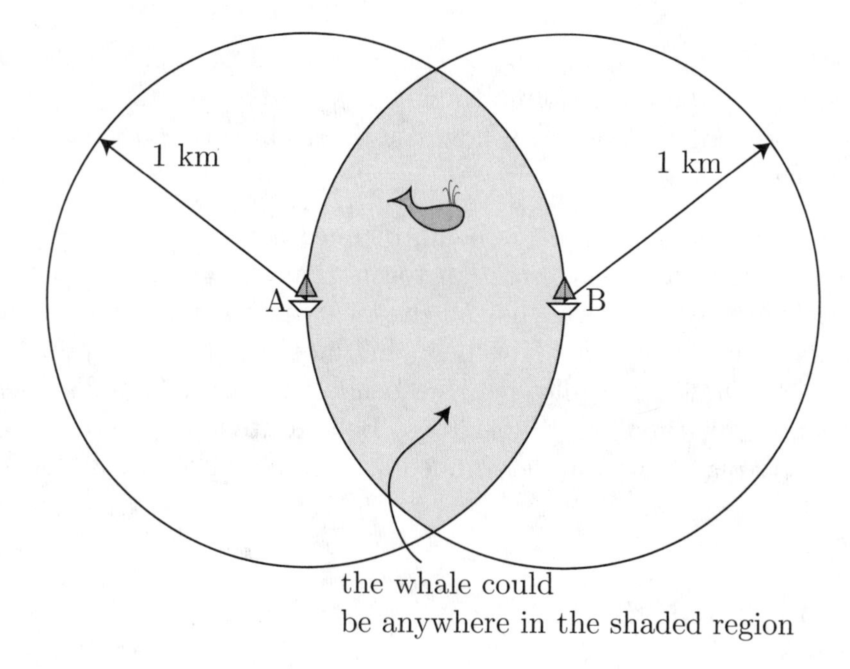

Figure 7.56: Two Boats and a Whale

a flat plane). This plane and the circle of locations where the airplane might be located either meet in a single point (unlikely) or in two points. So even with this information, we still can't pinpoint the location of the airplane, but we can narrow it down to two locations.

Problems for Section 7.3 on Circles and Spheres

1. A new mall is to be built to serve the towns of Sunnyvale and Frownietown, whose centers are 6 miles apart. The developers want to locate the mall not more than 3 miles from the center of Sunnyvale and also not more than 5 miles from the center of Frownietown. Draw a simple map showing Sunnyvale, Frownietown, and *all* potential locations for the new mall based on the given information. Be sure to show the scale of your map. Explain how you determined the possible locations for the mall.

2. A radio beacon indicates that a certain dolphin is less than 1 mile from boat A and at least $1\frac{1}{2}$ miles from boat B. Boats A and B are 2 miles

apart. Draw a simple map showing the locations of the boats and *all* the places where the dolphin might be located. Be sure to show the scale of your map. Explain how you determined all possible locations for the dolphin.

3. A new Giant Superstore is being planned somewhere in the vicinity of Kneebend and Anklescratch, towns that are 10 miles apart. The developers will only say that all the locations they are considering are more than 7 miles from Kneebend and more than 5 miles from Anklescratch. Draw a map showing Kneebend, Anklescratch, and all possible locations for the Giant Superstore. Be sure to show the scale of your map. Explain how you determined all possible locations for the Giant Superstore.

7.4 Triangles

In this section we will study some basic triangle terminology and one key fact about triangles: that the sum of the angles in a triangle is 180°.

A **triangle** is a closed shape in a plane consisting of three line segments. The term **closed** means that every endpoint of one of the line segments meets an endpoint of another line segment. Informally, we can say that a shape is closed if it has "no loose, dangling ends." Figure 7.57 shows examples of triangles and shapes that are not triangles.

Some kinds of triangles have special names. Figure 7.58 shows some examples of special kinds of triangles. A **right triangle** is a triangle that has a right angle (90°). In a right triangle, the side opposite the right angle is called the **hypotenuse**. A triangle that has three sides of the same length is called an **equilateral triangle**. A triangle that has as least two sides of the same length is called an **isosceles triangle**.

There are two especially famous facts about triangles. One is that the sum of the angles in a triangle is 180°, and the other is the Pythagorean Theorem. In this section, we will study the fact about the sum of the angles in a triangle. We will study the Pythagorean theorem in Chapter 9.

The Sum of the Angles in a Triangle

You probably already know the theorem that the angles in a triangle add to 180°. If you stop for a moment and think about this, it is really quite

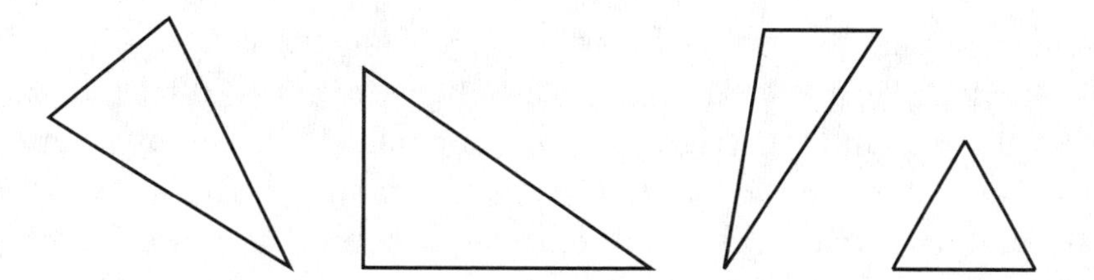

The shapes above are triangles.

The shapes below are not triangles.

This is not made out of line segments.

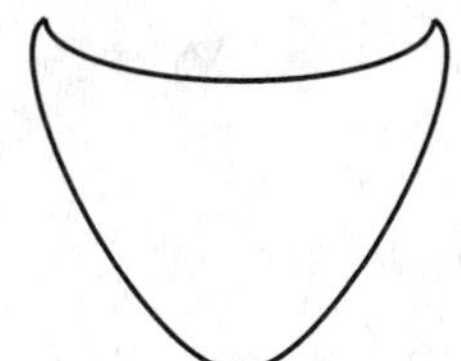

This is not closed.

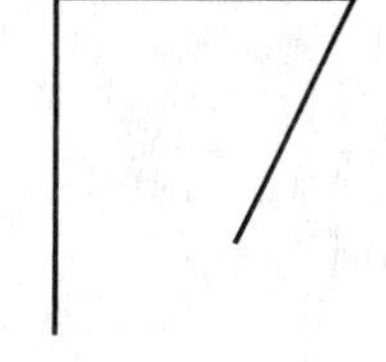

Figure 7.57: Triangles and Shapes That are Not Triangles

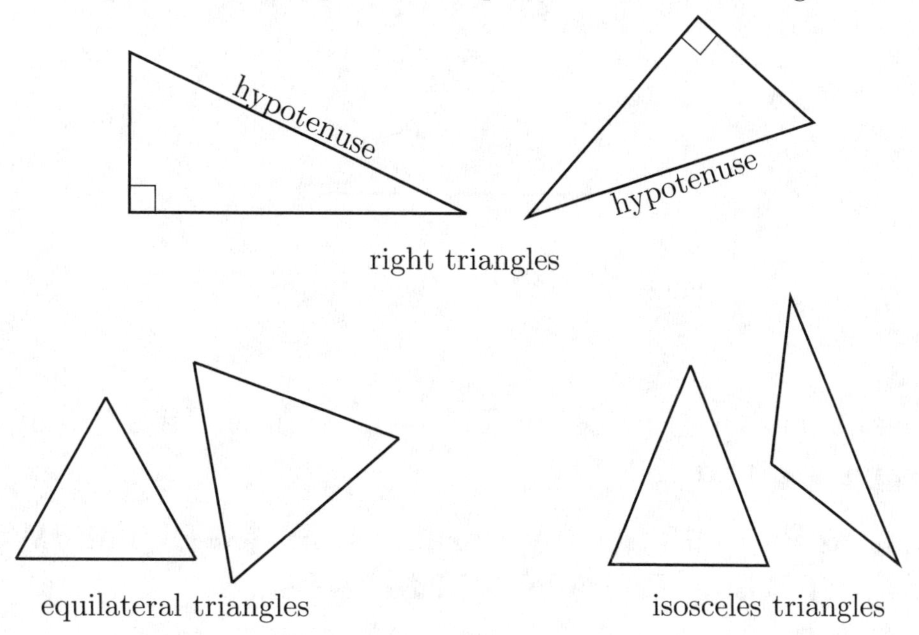

Figure 7.58: Special Kinds of Triangles

remarkable. No matter what the triangle, no matter how oddly shaped it may be—as long as it's a triangle—the angles *always* add to 180°. This theorem tells us about *all* triangles, and not only that, but we can explain why this theorem is true *without checking every single triangle individually.* Many people are attracted to mathematics because the powerful reasoning of mathematics enables us to deduce general truths.

What is a theorem? A **theorem** is a mathematical statement that has been proven to be true by using logical reasoning, based on previously known (or assumed) facts. The statement that the angles in a triangle add to 180° is a theorem.

Class Activities 7S, 5CA:SumAngTri2, and 7U will give you several ways to think about why the angles in a triangle always add to 180°. Here is how we commonly use this fact: If you know two of the angles in a triangle, you can figure out what the third angle must be. For example, in Figure 7.59 the triangle was constructed to have an angle of 40° and an angle of 70°. These two angles already add to 110°, so the third angle must be 70° so that all three will add to 180°.

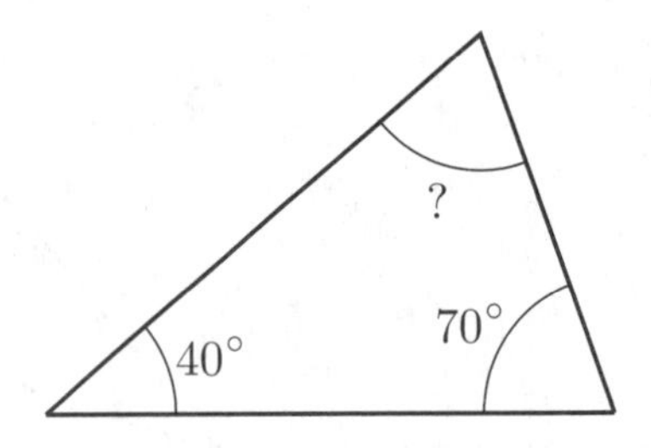

Figure 7.59: What is the Third Angle?

Class Activity 7S: Seeing That the Angles in a Triangle Always Add to 180°

Class Activity 7T: Using Parallel Lines to Explain Why the Angles in a Triangle Add to 180°

Class Activity 7U: Explaining Why the Angles in a Triangle Add to 180° by Walking and Turning

Class Activity 7V: Do the Angles of a Shape Inside Another Shape Add to Less?

Exercises for Section 7.4 on Triangles

1. For each of the triangles in Figure 7.60, determine the unknown angle.

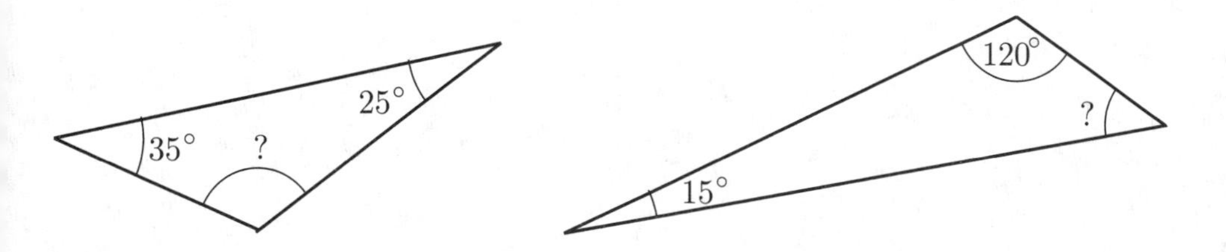

Figure 7.60: Find the Missing Angles

2. Use the "walking and turning" method of Class Activity 7U to explain why the angles in a triangle add to 180°.

3. Figure 7.61 shows a square inscribed in a triangle. Since the square is *inside* the triangle, how can it be that the angles in the square add up to more degrees (360°) than the angles in the triangle (which only add to 180°)?

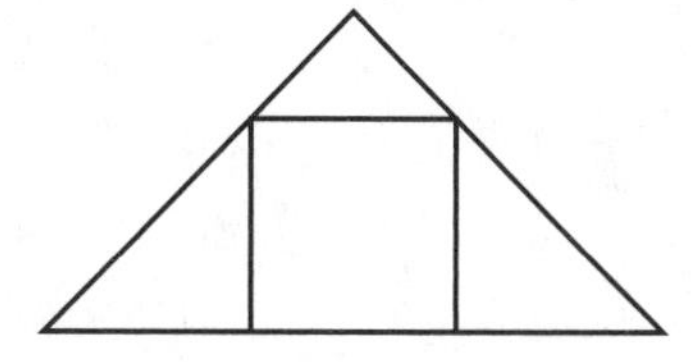

Figure 7.61: A Square in a Triangle

Answers to Exercises for Section 7.4

1. Left triangle: 120°. Right triangle: 45°.

2. As seen in Class Activity 7U, if a person were to walk around a triangle, returning to their original position, they would have turned a total of 360°. This means that $d+e+f=360°$, where d, e, f are the exterior angles, as shown in Figure 7.62. But since

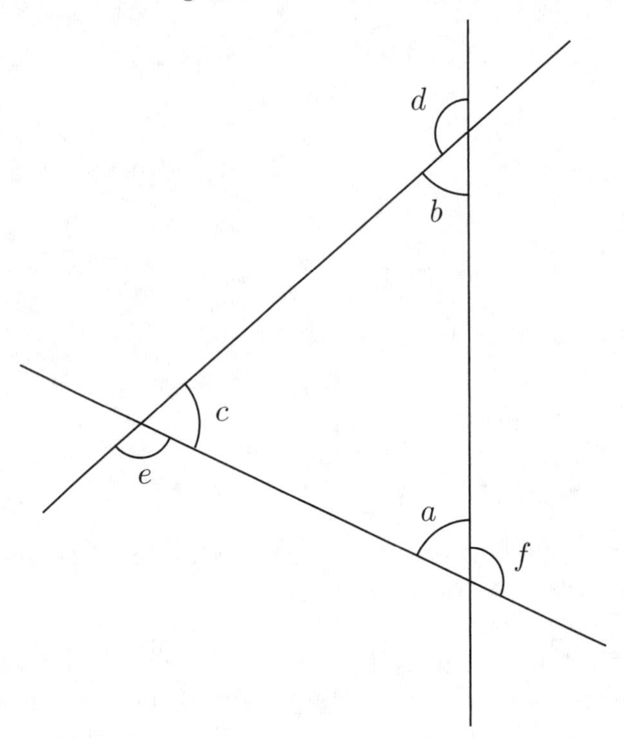

Figure 7.62: The Exterior Angles of a Triangle Add to 360°

$$\begin{aligned} a+f &= 180°, \\ b+d &= 180°, \text{ and} \\ c+e &= 180°, \end{aligned}$$

therefore, on the one hand,

$$\begin{aligned} (a+f)+(b+d)+(c+e) &= 180°+180°+180° \\ &= 540° \end{aligned}$$

while on the other hand,

$$\begin{aligned} (a+f)+(b+d)+(c+e) &= (a+b+c)+(d+e+f) \\ &= (a+b+c)+360°. \end{aligned}$$

Since $(a+f)+(b+d)+(c+e)$ is equal to both $540°$ and $(a+b+c)+360°$, therefore these last two expressions are equal to each other, i.e.,

$$(a + b + c) + 360° = 540°.$$

Therefore

$$a + b + c = 540° - 360° = 180°.$$

3. Unlike area, for example, the angles don't depend on the size of the square or the triangle. When we add the angles in the triangle, or the angles in the square, we are adding up amounts of turning at the corners. The amount of turning at the corners of the square is not related to the amount of turning at the corners of the triangle.

Problems for Section 7.4 on Triangles

1. Figure 7.63 shows a large triangle subdivided into three smaller triangles. If the angles in each of the three smaller triangles add to $180°$, then why don't the angles in the larger triangle add to

$$180° + 180° + 180° = 540°$$

instead of $180°$? Explain carefully.

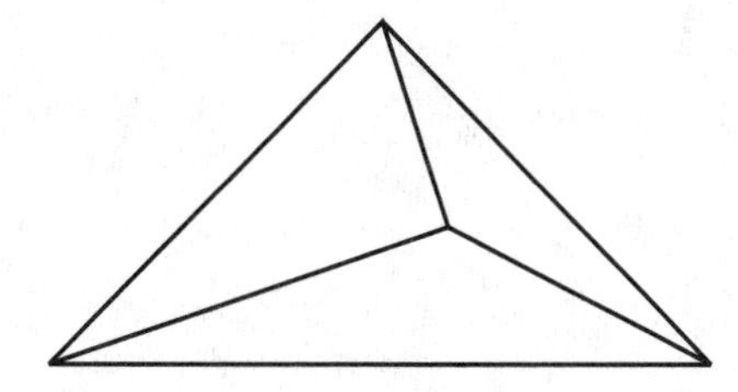

Figure 7.63: A Triangle Subdivided into Three Smaller Triangles

7.5 Quadrilaterals and Other Polygons

In this section we will investigate properties of some special kinds of quadrilaterals and we will see how various kinds of quadrilaterals are related. We'll also see that triangles and quadrilaterals are certain kinds of polygons. At the end of this section we will study Venn diagrams, which are a nice pictorial way of showing relationships between groups of things. Venn diagrams can be used to show how various kinds of shapes are related to each other.

A **quadrilateral** is a closed shape in a plane consisting of four line segments that do not cross each other. As before, the term **closed** means that every endpoint of one of the line segments meets an endpoint of another line segment. Figure 7.64 shows examples of quadrilaterals and shapes that are not quadrilaterals. The name *quadrilateral* makes sense because it means *four sided* (quad = four, lateral = side).

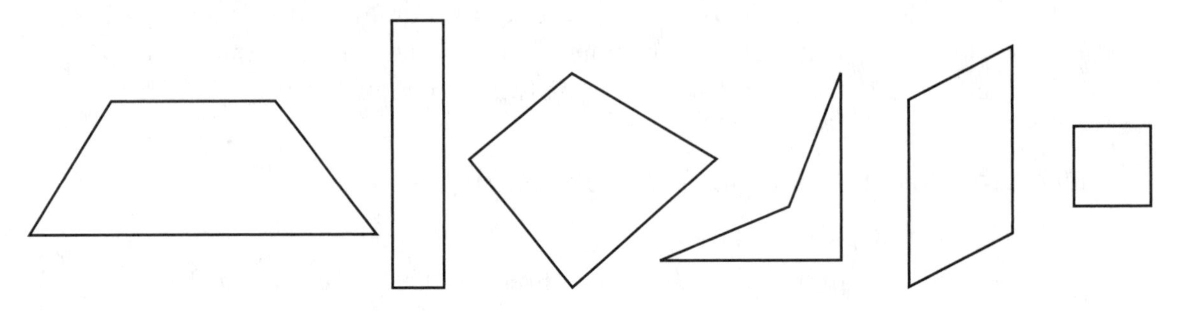

The shapes above are quadrilaterals.

The shapes below are not quadrilaterals.

This is not made out of line segments. This is not closed. This has sides that cross.

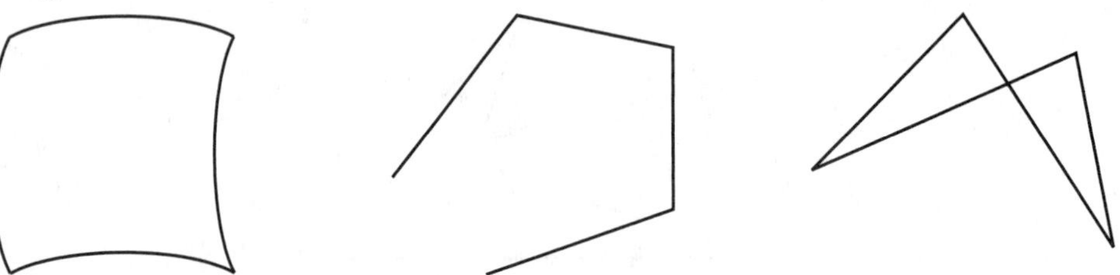

Figure 7.64: Quadrilaterals and Shapes That are not Quadrilaterals

Triangles and quadrilaterals are kinds of polygons. A **polygon** is a closed shape in a plane consisting of line segments that do not cross each other. Triangles are polygons made of 3 line segments, quadrilaterals are polygons made of 4 line segments, **pentagons** are polygons made of 5 line segments, **hexagons** are polygons made of 6 line segments, **octagons** are polygons made of 8 line segments. Figure 7.65 shows some examples. A polygon with n sides can be called an "n-gon." For example, a 13-sided polygon is a 13-gon.

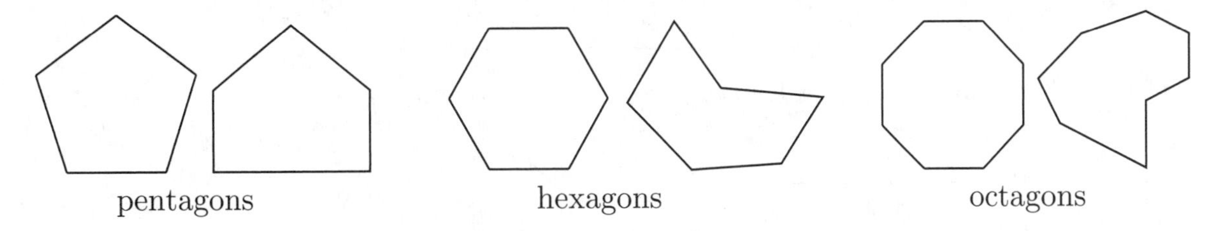

Figure 7.65: Pentagons, Hexagons, and Octagons

The name *polygon* makes sense because it means *many sided* (poly = many, gon = side). Similarly for the names *pentagon*, *hexagon*, and so on, since penta means 5 and hexa means 6, and so on. It would be perfectly reasonable to call triangles *trigons* and quadrilaterals *quadrigons*, although these terms are not conventionally used.

A polygon is called **regular** if all sides have the same length and all angles are equal. In Figure 7.65 the left-most pentagon, hexagon, and octagon are all regular polygons, whereas the pentagon, hexagon, and octagon on the right are not regular polygons.

We will now study some special kinds of quadrilaterals. The most familiar special quadrilaterals are squares and rectangles. Below is a list of definitions of some special quadrilaterals. See Figure 7.66 for examples.

square A quadrilateral with four right angles whose sides all have the same length.

rectangle A quadrilateral with four right angles.

rhombus A quadrilateral whose sides all have the same length. The name **diamond** is sometimes used instead of rhombus.

parallelogram A quadrilateral for which opposite sides are parallel.

trapezoid A quadrilateral for which at least one pair of opposite sides are parallel. (Some books define a trapezoid as a quadrilateral for which *exactly one* pair of opposite sides are parallel.)

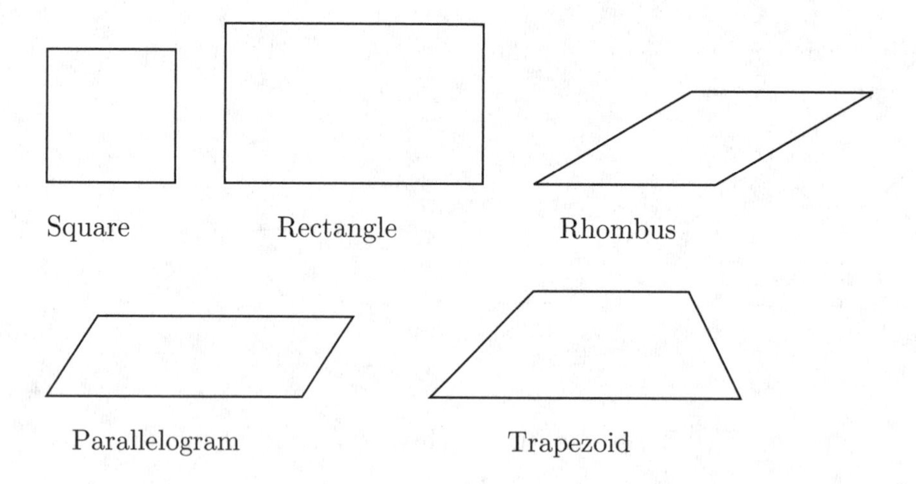

Figure 7.66: Some Special Quadrilaterials

Some of the following activities explore special properties of various kinds of quadrilaterals. Several investigate diagonals of quadrilaterals. The **diagonals** of a quadrilateral are line segments connecting opposite corner points, as shown in Figure 7.67.

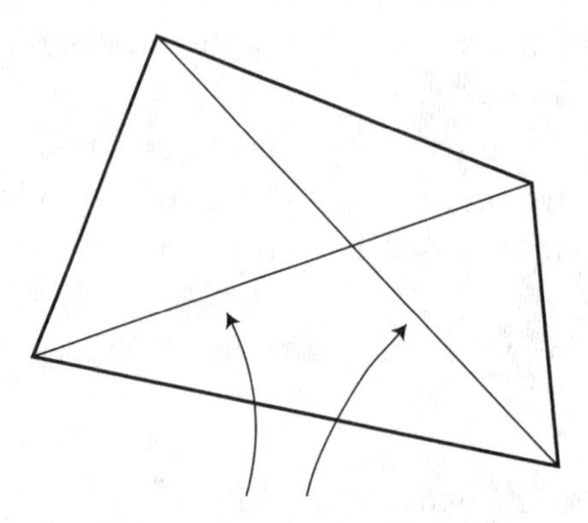

Figure 7.67: The Diagonals of a Quadrilateral

Class Activity 7W: Constructing Quadrilaterals with Geometer's Sketchpad

Class Activity 7X: Relating the Kinds of Quadrilaterals

Showing Relationships with Venn Diagrams

There is a nice way to use pictures called *Venn diagrams* to show relationships between collections of items. A **Venn diagram** is a picture that shows how certain sets are related. A **set** is a collection of objects.

Figure 7.68 shows a Venn diagram relating the set of mammals and the set of animals with four legs. The overlapping region represents mammals that have four legs because these animals fit in both categories: animals with four legs and mammals.

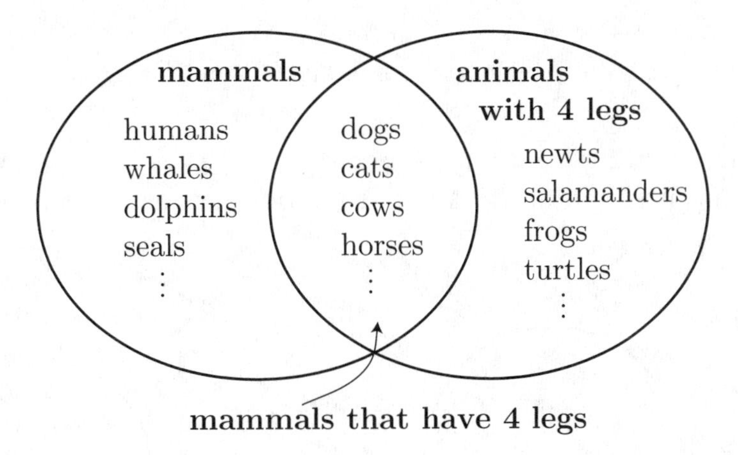

Figure 7.68: A Venn Diagram

Venn diagrams are not just used in mathematics. My son's third grade class used Venn diagrams in language arts. The children drew Venn diagrams comparing themselves to a character in a story. Figure 7.69 shows such an example. One set consists of the story character's attributes, the other set consists of Joey's attributes, and the overlap consists of attributes common to both the story character and to Joey.

A Venn diagram of two sets doesn't have to overlap the way the previous two examples did. Figure 7.70 shows two other kinds of Venn diagrams. The first shows a Venn diagram relating the set of boys and the set of girls. Of course, these sets do not overlap. The other Venn diagram in Figure 7.70 shows that the set of whole numbers is contained within the set of rational numbers. This is so because every whole number can also be expressed as a

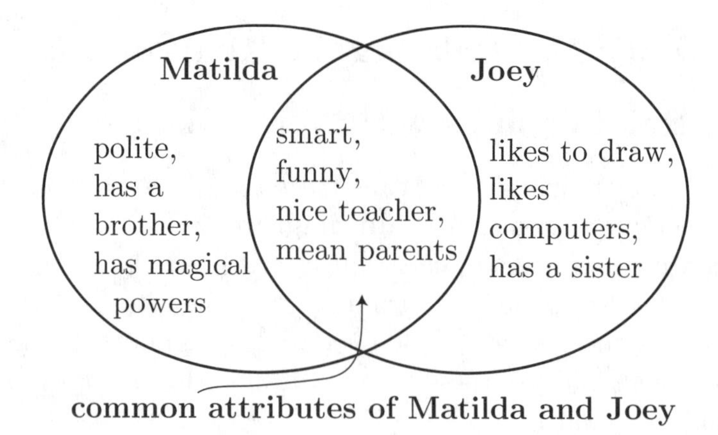

Figure 7.69: A Language Arts Venn Diagram

fraction (by "putting it over 1"). (Figure 2.6 on page 31 of volume I shows a more extensive Venn diagram relating various systems of numbers.)

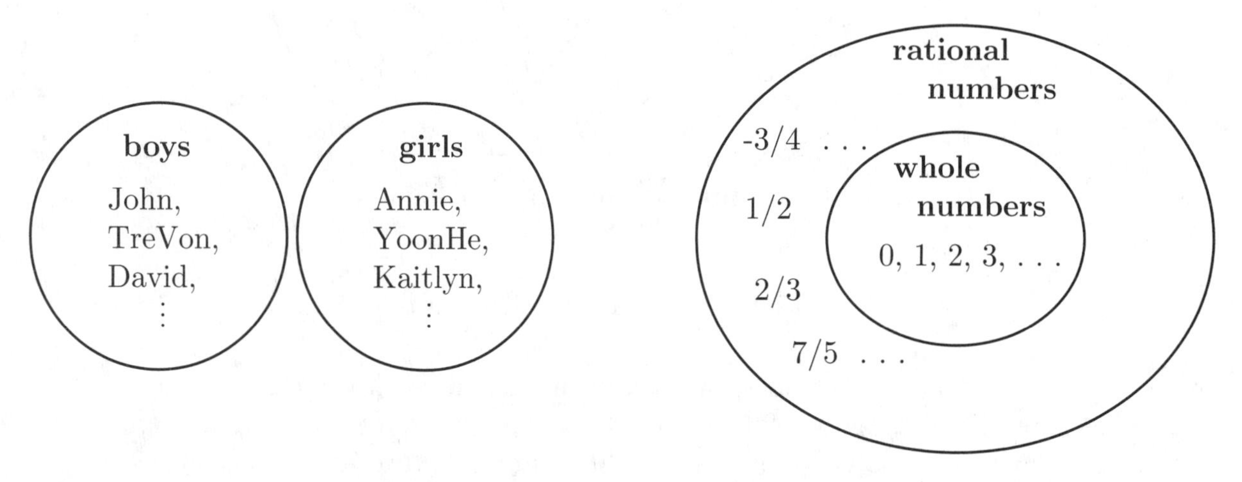

Figure 7.70: Other Kinds of Venn Diagrams

When three or more sets are involved, Venn diagrams can become quite complex. In general, there can be double overlaps, triple overlaps, or more (if more than three sets are involved). For example, Figure 7.71 shows a Venn diagram relating the set of warm blooded animals, the set of animals that lay eggs, and the set of carnivorous animals. There are three double overlaps and one triple overlap.

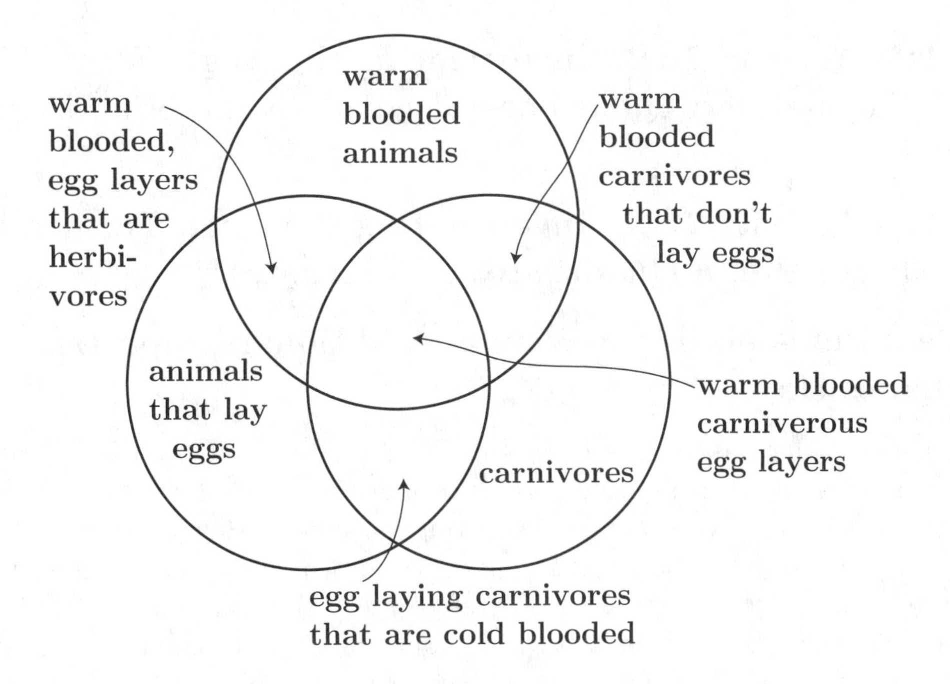

Figure 7.71: A Venn Diagram of Three Sets

Venn diagrams can be used to show how the various kinds of special quadrilaterals are related.

Class Activity 7Y: Venn Diagrams Relating Quadrilaterals

The following class activities will help you discover that some kinds of quadrilaterals have special properties that other quadrilaterals don't always have.

Class Activity 7Z: Investigating Diagonals of Quadrilaterals with Geometer's Sketchpad

Class Activity 7AA: Investigating Diagonals of Quadrilaterals (alternate)

Class Activity 7BB: Investigating the Angle Formed by a Diagonal of a Quadrilateral with Geometer's Sketchpad

Class Activity 7CC: Investigating the Angle Formed by a Diagonal of a Quadrilateral (alternate)

Definitions of Shapes Versus Additional Properties the Shapes Have

If you did the class activities in this section, then you probably noticed that shapes often have special properties that are not readily apparent from their definitions. For example, when you construct a rhombus and look at it carefully, you will notice that opposite sides of the rhombus appear to be parallel (in fact, they are). Therefore every rhombus should also be a parallelogram. But now look back at the definition of rhombus: a rhombus is a quadrilateral whose four sides all have the same length. There is nothing in this *definition* that tells us that opposite sides of a rhombus should be parallel. We need some kind of additional information in order to *deduce* that every rhombus is in fact also a parallelogram.

Here is another example. If you did activities investigating diagonals of quadrilaterals, then you may have noticed that the following seem to be true for every rhombus (in fact, they are true):

- the diagonals meet at a point that is half way across each diagonal,
- the diagonals are perpendicular,
- the diagonals have different lengths unless the rhombus is a square,
- a diagonal cuts an angle of a rhombus in half.

You may also have noticed that the following seem to be true for every rectangle (in fact, they are true):

- the diagonals meet at a point that is half way across each diagonal,
- the diagonals are only perpendicular when the rectangle is a square,
- the diagonals always have the same length.

Once again, these are examples of properties of shapes that we can't tell are true just from the definitions of these shapes—in fact, the definitions of rhombus and rectangle don't say anything at all about diagonals.

This leads straight to the heart of what theoretical mathematics is all about: the theoretical study of mathematics is about *starting with some assumptions and some definitions of objects and concepts, discovering additional properties that these objects or concepts must have, and then reasoning logically to deduce that the objects or concepts do indeed have those properties.* You probably already carried out such a logical deduction, for example, if you proved the Pythagorean Theorem. The *definition* of right triangle does not tell us that the square of the hypotenuse is equal to the sum of the squares of the lengths of the other two sides, but logical reasoning allows us to *deduce* this theorem. In the examples above, we took the first step of *discovering* additional properties that rhombuses and rectangles must have, beyond those properties given in the definition. In these cases, we won't carry out the next step of deducing logically that rhombuses and rectangles really do have these additional properties, but *it is possible to do so.* In other words, it is possible to go beyond observing that rhombuses and rectangles always seem to have certain properties, to explain why these shapes must always have those properties.

Exercises for Section 7.5 on Quadrilaterals

1. Define the following terms: quadrilateral, polygon, pentagon, hexagon, octogon, 13-gon, square, rectangle, rhombus, parallelogram, trapezoid.

2. Draw a Venn diagram showing how the set of rectangles and the set of rhombuses are related. Is there an overlap? If so, what does it consist of? Explain.

3. Draw a Venn diagram showing how the set of parallelograms and the set of trapezoids are related. Explain.

4. Some books define trapezoids as quadrilaterals that have *exactly one* pair of parallel sides. Draw a Venn diagram showing how the set of parallelograms and the set of trapezoids are related when this alternate definition of trapezoid is used. Explain.

5. Draw a Venn diagram showing how the set of rhombuses and the set of parallelograms are related. In doing this, can you rely only on the information given in the the definitions of rhombus and parallelogram, or do you need to rely on some additional information?

6. What special properties do diagonals of rhombuses have that diagonals of other quadrilaterals do not necessarily have?

7. What special properties do diagonals of rectangles have that diagonals of other quadrilaterals do not necessarily have?

Answers to Exercises for Section 7.5

1. See text.

2. See Figure 7.72. The shapes in the overlap are those shapes that have four right angles (as rectangles do) and have four sides of the same length (as rhombuses do). According to the definition, those are exactly the shapes that are squares.

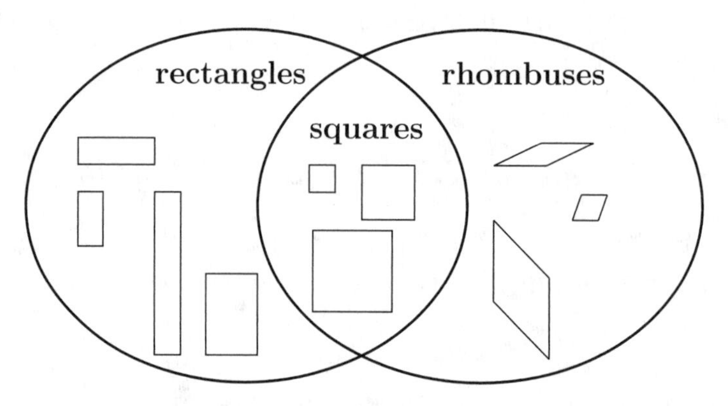

Figure 7.72: Venn Diagram of Rectangles and Rhombuses

3. See Figure 7.73. According to the definitions, every parallelogram is also a tapezoid because parallelograms have two pairs of parallel sides, so they can be said to have at least one pair of parallel sides.

4. See Figure 7.74. According to the alternate definition, no parallelogram is a trapezoid because parallelograms have two pairs of parallel sides, so

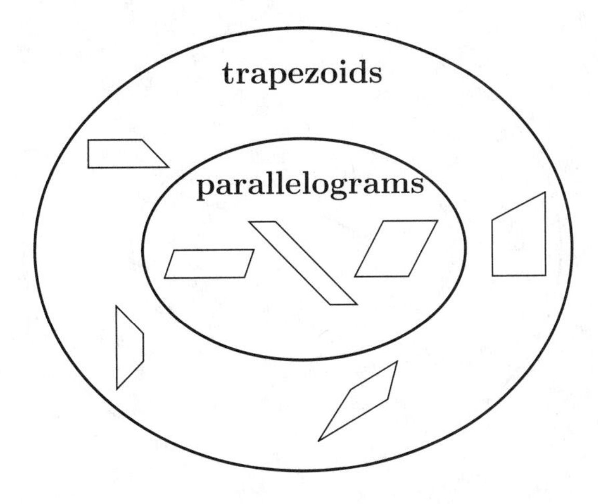

Figure 7.73: Venn Diagram of Parallelograms and Trapezoids

they don't have exactly one pair of parallel sides. Therefore, with this alternate definition, the set of parallelograms and the set of trapezoids do not have any overlap.

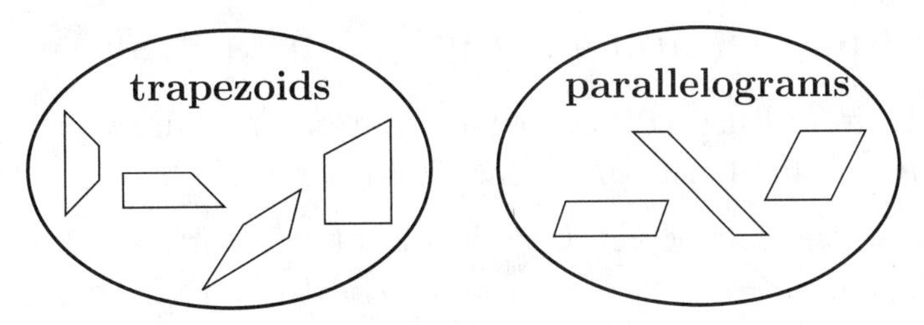

Figure 7.74: Venn Diagram of Parallelograms and Trapezoids Using Alternate Definition

5. See Figure 7.75. Looking at rhombuses, it appears that opposite sides are parallel and therefore that every rhombus is also a parallelogram. Since the definition of rhombus does not say anything about opposite sides being parallel, some additional information would be needed in order to explain why rhombuses really are parallelograms.

6. The diagonals of a rhombus are perpendicular. The two diagonals of a rhombus meet at a point that is half way across each diagonal. A diagonal of a rhombus cuts both angles it passes through in half.

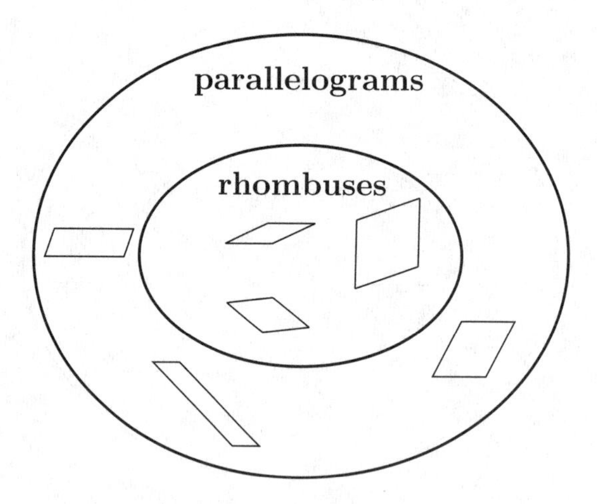

Figure 7.75: Venn Diagram of Rhombuses and Parallelograms

7. The two diagonals of a rectangle have the same length and meet at a point that is half way across each diagonal.

Problems for Section 7.5 on Quadrilaterals

1. (a) Draw at least 3 different quadrilaterals and in each case, show how to subdivide the quadrilateral into 2 triangles.
   (b) Use the fact that every quadrilateral can be subdivided into 2 triangles to determine the sum of the angles in a quadrilateral. Explain your reasoning clearly.
   (c) Draw at least 3 different pentagons and in each case, show how to subdivide the pentagon into 3 triangles.
   (d) Use the fact that every pentagon can be subdivided into 3 triangles to determine the sum of the angles in a pentagon. Explain your reasoning clearly.
   (e) Based on your work above, what should the sum of the angles in a hexagon be? In general, what should the sum of the angles in an n-gon be? Explain briefly.

2. (a) Use the reasoning of Class Activity 7U and the answer to exercise 2 on page 56 for triangles to determine the sum of the exterior angles of the quadrilateral in Figure 7.76. In other words, determine

$e + f + g + h$. Measure with a protractor to check that your formula is correct for this quadrilateral.

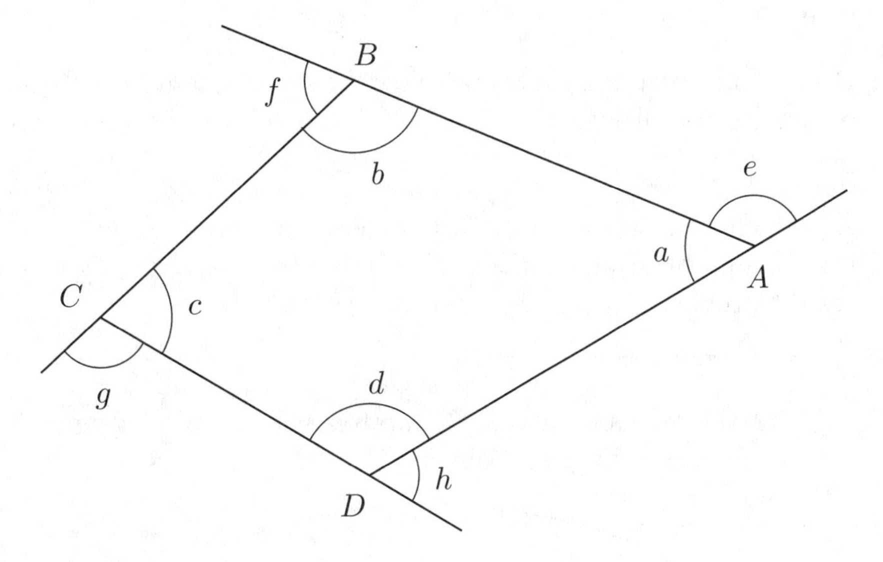

Figure 7.76: A Quadrilateral

(b) Will there be a similar formula for the sum of the exterior angles of pentagons, hexagons, 7-gons (a.k.a. heptagons), octogons, etc.? Explain briefly!

(c) Using your formula for the sum of the exterior angles of a quadrilateral, deduce the sum of the interior angles of the quadrilateral, in other words find $a + b + c + d$, as pictured in Figure 7.76. Explain your reasoning. Measure with a protractor to verify that your formula is correct for this quadrilateral.

(d) Based on the class activity and your work above, what formula would you expect to be true for the sum of the interior angles of a pentagon? What about for hexagons? What about for a polygon with n sides?

3. Draw a Venn diagram showing how the set of rectangles and the set of parallelograms are related. In doing this, can you rely only on the information given in the *definitions* of parallelogram and rectangle, or do you need to rely on some additional information? Give a careful, thorough explanation for why you can draw the Venn diagram the way

you do. Hint: Show that if opposite sides of a rectangle were not parallel they would be part of a triangle whose angles would add to more than 180°.

4. Draw a Venn diagram that shows how the sets of quadrilaterals, squares, rectangles, parallelograms, rhombuses, and trapezoids are related. Explain.

5. A problem that was given to 5th graders is shown in Figure 7.77. Criticize the problem on *mathematical grounds* and rewrite it to make a correct problem.

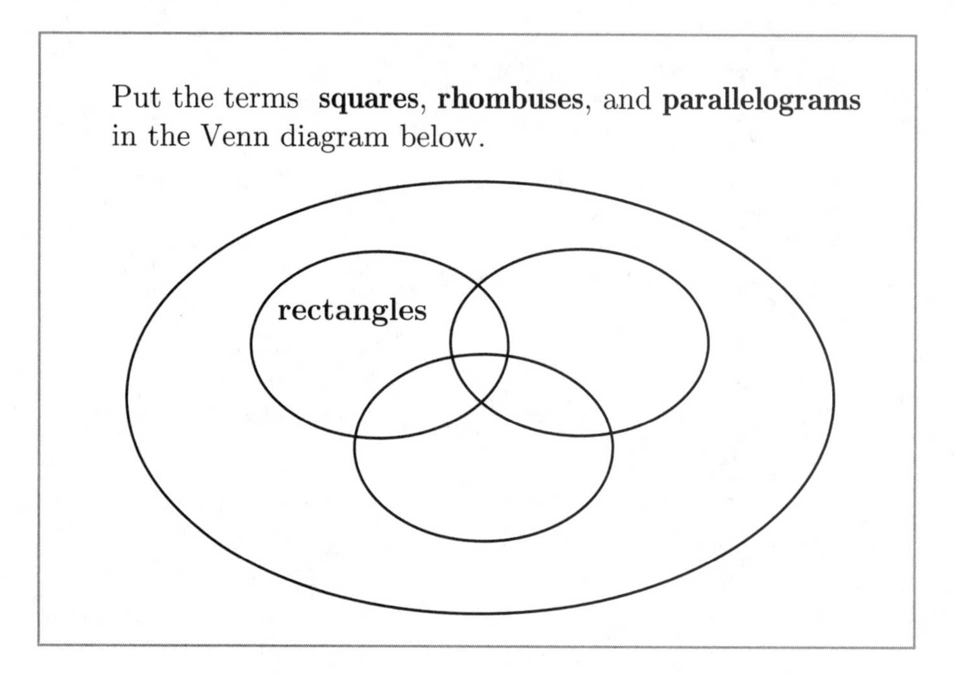

Figure 7.77: A Problem Given to 5th Graders

6. Before builders lay the foundation for a rectangular house, they usually pound stakes into the ground where the 4 corners of the house will be. The builders then measure the diagonals of the quadrilateral formed by the four stakes. Suppose that these diagonals turn out to have different lengths. In this case, based on the material in this section about diagonals of quadrilaterals, what can you conclude about the

angles at the corners? Explain clearly. It is important for the corners of the foundation to be right angles, so that the walls above will be structurally stable. Should the builders be satisfied with the proposed foundation, or should they make adjustments to it?

7.6 Constructions With Straightedge and Compass

In this section we will explore some fun and easy constructions that can be done with a compass and a straightedge. We will see that these constructions work because of special properties of rhombuses. A **straightedge** is just a "straight edge," namely a ruler, except that it need not have any markings.

Constructions with straightedge and compass have their origins in the desire to make precise drawings using only simple, reliable tools. Think back to the times before there were computers. If you were studying shapes in a plane and if you wanted to discover additional properties these shapes have beyond those properties given in the definitions, then you would need some way to make precise drawings. A sloppy drawing could hide interesting features, or worse, could seem to show features that aren't really there. Nowadays we have computer technology that can help us make accurate drawings easily, but we can still benefit from learning some of the old "hands on" techniques.

Dividing a Line Segment in Half and Constructing a Perpendicular Line

Starting with any line segment, here is a procedure to construct a line that is perpendicular to the given line segment *and* that passes through the point half way across the original line segment. The point half way across a line segment is often called a **midpoint** of the line segment. So this construction kills two birds with one stone: it constructs a perpendicular line and it also finds a midpoint of a line segment.

1. Starting with a line segment AB, with endpoints A and B, as shown in Figure 7.78, open a compass to at least half the length of the line segment (other than that, it doesn't matter how wide you open it).

2. Stick the point of the compass at point A and draw a circle. You really only need to draw part of the circle, as shown in Figure 7.79.

Figure 7.78: Line Segment AB

3. Without changing how wide the compass is opened, stick the point of the compass at point B and draw (part of) a circle, as shown in Figure 7.80.

4. Draw a line through the two points where the two circles you have just drawn meet, as shown in Figure 7.81. This new line is perpendicular to the original line segment AB, and the point where this perpendicular line meets the line segment AB is the midpoint of AB.

Dividing an Angle in Half

Starting with two rays (or line segments) that meet at a point P, here is a construction to divide the angle formed by the rays in half. In other words, this construction produces another ray, also starting at P, that is half way between the two original rays.

1. Starting with two rays that meet at a point P, as shown in Figure 7.82, open a compass to any width.

2. Stick the point of the compass at the point P, and draw a circle as shown in Figure 7.83 (you actually only need a part of a circle). Let Q and R be the names of the two points where the circle meets the two rays.

3. Keep the compass opened to the same width. The remainder of the construction is identical to construction of a perpendicular line through the midpoint of the line segment connecting Q and R. Stick the point of the compass at Q and draw (part of) a circle, as shown in Figure 7.84.

4. Without changing how wide the compass is opened, stick the point of the compass at R and draw (part of) a circle, as shown in Figure 7.85.

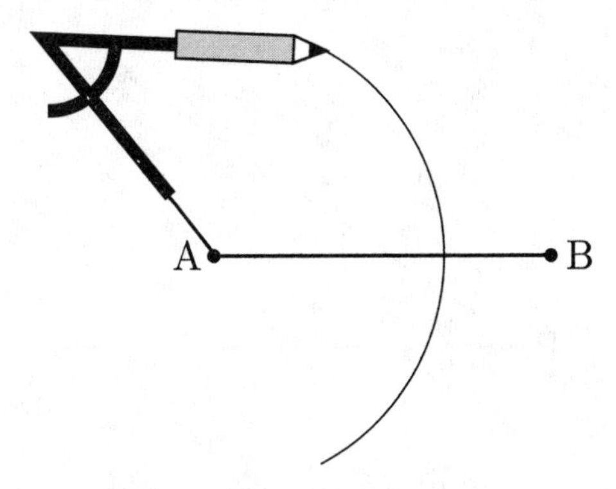

Figure 7.79: Draw a Circle with Center A

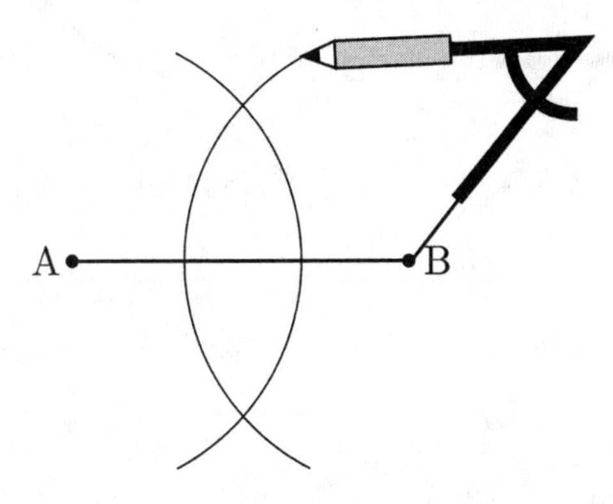

Figure 7.80: Draw a Circle with Center B

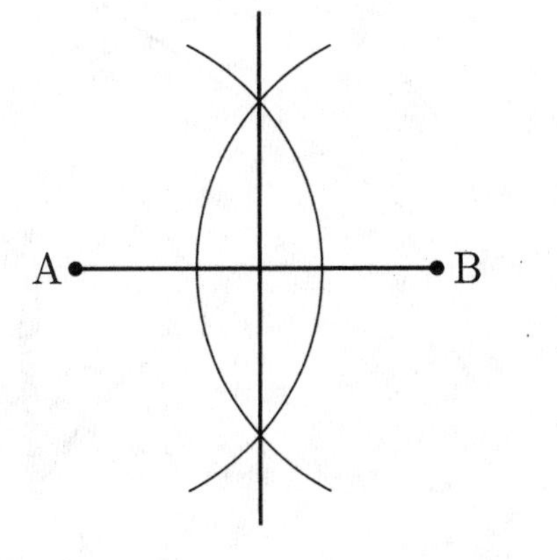

Figure 7.81: Draw a Line Through the Points Where the Circles Meet

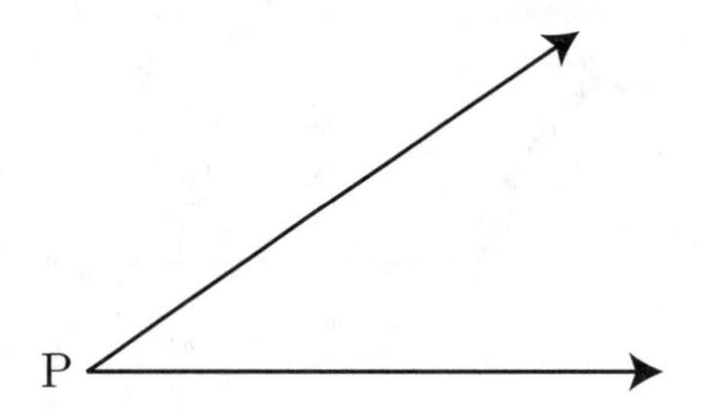

Figure 7.82: An Angle

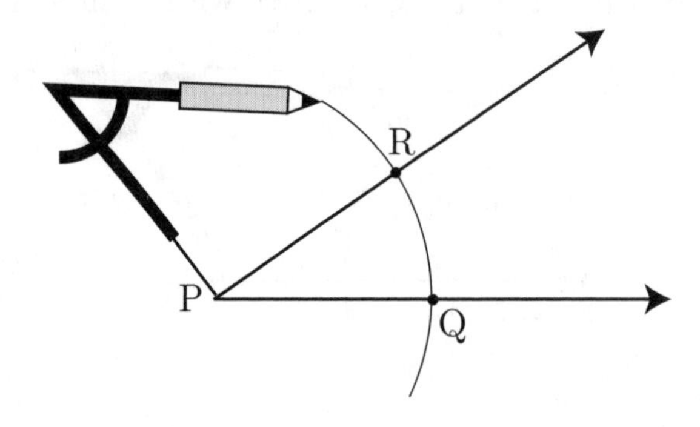

Figure 7.83: Draw a Circle Centered at P

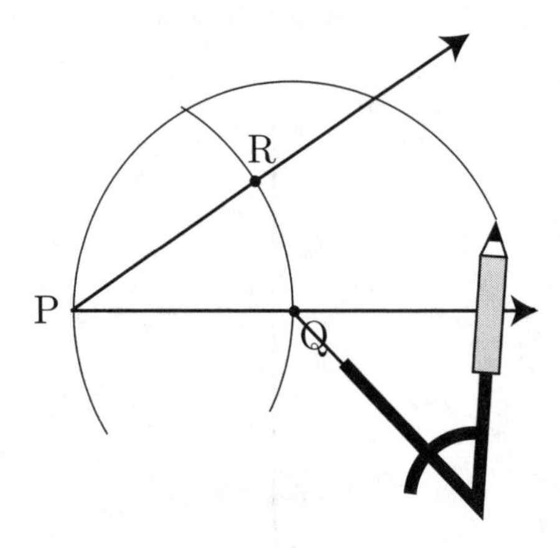

Figure 7.84: Draw a Circle Centered at Q

5. Draw a line through P and the other point where the circles meet, as shown in Figure 7.86.

Class Activity 7DD: Constructing a Square and an Octagon with Straightedge and Compass

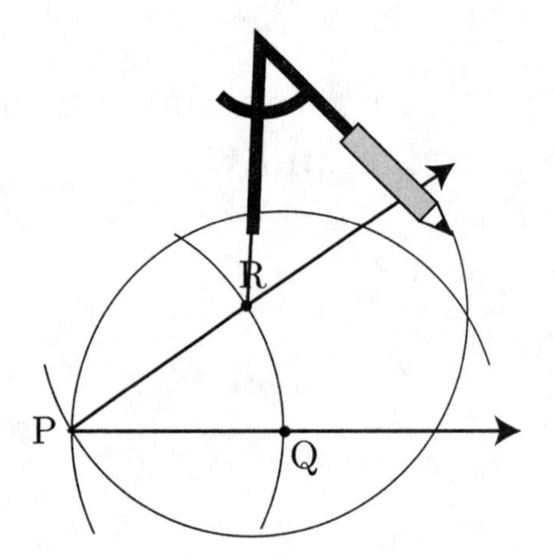

Figure 7.85: Draw a Circle Centered at R

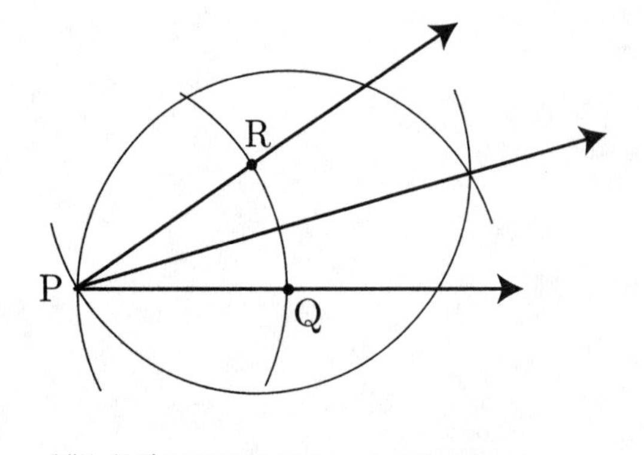

Figure 7.86: Draw a Line

Class Activity 7EE: Relating the Constructions to Properties of Rhombuses

Exercises for Section 7.6 on Constructions

1. Draw several line segments. Use a straightedge and compass to construct lines that are perpendicular to your line segments and divide your line segments in half.

2. Draw several angles. Use a straightedge and compass to divide your angles in half.

3. Starting with any line segment, use a straightedge and compass to construct an equilateral triangle (a triangle for which all three sides have the same length) which has your line segment as one of its sides. It will help you to remember what the definition of a circle is.

4. Use facts about rhombuses from Section 7.5 (see page 64) to explain why the construction described on page 71 really does construct a line that is perpendicular to a given line segment and that passes through the midpoint of the given line segment. If you say that some shape is a rhombus, be sure to explain why it really is a rhombus.

Answers to Exercises for Section 7.6

3. Draw your starting line segment. Let's call the endpoints A and B. Figure 7.87 shows the following construction. Open your compass to the width of your line segment, put the point at A and draw a circle (or part of a circle). Do the same with the point of the compas at B. Pick either one of the two points where the two circles meet, let's call the chosen point C. Draw line segments connecting A to C and B to C. These three segments form an equilateral triangle because of the way it was constructed with circles. The first circle consists of all points that are the same distance from A as B is; since C is on this circle, the distance from C to A is the same as the distance from B to A. The second circle consists of all points that are the same distance from B as A is; since C is also on this circle, the distance from C to B is the same as the distance from A to B.

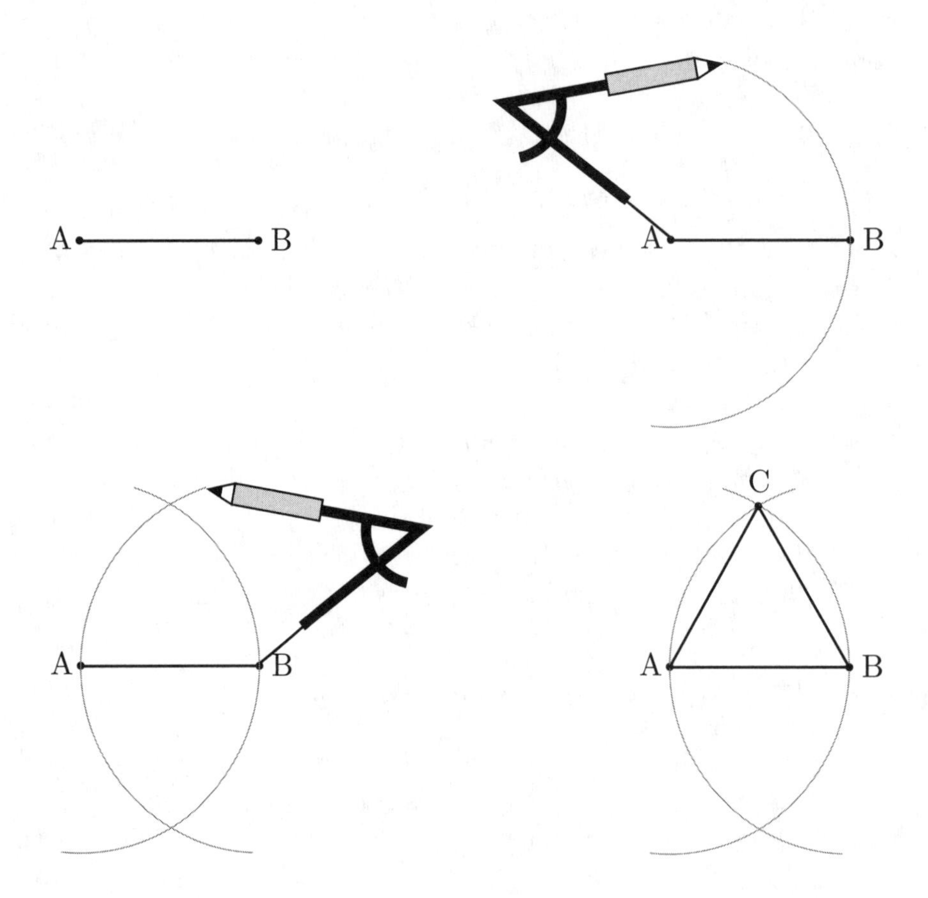

Figure 7.87: Constructing an Equilateral Triangle

4. Starting with the finished construction in Figure 7.81, let C and D be the points in the construction where the two circles meet. Form the four line segments AC, BC, AD, and BD, as shown in Figure 7.88. Notice that AC and AD are radii of the circle centered at A, therefore AC and AD have the same length. Similarly, BC and BD have the same length because they are radii of the circle centered at B. But the circles centered at A and at B have the same radius because they were constructed that way (without changing the width of the compass). Therefore all four line segments AC, BC, AD, and BD have the same length. By definition, this means that the quadrilateral ACBD is a rhombus. According to the list of properties of rhombuses on page 64, the diagonals of a rhombus meet at a point that is half way across each diagonal, and the diagonals of a rhombus are perpendicular. This explains why the construction described on page 71 really does construct a line that is perpendicular to the line segment AB and that passes through the midpoint of AB.

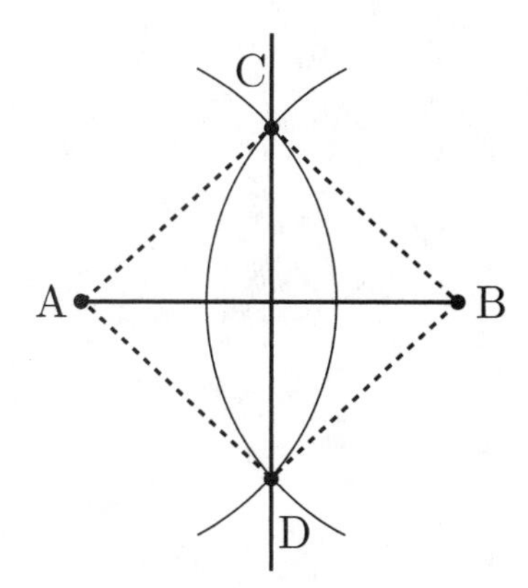

Figure 7.88: The Construction for a Perpendicular Line Forms a Rhombus

Problems for Section 7.6 on Constructions

1. Describe how to use a compass to construct the pattern of three circles shown in Figure 7.89 so that the triangle shown inside the circles is an equilateral triangle. Then explain why the triangle really is an equilateral triangle.

Figure 7.89: A Pattern of Three Circles

2. Use a straightedge and compass to carefully construct a square.

3. Use a straightedge and compass to carefully construct a regular hexagon. Explain why your method of construction guarantees that all six sides of your hexagon have the same length. It may help you to examine Figure 7.90.

Figure 7.90: A Pattern of Circles

4. Use a straightedge and compass to carefully construct a regular 12-gon. Explain why your method of construction guarantees that all twelve sides of your 12-gon have the same length. It may help you to examine Figure 7.90.

5. Draw a large circle on a piece of paper. This circle will represent a pie. Use a straightedge and compass to do a careful construction that divides your pie into 6 equal pie pieces. Explain why your method of construction guarantees that all 6 pie pieces are of equal size. It may help you to examine Figure 7.90.

6. Draw a circle on a piece of paper. This circle will represent a pie. Use a straightedge and compass to do a careful construction that divides your pie into 12 equal pieces. Explain why your method of construction guarantees that all 12 pieces are of equal size. It may help you to examine Figure 7.90.

7. Use facts about rhombuses from Section 7.5 (see page 64) to explain why the construction described on page 72) really does divide a given angle in half. If you say that some shape is a rhombus, be sure to explain why it really is a rhombus.

7.7 Polyhedra and Other Solid Shapes

So far in this chapter we have studied a number of plane shapes: circles, triangles, quadrilaterals, and other polygons. All of these shapes are flat because they lie in a plane. But plane shapes can be used to make a variety of interesting three-dimensional, solid shapes. In this section we will study some special kinds of solid shapes that can be made out of polygons as well as some related solid shapes.

A closed shape in space that is made out of polygons is called a **polyhedron**. The plural of polyhedron is **polyhedra**. Figure 7.91 shows some polyhedra and a shape that is not a polyhedron. The polygons that make up the surface of the polyhedron are called the **faces** of the polyhedron. The place where two faces come together is called an **edge** of the polyhedron. A corner point where several faces come together is called a **corner** or **vertex** of the polyhedron. The plural of vertex is **vertices**.

The name *polyhedron* comes from the Greek; it makes sense because *poly* means *many* and *hedron* means *face*, so that *polyhedron* means *many faces*.

Prisms, Cylinders, Pyramids, and Cones

One special type of polyhedron is a prism. Some people have glass or crystal prisms hanging in their windows to bend the incoming light and cast rainbow patterns around the room.

From a mathematical point of view, a *right prism* is a polyhedron that roughly speaking can be thought of as "going straight up over a polygon," as shown in Figure 7.92. We can think of **right prisms** as formed in

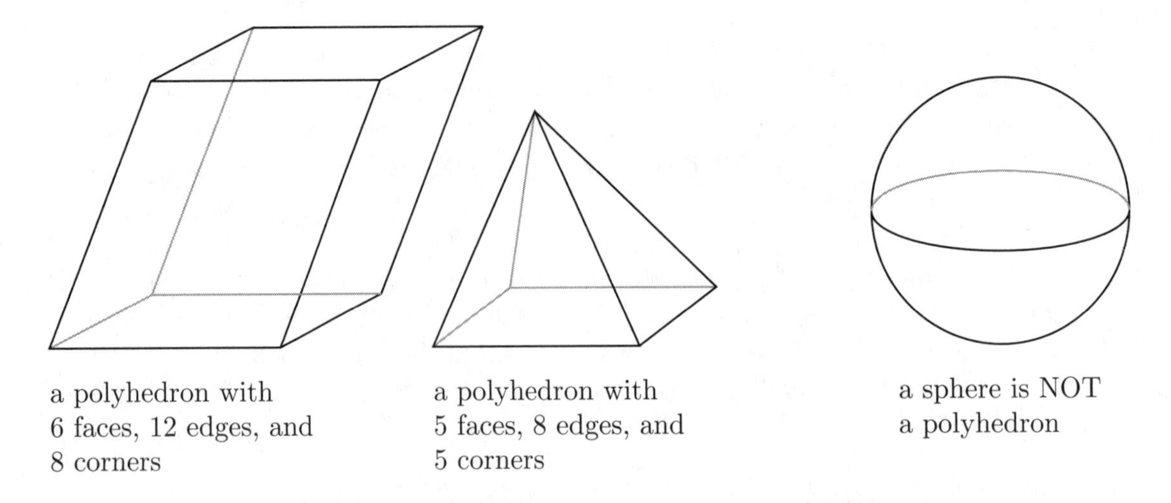

Figure 7.91: Polyhedra and a Shape That is Not a Polyhedron

the following way. Take two paper copies of any polygon and lay both flat on a table, one on top of the other so that they match up. Move the top polygon *straight up* above the bottom one. If vertical rectangular faces are now placed so as to connect corresponding sides of the two polygons, then the shape formed this way is a right prism. The two polygons that you started with are called the **bases** of the right prism.

By modifying the previous description, we get a description of *all* prisms, not just *right* prisms. As before, start with two paper copies of any polygon and lay both flat on a table, one on top of the other so that they match up.

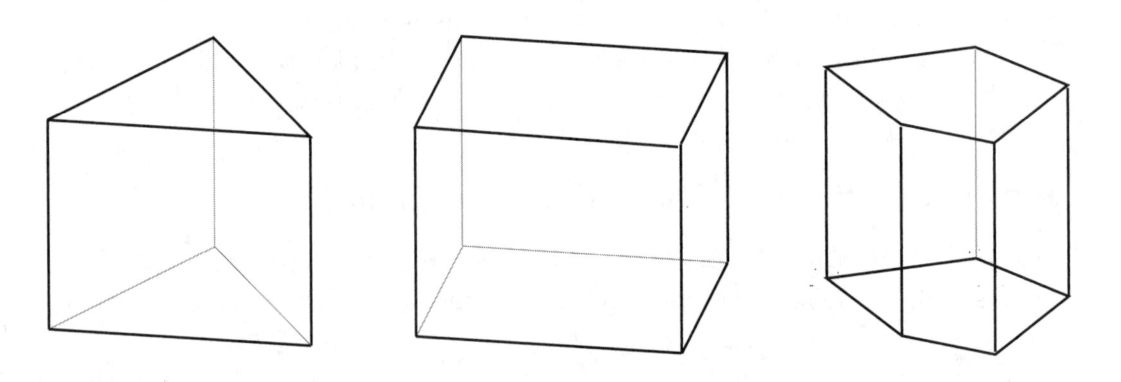

Figure 7.92: Some Right Prisms

Move the top polygon up without twisting, away from the bottom polygon, keeping the two polygons parallel. This time, the top polygon does not need to go *straight up* over the bottom polygon, but can go off to one side, as long as it is not twisted, and remains parallel to the bottom polygon. Once again, if faces are now placed so as to connect corresponding sides of the two polygons, then the shape formed this way is a **prism** (this time the faces will be parallelograms). Every right prism is a prism, but Figure 7.93 shows prisms that are not right prisms. A prism that is not a right prism can be called an **oblique prism** . As before, the two polygons that you started with are called the **bases** of the prism.

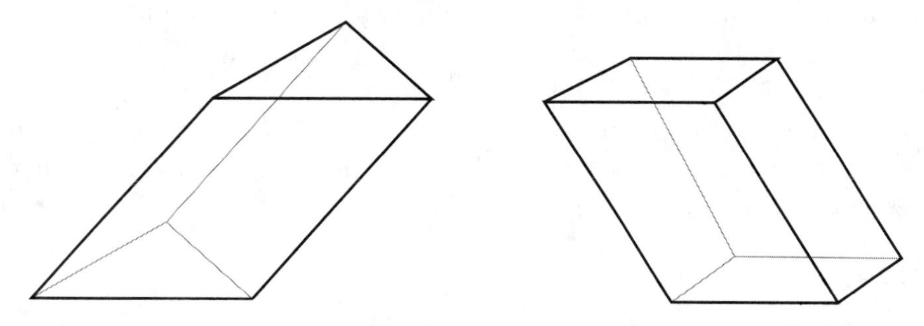

Figure 7.93: Oblique Prisms

If a prism is moved to a different orientation in space, it is still a prism. So, for example, the polyhedra shown in Figure 7.94 are prisms, although you may need to rotate them mentally to be convinced that they really are prisms.

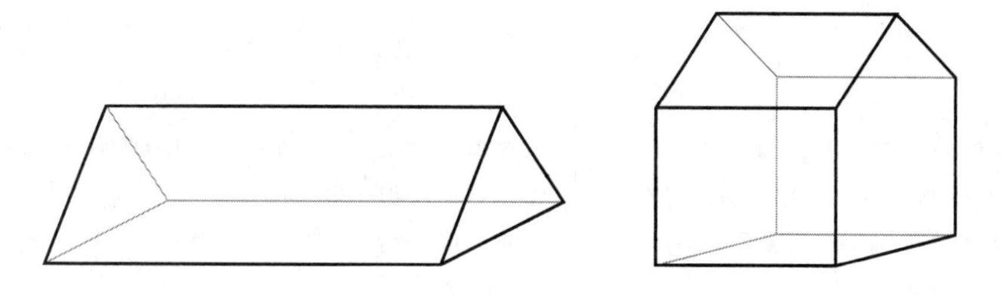

Figure 7.94: A Prism Turned Sideways is Still a Prism

Prisms are often named according to the kind of polygon that make the bases of the prism. So, for example, a prism with a triangle bases can be called a **triangular prism**, while a prism with a rectangle base can be called a **rectangular prism**.

A cylinder is a kind of shape that is related to prisms. Roughly speaking, a cylinder is a tube-shaped object, as shown in Figure 7.95. The tube inside a roll of paper towels is an example of a cylinder. **Cylinders** can be described in the following way. Draw a closed curve on paper (such as a circle or an oval), cut it out, make a copy of it, and lay the the two copies on a table, one on top of the other, so that they match. Now take the top copy and move it up without twisting, away from the bottom copy, keeping the two copies parallel. Now imagine paper or or some other kind of material that connects the two curves in such a way that every line between corresponding points on the curves lies on this paper or material. The shape formed by this paper or other material is a cylinder. The regions formed by the two starting curves can again be called the **bases** of the cylinder. In some cases, one wants to consider the two bases as part of the cylinder, in other cases not. Both kinds of shapes can be called cylinders.

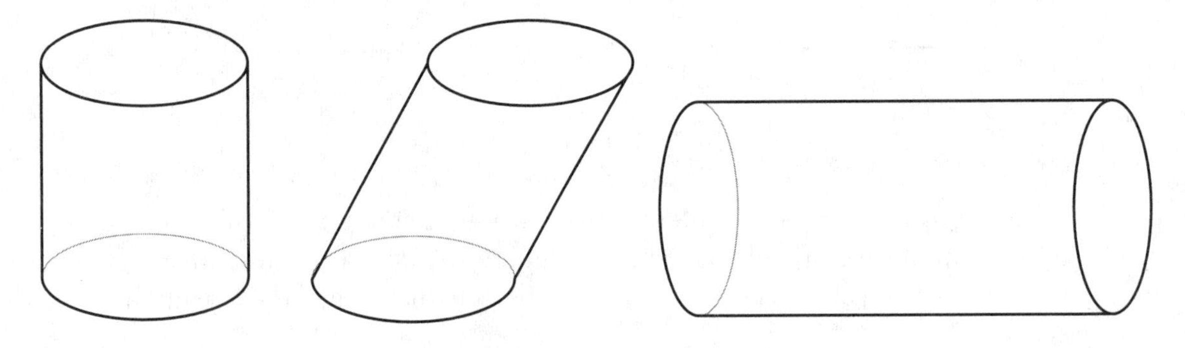

Figure 7.95: Cylinders

As with prisms, a cylinder can be a **right cylinder** or an **oblique cylinder**, according as one base is "straight up over" the other or not. The outer two cylinders in Figure 7.95 are right cylinders, whereas the middle cylinder is an oblique cylinder.

In addition to prisms, another special type of polyhedron is a *pyramid*. You have probably seen pictures of the famous pyramids in Egypt. Mathematical pyramids include shapes like the Egyptian pyramids as well as variations on this kind of shape, as shown in Figure 7.96. **Pyramids** can be described as follows. Start with any polygon and a separate single point that does not lie in the plane of the polygon. Now use additional polygons to connect the point to the original polygon. This should be done in such a way that all the lines that connect the point to the polygon lie on these additional polygons. These new polygons together with the original polygon form a pyramid. As usual, the original polygon is called the **base** of the pyramid.

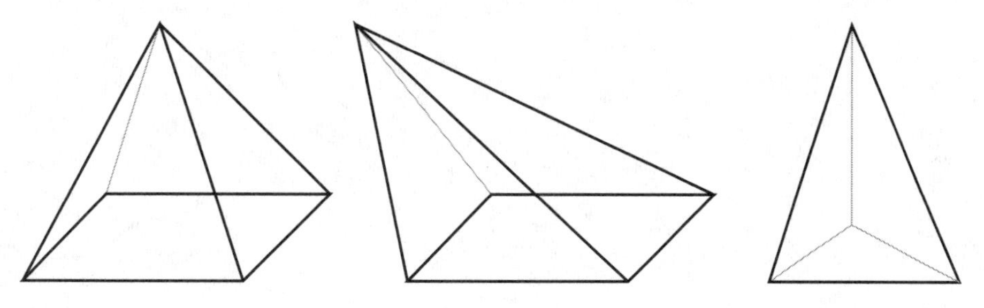

Figure 7.96: Pyramids

As with prisms and cylinders, certain kinds of pyramids are called right pyramids. A **right pyramid** is a pyramid for which the point lies "straight up over" the center of the base. A pyramid that is not a right pyramid can be called an **oblique pyramid**. In Figure 7.96 the outer two pyramids are right pyramids, whereas the middle pyramid is an oblique pyramid. Figure 7.97 shows an oblique pyramid decorating the top of a buidling.

Figure 7.97: An Oblique Pyramid on a Building

Just as cylinders are shapes that are related to prisms, *cones* are shapes that are related to pyramids. Roughly speaking, cones are objects like ice-cream cones or cone-shaped paper cups, as well as related objects, as shown in Figure 7.98. As with pyramids, **cones** can be described by starting with

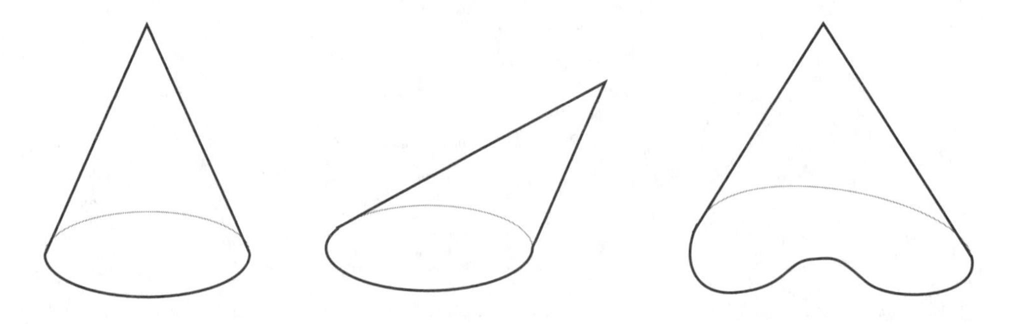

Figure 7.98: Cones

a closed curve in a plane and a separate point that does not lie in that plane. Now imagine paper or some other kind of material that joins the point to the curve in such a way that all the lines that connect the point to the curve lie on that paper (or other material). That paper (or other material), together with the original curve, form a cone. As usual, the starting curve and the region inside it is called the **base** of the pyramid. Sometimes the base of a cone is considered a part of the cone, and sometimes it isn't. Either shape (with or without the base) can be called a cone.

As with prisms, cylinders, and pyramids, certain kinds of cones called right cones. A **right cone** is a cone whose point lies directly over the center of its base. A cone that is not a right cone can be called a **oblique cone** In Figure 7.98, the outer two cones are right cones, whereas the cone in the middle is an oblique cone.

Class Activity 7FF: Making Prisms, Cylinders, Pyramids, and Cones

Class Activity 7GG: Analyzing Prisms, Cylinders, and Cones

The Platonic Solids

Five special polyhedra are called the **Platonic solids**, in honor of Plato, who thought of these solid shapes as associated with earth, fire, water, air, and the whole universe. These are the five Platonic solids:

Tetrahedron is made of 4 equilateral triangles, with 3 triangles coming together at each corner.

Cube is made of 6 squares, with 3 squares coming together at each corner.

Octahedron is made of 8 equilateral triangles, with 4 triangles coming together at each corner.

Dodecahedron is made of 12 regular pentagons, with 3 pentagons coming together at each corner.

Icosahedron is made of 20 equilateral triangles, with 5 triangles coming together at each corner.

The Platonic solids are special because each one is made of only one kind of polygon, and the same number of polygons come together at each corner. The names of the Platonic solids make sense because *hedron* comes from the Greek for *faces*, while *tetra*, *octa*, *dodeca*, and *icosa* mean 4, 8, 12, and 20, respectively. So for example, *icosahedron* means *20 faces*, which makes sense because an icosahedron is made out of 20 triangles.

While cubes are commonly found in daily life (such as boxes and ice cubes), the other Platonic solids are not commonly seen. However, these shapes do occur in nature occasionally. For example, the mineral pyrite can form a crystal in the shape of a dodecahedron (this crystal is often called a pyritohedron). See the website

`http://www.minerals.net/mineral/sulfides/pyrite/pyrite2.htm`

for a neat picture. The mineral fluorite can form a crystal in the shape of an octahedron, and although rare, gold can too. See the bottom picture of

`http://home.pacifier.com/~leopard/minpix9.htm`

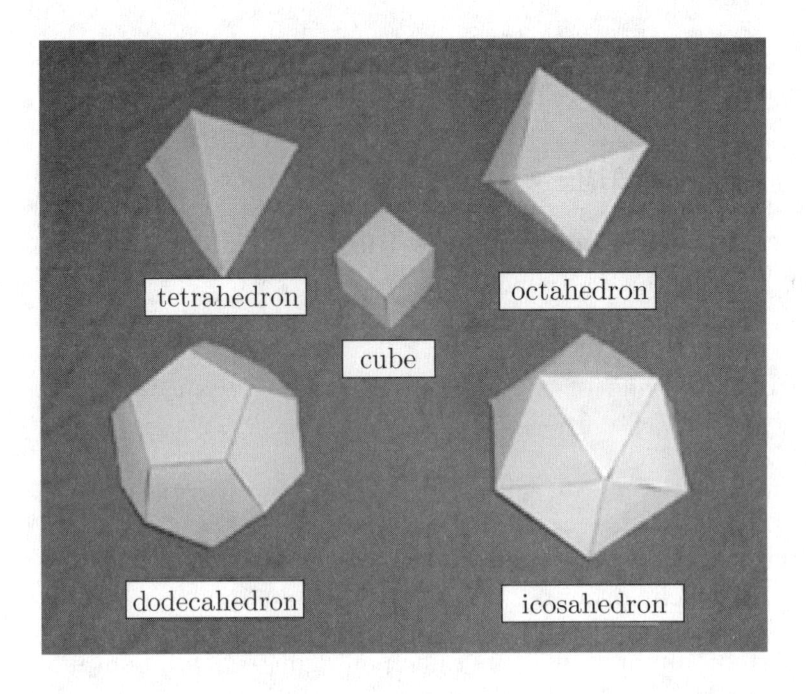

Figure 7.99: The Platonic Solids

for a picture of fluorite in octahedron form.

Some viruses are shaped like an icosahedron. See

`http://www.tulane.edu/~dmsander/WWW/335/335Structure.html`

(scroll down) and

`http://www.bocklabs.wisc.edu/Herpesvirus.html`

for some pictures.

Figure 7.100 shows a picture of a dodecahedron sculpture in front of a school in Jesup, Georgia.

The real attraction of the Platonic solids is their beautiful perfection and symmetry. You can't really appreciate the Platonic solids unless you make models of these wonderful shapes, so I recommend that you construct models of them—see Class Activity 7HH and exercise 5.

Class Activity 7HH: Making Platonic Solids with Toothpicks and Marshmallows

Figure 7.100: A Dodecahedron Sculpture in Jesup, Georgia

Class Activity 7II: Why Are There No Other Platonic Solids?

Class Activity 7JJ: What's Inside the Magic 8 Ball?

The following class activity applies to all polyhedra, not just the Platonic solids.

Class Activity 7KK: Relating the Numbers of Faces, Edges, and Corners of Polyhedra

Exercises for Section 7.7 on Polyhedra

1. Find examples of objects in the "real world" that are prisms, cylinders, pyramids, and cones.

2. Examine the small patterns in Figure 7.101. Try to visualize: if you were to cut these patterns out on the heavy lines, fold them on the dotted lines, and tape various sides together, what kinds of polyhedra

would they make?

Now cut out the large versions of these patterns on page 373 and 375 along the heavy lines. Fold up the shapes and tape the sides together to make polyhedra. Were your predictions correct? If not, undo your polyhedra and try to visualize again how they turn into their final shapes.

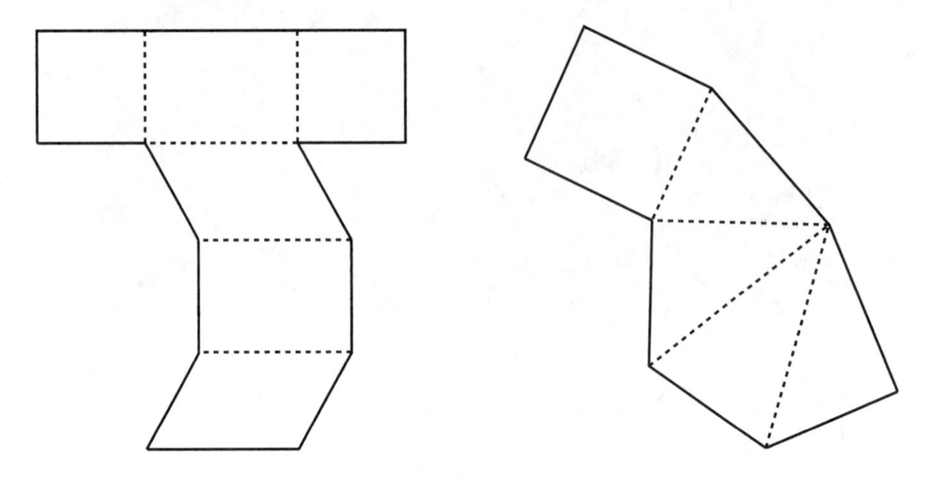

Figure 7.101: Small Patterns for Two Polyhedra

3. Make a pattern for a prism whose two bases are congruent to the triangle in Figure 7.102. Include the bases in your pattern (use a straightedge and compass to make a copy of the triangle in Figure 7.102). Label all sides that have length a, b, and c on your pattern.

4. Make a pattern for a pyramid whose base is congruent to the triangle in Figure 7.102. Include the base in your pattern (use a straightedge and compass to make a copy of the triangle in Figure 7.102). Label all sides that have length a, b, and c on your pattern.

5. Referring to the descriptions of the Platonic solids on page 87, construct the five Platonic solids by cutting out the triangles, squares, and rectangles on pages 377, 379, and 381, and taping these shapes together. You may wish to copy pages 377, 379, and 381 onto card stock in order to make sturdier models that are easier to work with.

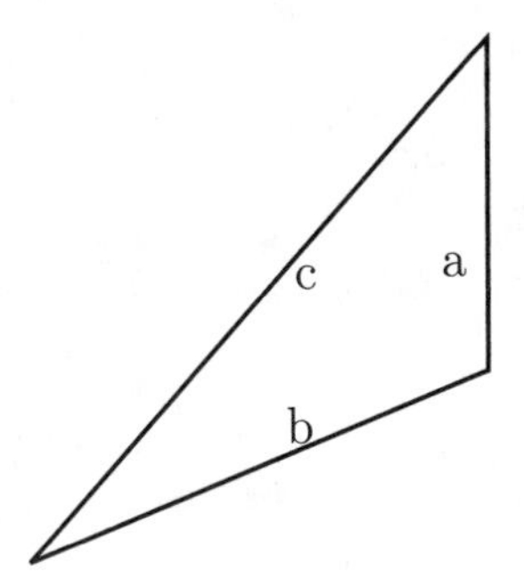

Figure 7.102: A Base for a Prism and a Pyramid

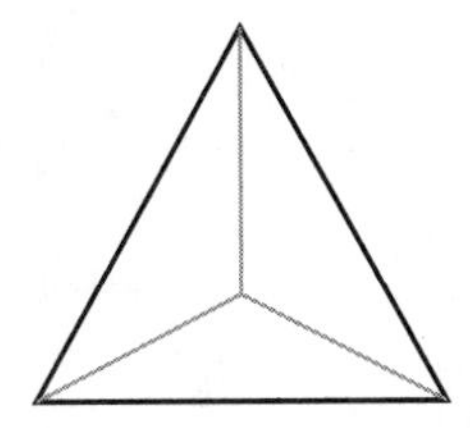

Figure 7.103: A Tetrahedron

6. What other name can you call a tetrahedron? (See Figure 7.103.)

7. What happens if you try to make a polyhedron whose faces are all equilateral triangles and for which 6 triangles come together at every corner?

8. Is it possible to make a polyhedron so that 7 or more equilateral triangles come together at every point and the polyhedron has no indentations or protrusions? Explain.

Answers to Exercises for Section 7.7

3. & 4. See Figure 7.104.

6. A tetrahedron is a special kind of pyramid with a triangle base.

7. If you put 6 equilateral triangles together so that they all meet at one point, then you will find that they make a flat hexagon, as seen in Figure 7.105. It makes sense that 6 triangles put together at a point

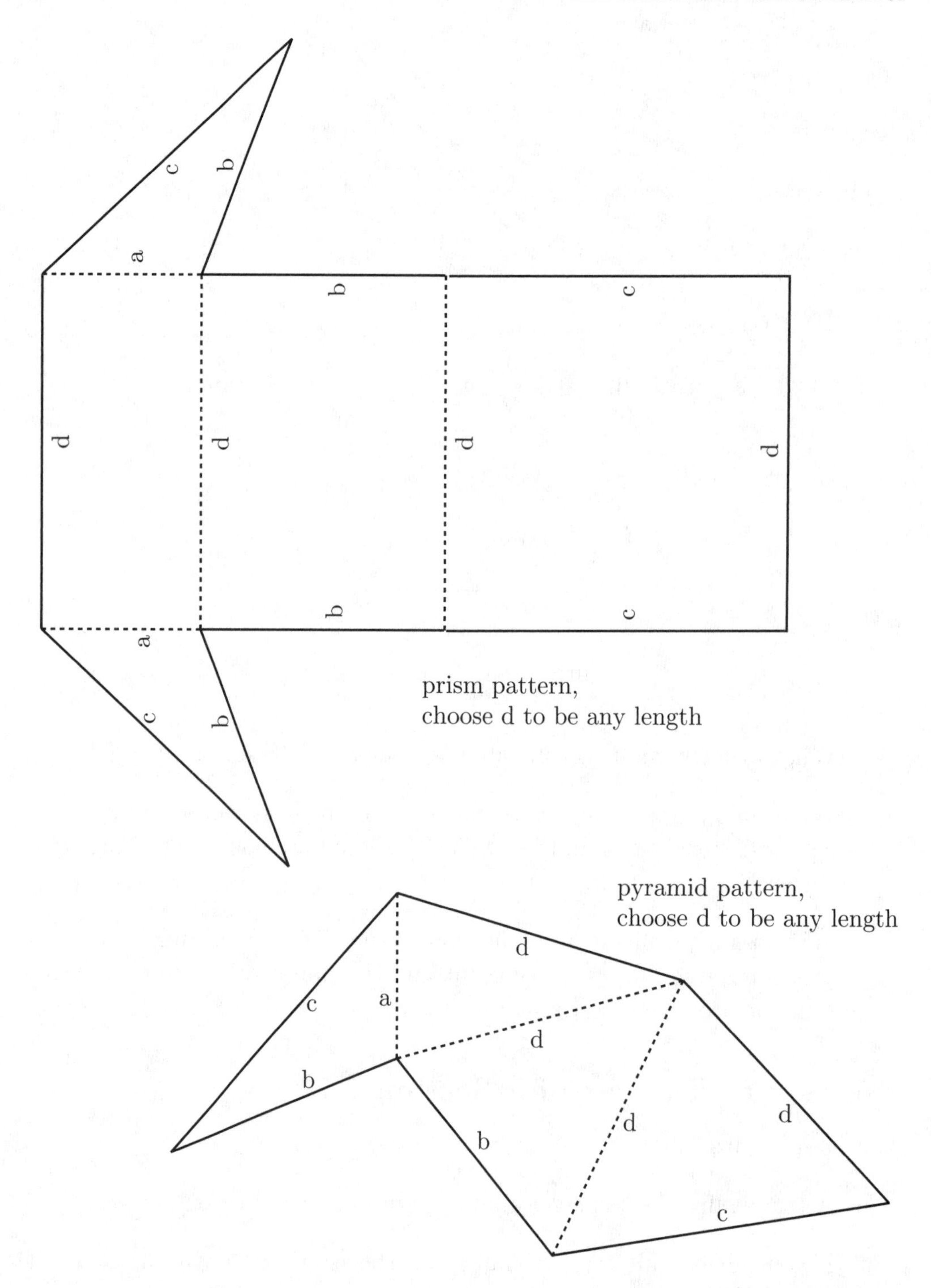

Figure 7.104: Patterns for a Prism and a Pyramid with Triangle Bases

make a flat shape because the angle at a corner of an equilateral triangle is 60°, so 6 of these angles side by side will make a full 360°. So if you tried to make a polyhedron in such a way that 6 equilateral triangles came together at every corner, you could never get the polyhedron to "close up"—you could only get a flat shape this way.

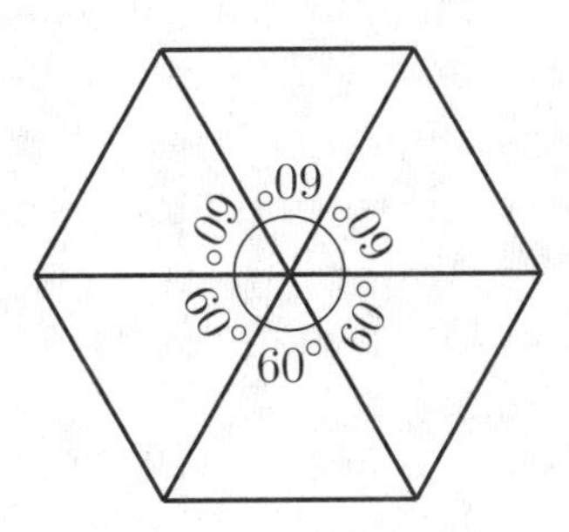

Figure 7.105: Six Triangles Put Together at a Point

8. If you put 7 or more equilateral triangles together so that they all meet at one point, then you will find that some triangles will have to slant up and others will have to slant down—thus you will have to create an indentation when you put triangles together in this way. It makes sense that 7 triangles meeting at a point will do this because the angle at a corner of an equilateral triangle is 60°, so 7 or more of these angles side by side will make more than 360°. This forces some triangles to slant up and others to slant down in order to put so many triangles together at a point.

Problems for Section 7.7 on Polyhedra

1. Use a straightedge and compass to make a pattern for a prism whose two bases are congruent to the triangle in Figure 7.102. Make your pattern different from the one for a prism given in Figure 7.104. Include the bases in your pattern (use a straightedge and compass to make a copy of the triangle in Figure 7.102). Label all sides that have length a, b, and c on your pattern.

2. Use a straightedge and compass to make a pattern for a pyramid whose base is congruent to the triangle in Figure 7.102. Make your pattern

different from the one for a pyramid given in Figure 7.104. Include the base in your pattern (use a straightedge and compass to make a copy of the triangle in Figure 7.102). Label all sides that have length a, b, and c on your pattern.

3. Make a pattern for an oblique cylinder with a circular base. You may leave the bases off your pattern. (Advice: be willing to experiment first! You might start by making a pattern for a right cone and modifying it.) The sleeve of mosts shirts and blouses are more or less in the shape of an oblique cylinder. If your pattern were to make a sleeve, what part of your pattern would be at the shoulder? What part of your pattern would be at the armpit?

4. Make a pattern for an oblique cone with a circular base. You may leave the base off your pattern. (Advice: be willing to experiment first! You might start by making a pattern for a right cone and modifying it.)

5. Make a pattern for the "bottom portion" of a right cone, as pictured in Figure 7.106. What article of clothing is often shaped like this?

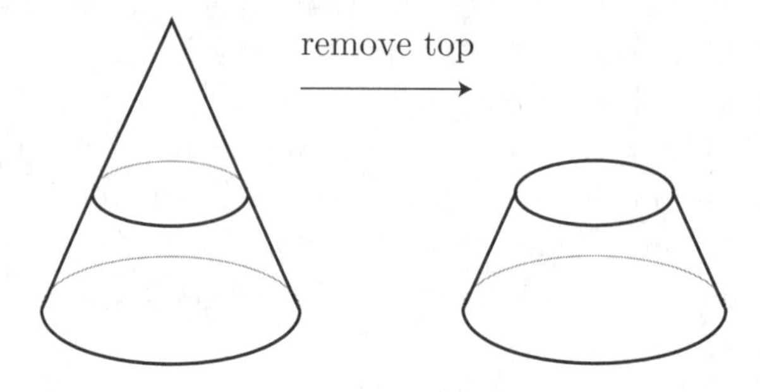

Figure 7.106: The Bottom Portion of a Cone

6. Recall that an n-gon is a polygon with n sides. For example, a polygon with 17 sides can be called a 17-gon. A triangle could be called a 3-gon. Find formulas (in terms of n) for the following, and explain why your formulas are valid:

   (a) the number of faces on a prism that has n-gon bases;

   (b) the number of edges on a prism that has n-gon bases;

(c) the number of corners on a prism that has n-gon bases;

(d) the number of faces on a pyramid that has an n-gon base;

(e) the number of edges on a pyramid that has an n-gon base;

(f) the number of corners on a pyramid that has an n-gon base;

7. This problem goes with Class Activity 7JJ on the Magic 8 Ball. Make a closed, three dimensional shape out of some or all of the triangles in Figure A.10. Make any shape you like—feel free to be creative! Answer the following:

   (a) Do you think your shape could be the one that is actually inside the Magic 8 Ball? Why or why not?

   (b) If you were going to make your own advice ball but with a (possibly) different shape inside, what would be the advantages or disadvantages of using your shape?

8. Two gorgeous polyhedra can be created by *stellating* an icosahedron and a dodecahedron. **Stellating** means *making star-like.* Imagine turning each face of an icosahedron into a "star point," namely a cone whose base is a triangular face of the icosahedron. Likewise, imagine turning each face of a dodecahedron into a "star point," namely a cone whose base is a pentagonal face of the dodecahedron. So a stellated icosahedron will have 20 "star points," while a stellated dodecahedron will have 12 "star points."

   (a) Make 20 copies of Figure 7.107 on card stock. Cut, fold, and tape them to make 20 triangle-based pyramid "star points." (The pattern makes star point pyramids that don't have bases.) Tape the star points together as though you were making an icosahedron out of their (open) bases.

   (b) Make 12 copies of Figure 7.108 on card stock. Cut, fold, and tape them to make 12 pentagon-based pyramid "star points." (The pattern makes star point pyramids that don't have bases.) Tape the star points together as though you were making a dodecahedron out of their (open) bases.

9. If you try to make a polyhedron whose faces are all regular hexagons, what will happen? Explain.

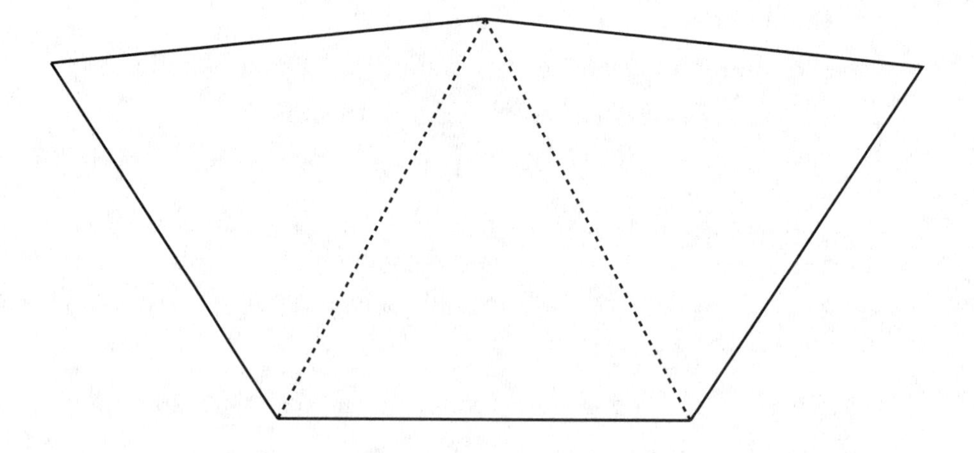

Figure 7.107: Pattern for a Star Point of an Icosahedron

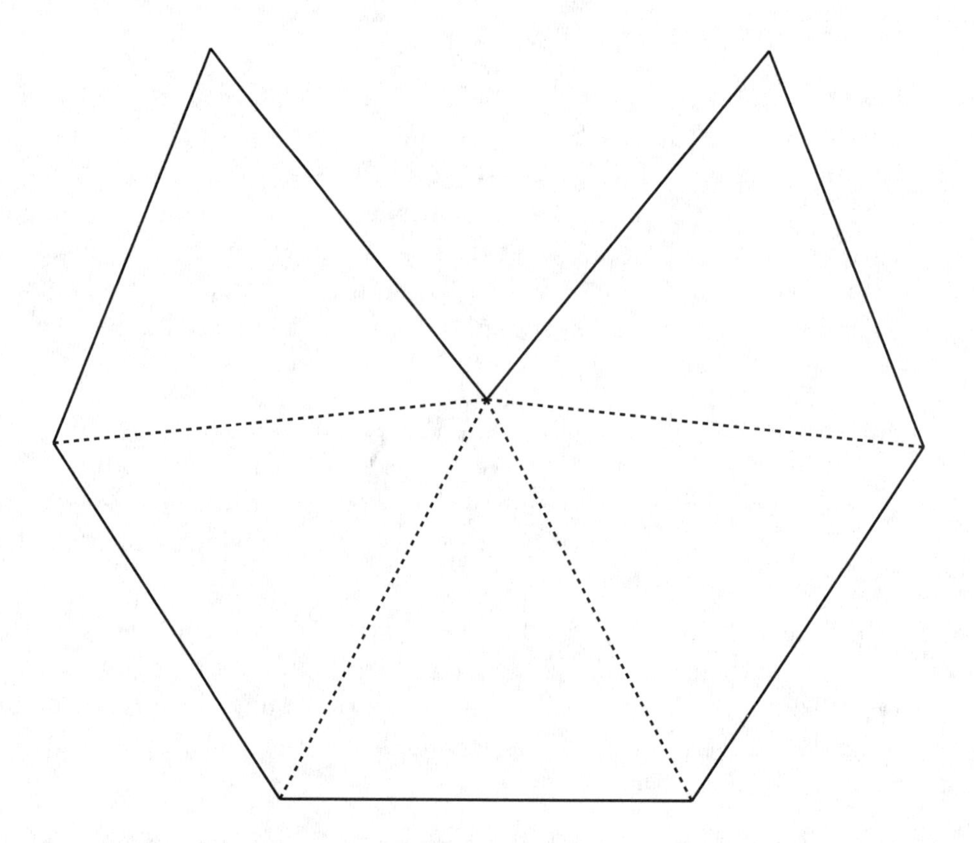

Figure 7.108: Pattern for a Star Point of a Dodecahedron

Chapter 8

Geometry of Motion and Change

In the last chapter we studied shapes from a static point of view, that is, we did not consider moving shapes around, or changing their sizes. In this chapter, we bring movement and size-change into the study of shapes. The study of movements of shapes leads to the study of symmetry: by moving shapes in different ways, we create different kinds of symmetry. We will also study the extent to which a given shape can be changed. Certain shapes are rigid and inflexible, while others are "floppy" and moveable—this relates to construction practices. Finally, we will study what happens when shapes or objects change size, but otherwise remain the same. This theory has widely used practical applications, such as the measurement of distances in land surveying.

8.1 Reflections, Translations, and Rotations

In the last chapter, we saw various special properties that shapes in a plane can have. One especially attractive property that some shapes and designs have is that of symmetry. In order to study symmetry we will first study certain transformations of planes, namely translations, reflections, and rotations. We will use translations, reflections, and rotations to define what we mean by symmetry, as well as to create designs that have symmetry. From this point of view, translations, reflections, and rotations are the "building blocks" of symmetry.

Roughly speaking, a **transformation** of a plane is just what the name sounds like: an action that changes or transforms a plane. We will be interested in transformations that start with a plane and change it back into the same plane. Even though we will start and end with the same plane, the transformation will usually have caused most or all of the individual points on the plane to change location. The three kinds of transformations that we will study are *rotations*, *reflections*, and *translations*.

A **reflection** (or **flip**) across a line is one kind of transformation of a plane. The line involved in a reflection is called the **line of reflection**. One informal way to see the effect of a reflection is by drawing a point on paper in wet ink or paint and quickly folding the paper along the desired line of reflection: the location where the wet ink rubs off onto the paper shows the final position of the point after reflecting. Another way to see the effect of a reflection is by drawing a point on a semi-transparent piece of paper and flipping the paper upside down by twirling it around the desired line of reflection: you will see the location of the reflected point on the other side of the paper.

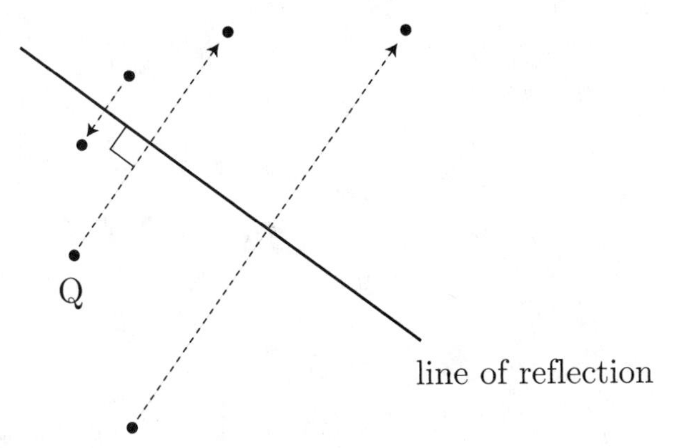

Figure 8.1: The Effect of a Reflection on a Point Q and on Other Points

Another way to think about reflections is that under a reflection, each point in the plane goes to its mirror image on the other side of the line of reflection (hence the name reflection). More precisely, to see where a point Q in the plane goes, draw a line through Q that is perpendicular to the given line of reflection. The point Q will go to the other point on this new line that is the same distance from the line of reflection but is on the other side of the line of reflection, as shown in Figure 8.1.

A **translation** (or **slide**) by a given distance in a given direction is another kind of transformation of a plane. In a translation, each point in the plane is moved the given distance in the given direction. To illustrate a translation, put a flat piece of paper on a table top and slide the piece of paper in some direction (without rotating it). Now imagine doing this with a whole plane instead of a piece of paper: this would produce a translation of the plane. Figure 8.2 indicates initial and final positions of various points under a translation.

A **rotation** (or **turn**) about a point through a given angle is a third kind of transformation of a plane. To illustrate a rotation about a point, put a flat piece of paper on a table top, stick a pin through the point so as to hold that point at a fixed point on the table, and rotate the piece of paper about that point. Now imagine doing this with an infinite plane instead of a piece of paper: this would would produce a rotation of the plane about the point that is fixed. Figure 8.3 shows the initial and final positions of some points

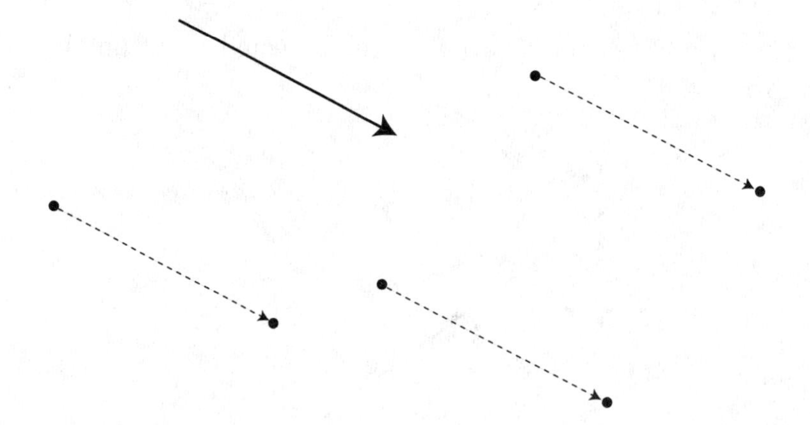

Figure 8.2: The Effect of a Translation on Various Points

in a plane under a rotation about the point P. Notice that points that are farther from the point P move a greater *distance* than points that are closer to P, even though all points rotate through the same *angle*.

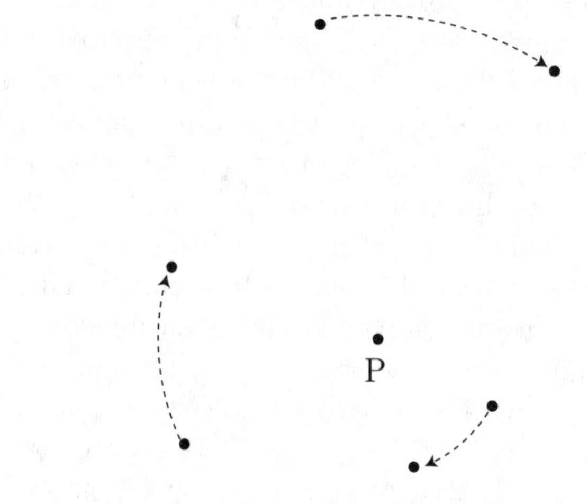

Figure 8.3: Where Various Points Go Under a Rotation About Point P

Class Activity 8A: Exploring Reflections with Geometer's Sketchpad

Class Activity 8B: Exploring Translations with Geometer's Sketchpad

Class Activity 8C: Exploring Rotations with Geometer's Sketchpad

Class Activity 8D: Exploring Rotations (alternate)

Class Activity 8E: Exploring Reflections (alternate)

Exercises for Section 8.1 on Rotations, Reflections, and Translations

1. Practice the class activities of this section.

2. Match the specified transformations in A, B, C, and D of Figure 8.4 to the effects shown in 1, 2, 3, and 4.

3. For each picture in Figure 8.5, determine what kind of transformation (reflection, translation, or rotation) will take the initial shape to the final shape.

Answers to Exercises for Section 8.1

2. A–4, B–1, C–2, D–3.

3. See Figure 8.6

Problems for Section 8.1 on Rotations, Reflections, and Translations

1. Investigate the following, either with Geometer's Sketchpad (preferable) or by making drawings on paper. What is the *net effect* if you rotate a shape or design about some point by 180° and then rotate the resulting shape by 180° *about some other point*? What *single* transformation (reflection, translation, or rotation) will take your initial shape

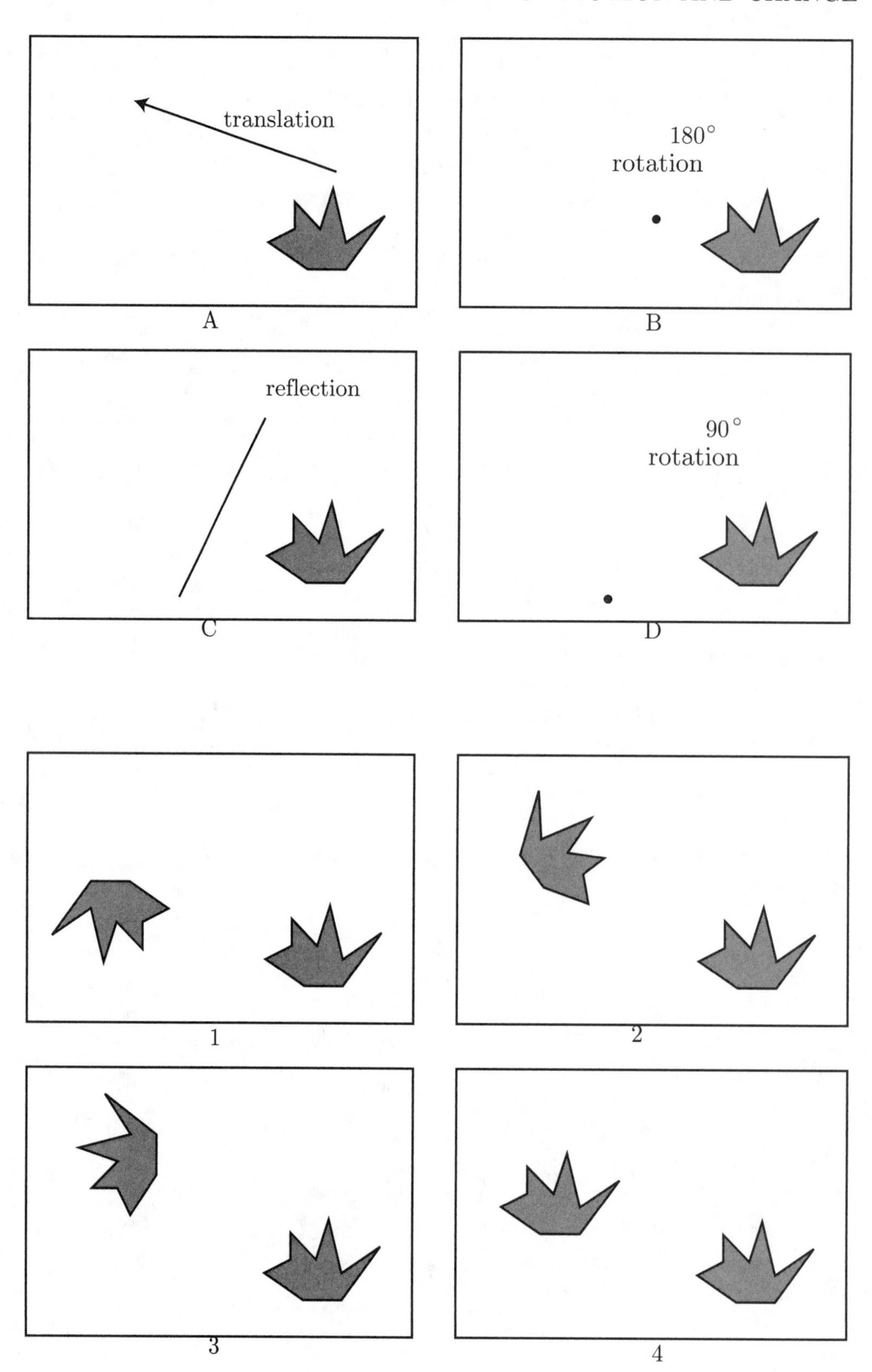

Figure 8.4: Various Transformations and Their Effects

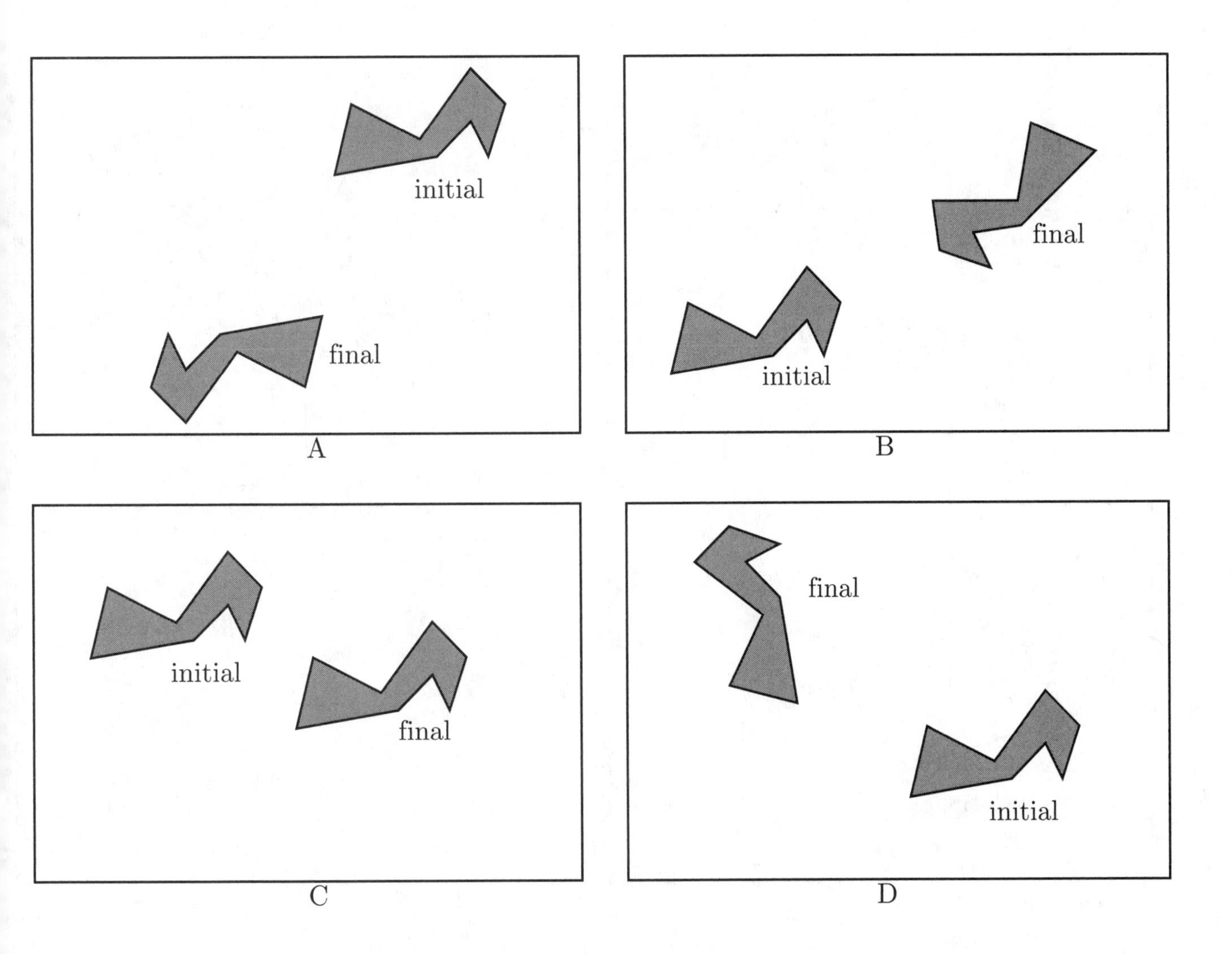

Figure 8.5: What Kinds of Transformations?

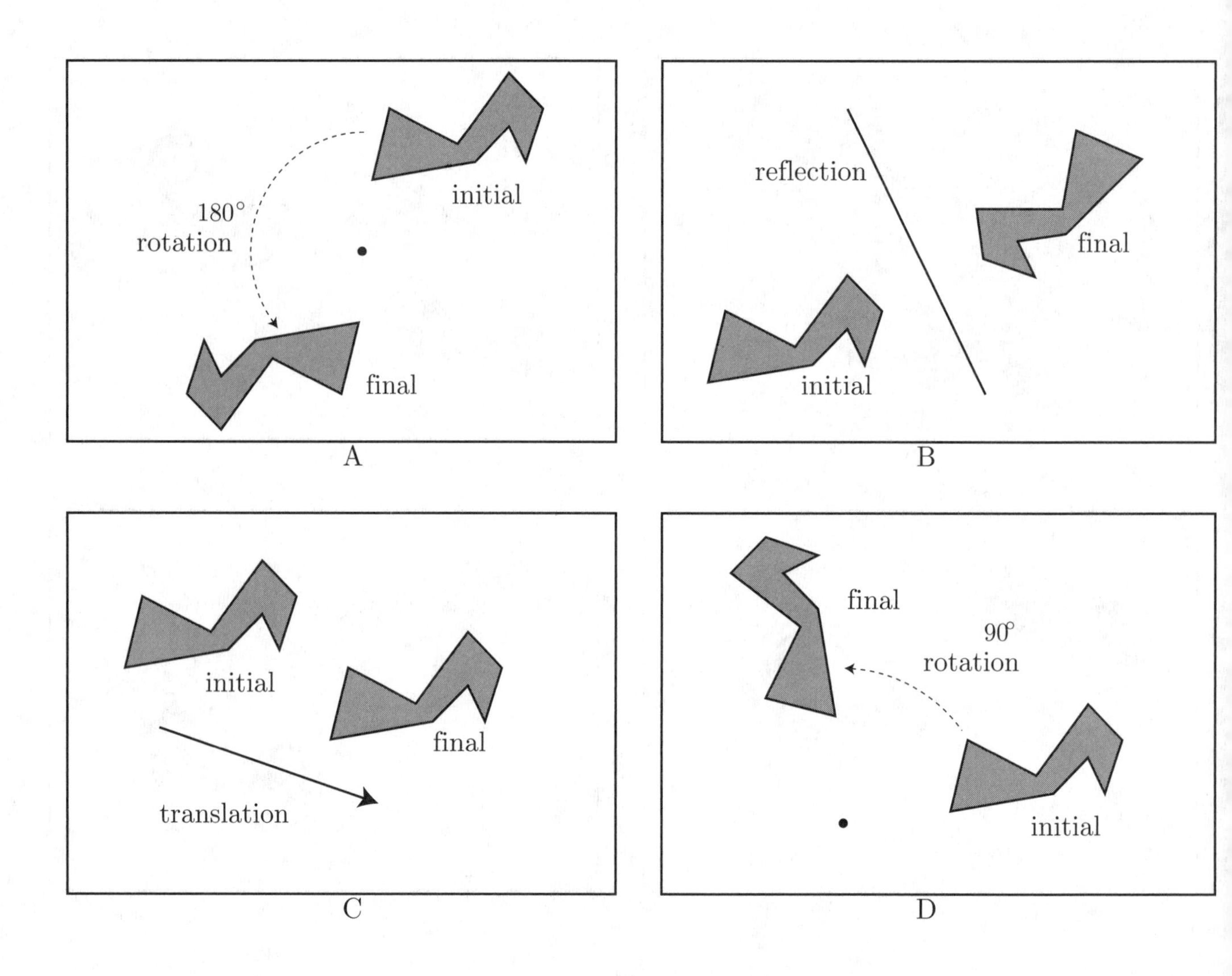

Figure 8.6: The Effects of Various Transformations on a Shape

to your final shape? (Be sure to use a "crazy" asymmetrical shape when investigating this.)

2. Investigate the following, either with Geometer's Sketchpad (preferable) or by making drawings on paper. What is the *net effect* if you reflect a shape or design across a line and then reflect the resulting shape across another line that is perpendicular to the first line? What *single* transformation (reflection, translation, or rotation) will take your initial shape to your final shape? (Be sure to use a "crazy" asymmetrical shape when investigating this.)

3. Investigate the following, either with Geometer's Sketchpad or by making drawings on paper. Draw a "crazy" asymmetrical shape, draw a separate line, and draw a point on the line.

   (a) If you reflect the shape across the line and then rotate the reflected shape 180° about the point, will the final position of the shape be the same as if you had *first* rotated the shape 180° and *then* reflected the rotated shape across the line?

   (b) If you reflect the shape across the line and then rotate the reflected shape 90° counterclockwise about the point, will the final position of the shape be the same as if you had *first* rotated the shape 90° counterclockwise and *then* reflected the rotated shape across the line?

8.2 Symmetry

Symmetry is an area shared by mathematics, the natural world, and art, and so it offers opportunities for cross-disciplinary study. There is something about symmetry that is deeply appealing to most people. Why is that so? Maybe it is because objects with symmetry sometimes seem more perfect than objects that don't have symmetry. Or maybe it is because objects with symmetry involve repetition. Is there something about human nature that causes us to enjoy repetition? For example, almost all music involves repetition of themes—often several different themes are repeated throughout a piece of music in a certain pattern. Young children love to have the same story read to them over and over again (parents sometimes dread having to read a favorite book for the umteenth time!). Maybe we like repetition because it helps us learn and understand, and maybe this plays a role in our enjoyment of symmetrical designs.

When we look at natural objects in the world around us, we find a mix of symmetry and asymmetry. Most creatures are symmetrical. Plants typically have symmetrical parts even if they are not symmetrical over all: leaves and flowerheads are usually symmetrical. On the other hand, geological features are rarely symmetrical. It would be surprising to see a mountain, a lake, a river, or a rock that we would describe as symmetrical, even though they are usually made up of smaller, repeated parts. On the other hand, some volcanoes, beaches, river stones, and waves are symmetrical, or nearly so.

Just as the concept of *circle* has both an informal or artistic interpretation as well as a mathematical definition, the notion of symmetry also has both an informal interpretation as well as a specific mathematical definition that applies to shapes in a plane or in space. We will now turn our attention to the mathematical definition of symmetry, which will require us to draw upon the transformations that we have just studied: rotations, reflections, and translations. There are three ways that we will use these transformations in our study of symmetry. First, we will use rotations, reflections, and translations to *define* symmetry. In other words, we will use these transformations to say what it means for a shape or a design to have symmetry. Second, we will use rotations, reflections, and translations to *create* shapes and designs that have symmetry. Finally, we will *analyze* symmetrical designs by looking for parts of the design that can be used to create the whole design by applying rotations, reflections, and translations.

Defining Symmetry

There are four kinds of symmetry that a shape or design in a plane can have: reflection symmetry, translation symmetry, rotation symmetry, and glide-reflection symmetry. We will use reflections, translations, and rotations to say what it means for a shape or design to have symmetry of one of these types.

A shape or design in a plane has **reflection symmetry** or **mirror symmetry** if there is a line in the plane such that the shape or design as a whole is located in the same position both before and after reflecting across the line. This line is called a **line of symmetry**. For example, Figure 8.7 shows a design and a line of symmetry of the design. Notice that reflecting across the line of symmetry causes most points on the design to

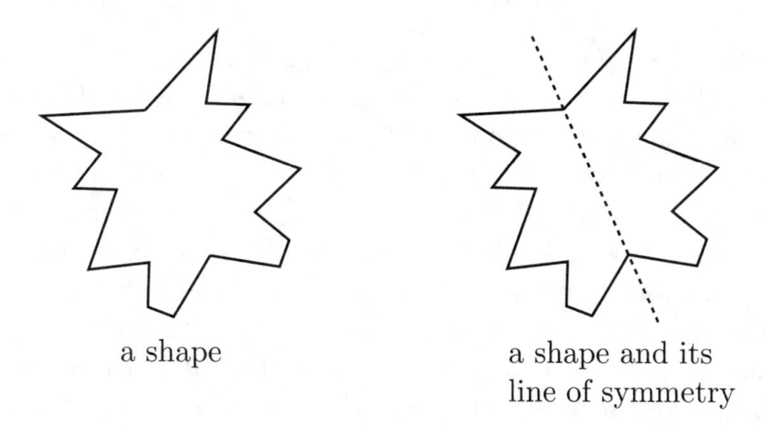

Figure 8.7: A Shape with Reflection Symmetry

swap locations with another point on the design, but the design *as a whole* remains in the same location. Shapes or designs can have more than one line of symmetry. For example, see Figure 8.8.

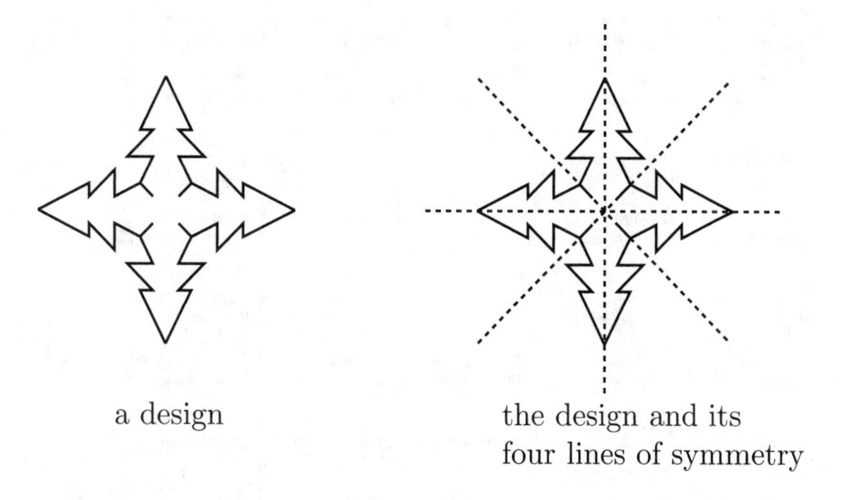

Figure 8.8: A Design with Several Lines of Symmetry

To check if a design or shape drawn on a semi-transparent piece of paper has reflection symmetry with respect to a certain line on the paper, fold the paper along that line. If you can see that the parts of the design on the two parts of the paper match up, then the design has reflection symmetry, and the line you folded along is a line of symmetry. Another way to check if a design or shape has reflection symmetry is to use a mirror that has a straight edge. Place the straight edge of the mirror along the line that you think might be a line of symmetry, and hold the mirror so that it is perpendicular to the design. When you look in the mirror, does the design look like it did without the mirror in place? If so, then the design has reflection symmetry. These methods may help you get a better feel for reflection symmetry when you are first learning about it, but the goal is to be able to use *visualization* to determine whether or not a design has reflection symmetry.

A design or pattern in a plane has **translation symmetry** if there is a translation of the plane such that the design or pattern *as a whole* is located in the same place in the plane both before and after the translation is applied. True translation symmetry only occurs in designs or patterns that take up an infinite amount of space. So when we draw a picture of a design or pattern that has translation symmetry, we can only show a small portion of it; we must imagine the pattern continuing on indefinitely. Figure 8.9 shows a pattern with translation symmetry. In fact, notice that this pattern

Figure 8.9: A Wallpaper Pattern With Translation Symmetry

has two independent translations that take the pattern as a whole to itself: one is "shift right", another is "shift up." Wallpaper patterns generally have translation symmetry with respect to translations in two independent directions.

Frieze patterns—often seen on narrow strips of wallpaper up around the top of the walls of a room—provide additional examples of designs with translation symmetry. Figure 8.10 shows a frieze pattern that has translation symmetry. Unlike a wallpaper pattern, a frieze pattern will only have tranlation symmetry in one direction (and its "reverse") instead of in two independent directions.

pattern continues forever to the right and to the left

Figure 8.10: A Frieze Pattern With Translation Symmetry

A shape or design in a plane has **rotation symmetry** if there is a rotation of the plane, of more than 0° but less than 360°, such that the shape or design *as a whole* is located in the same place in the plane both before and after the rotation is applied. For example, see Figure 8.11. Rotating

Figure 8.11: A Design with 5-fold Rotation Symmetry

72° about the center of the design takes the design as a whole to the same location in the plane, even though each individual curlicue in the design moved over to the position of another culycue. The design of Figure 8.11 is said to have **5-fold rotation symmetry** because by applying the 72° rotation about the center of the design 5 times, all points on the design are taken back to their initial starting positions. Generally, shapes or designs can have 2-fold, 3-fold, 4-fold, etc. rotation symmetry. A design has **n-fold rotation symmetry** provided that there is a rotation of more than 0° but less than 360° that takes the design as a whole to the same location and such that applying this rotation n times takes every point on the design back to its initial position. This is more complicated to say than to see! Figure 8.12 shows some examples of designs with various rotation symmetries.

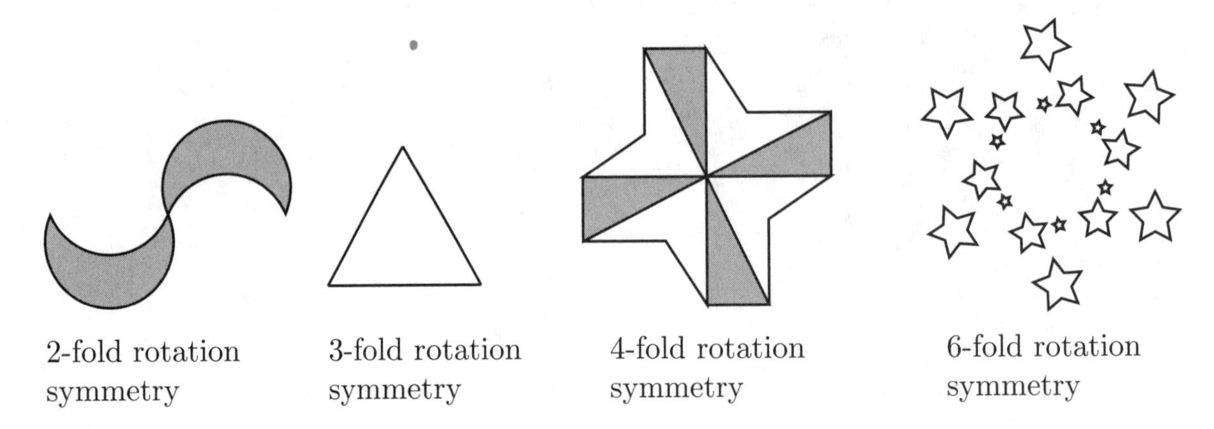

Figure 8.12: Designs with Rotation Symmetry

Here is how you can check if a design or shape drawn on a piece of paper has rotation symmetry: make a copy of the design on semi-transparent paper. Lay the semi-transparent paper over the original design so that the two designs match up. If you can rotate the semi-transparent paper so that the two designs match up again, then the design has rotation symmetry. If you can keep rotating by the same amount for a total of 2, 3, 4, 5, etc., times until the design on the semi-transparent paper is back to its initial position, then the design has 2-fold, 3-fold, 4-fold, 5-fold, etc., rotation symmetry, respectively. When you are first learning about rotation symmetry it may help you to physically rotate designs, however, the goal is to use *visualization* to determine if a design has rotation symmetry.

Finally, a design or pattern in a plane has **glide-reflection symmetry** if there is a reflection and a translation such that after applying the reflection followed by the translation, the *design as a whole* is in the same location. Glide reflection symmetry is often seen on frieze patterns, for example, Figure 8.13 shows an example of a frieze pattern with glide-reflection symmetry. If this design is reflected across a horizontal line through its middle, and then

Figure 8.13: A Frieze Pattern With Glide-Reflection Symmetry

translated to the right (or left), the design as a whole will be in the same location as it was originally (see Figure 8.14).

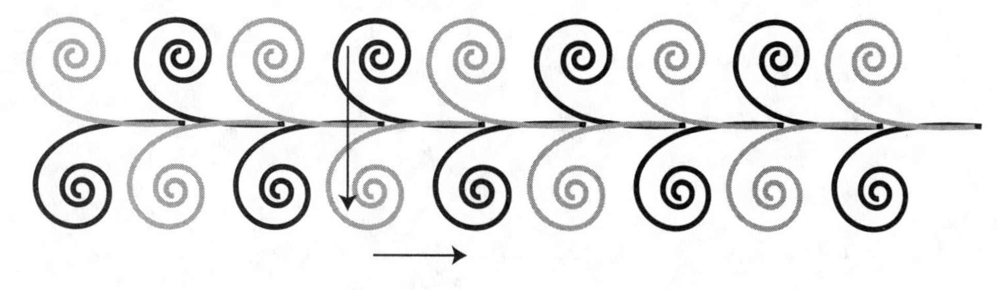

Figure 8.14: Understanding Glide-Reflection Symmetry

Notice that the frieze pattern in Figure 8.13 also has translation symmetry. Many designs have more than one type of symmetry. For example, the design shown in Figure 8.15 has both 2-fold rotation symmetry as well as reflection symmetry with respect to two lines: one horizontal and one vertical.

Figure 8.15: A Design on Egyptian Fabric

Class Activity 8F: What Kind of Symmetry?

Creating Symmetry

One way to create symmetrical designs is to start with any design and then either reflect it, repeatedly rotate it, or repeatedly translate the design, keeping the original and all copies of the design produced in the process. The new design, consisting of the original and all copies, will have reflection symmetry, rotation symmetry, or translation symmetry, respectively. Figure 8.16 indicates the process of creating symmetrical designs this way.

To create a design with 2-fold rotation symmetry, start with any design and rotate it $\frac{360}{2}^\circ = 180^\circ$ about any point in the plane. The original design together with the rotated design form a new design that has 2-fold rotation symmetry. To create a design with 6-fold rotation symmetry, start with any design and rotate it $\frac{360}{6}^\circ = 60^\circ$ five times, keeping the original design and all the rotated designs. All six designs together form a design with 6-fold rotation symmetry. In general, to create a design with n-fold rotation symmetry, start with any design and rotate it $\frac{360}{n}^\circ$ repeatedly until the original design gets back to its initial position. Thereafter, if you rotate the whole design, the constituent designs will cycle around, but the design as a whole moves to the same location in the plane. Since n rotations of $\frac{360}{n}^\circ$ produce a total of 360°, which is a full rotation, therefore the design produced this way has n-fold rotation symmetry.

Class Activity 8G: Creating Symmetrical Designs with Geometer's Sketchpad

Class Activity 8H: Creating Symmetrical Designs (alternate)

The artist M. C. Escher (1898–1972) created many interesting symmetrical designs involving congruent, interlocking shapes—something like Figure 8.17. Some of Escher's artwork can be seen at

`http://www.worldofescher.com/gallery/`

There is software called TesselMania! that can be used to create Escher-type designs (it is suitable for use even by young children). However, as you'll see if you do the next class activity, you can also use Geometer's Sketchpad to

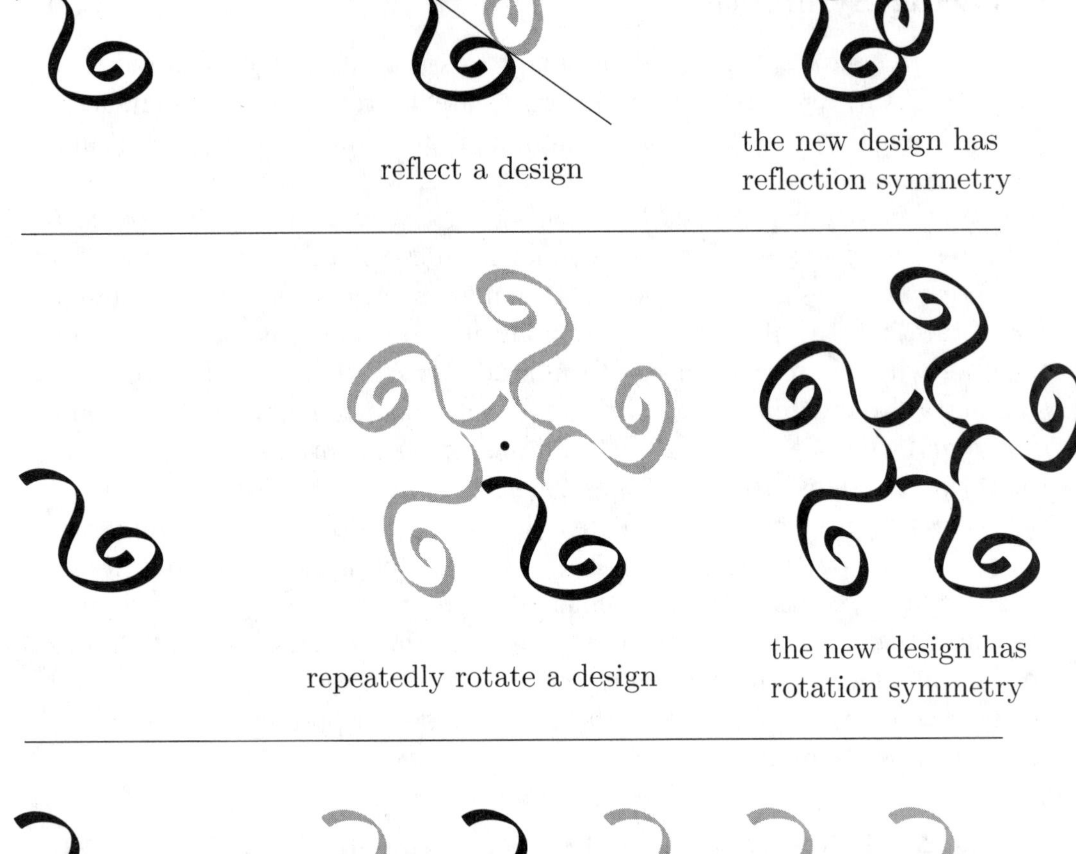

Figure 8.16: Creating Symmetrical Designs

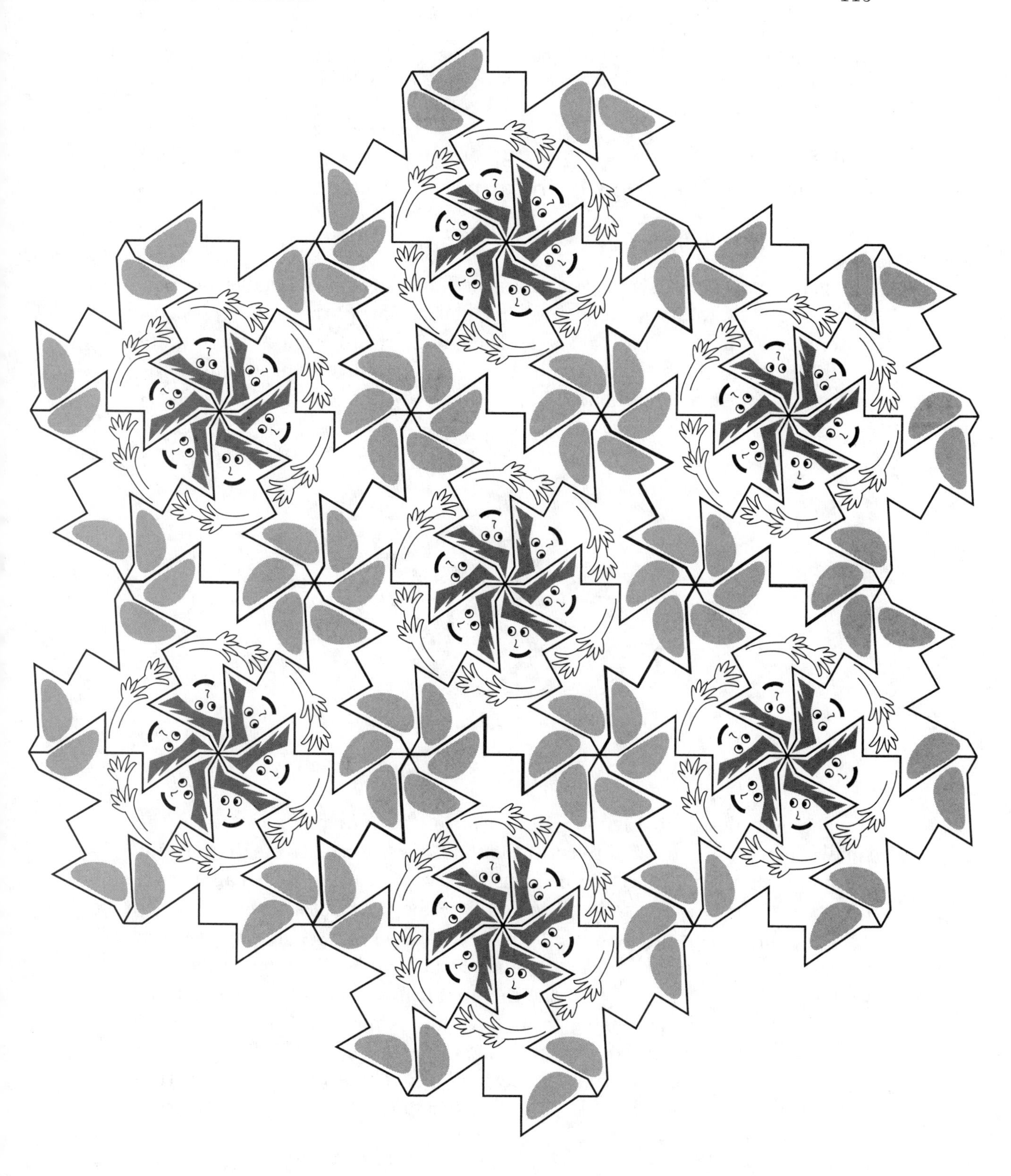

Figure 8.17: One Kind of Escher-type Design

create and modify Escher-type designs. By using Geometer's Sketchpad, you can get a feel for what makes these fascinating designs work.

Class Activity 8I: Creating Escher-type designs with Geometer's Sketchpad (for fun)

Analyzing Designs

By their very nature, symmetrical designs are made up of smaller designs, put together by rotating, reflecting, and/or translating the smaller designs. In some cases it's not hard to see how a symmetrical design is made up of smaller parts, such as the design in Figure 8.18. This design was made by repeatedly rotating a "curlicue."

Figure 8.18: A Simple Symmetrical Design

The quilt design in Figure 8.19 is a little more complex. Copies of the small square shown at the top left of Figure 8.20 were used to create the quilt design by repeatedly reflecting or translating and rotating the small square. Figure 8.20 shows a way to analyze and describe the quilt's design.

For some designs or patterns, it can be a challenge to see how the pattern is made up of copies of smaller designs. For example, Figure 8.21 shows a wallpaper pattern at the top of the picture. (Maybe too wild a pattern for most of us!) If you saw this pattern on a wall, it might not be clear how it was created out of a smaller design. The middle of Figure 8.21 shows a way of subdividing the pattern into smaller designs. The bottom of the picture shows how the pattern is broken into smaller parts. There are many other

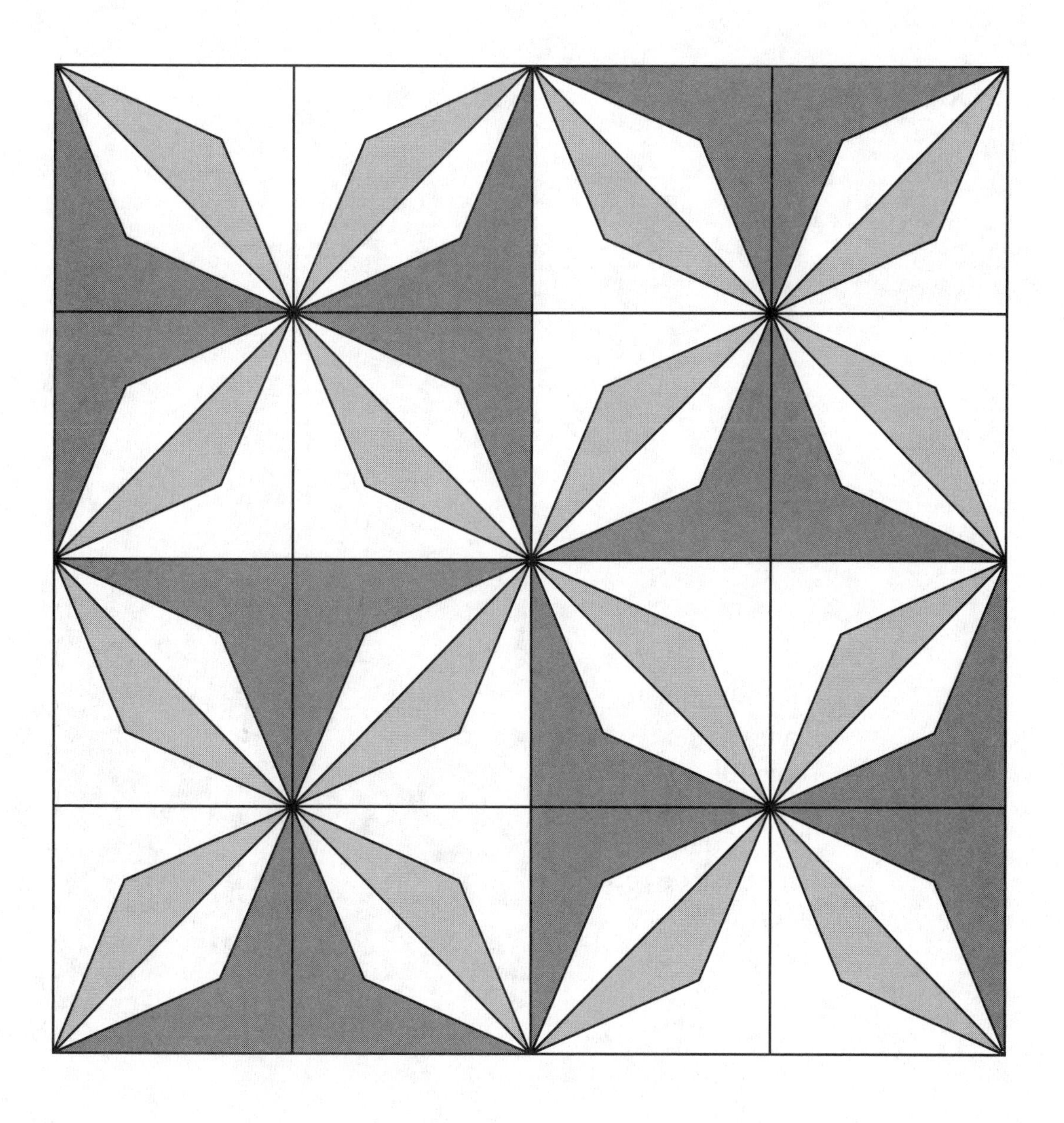

Figure 8.19: A Quilt

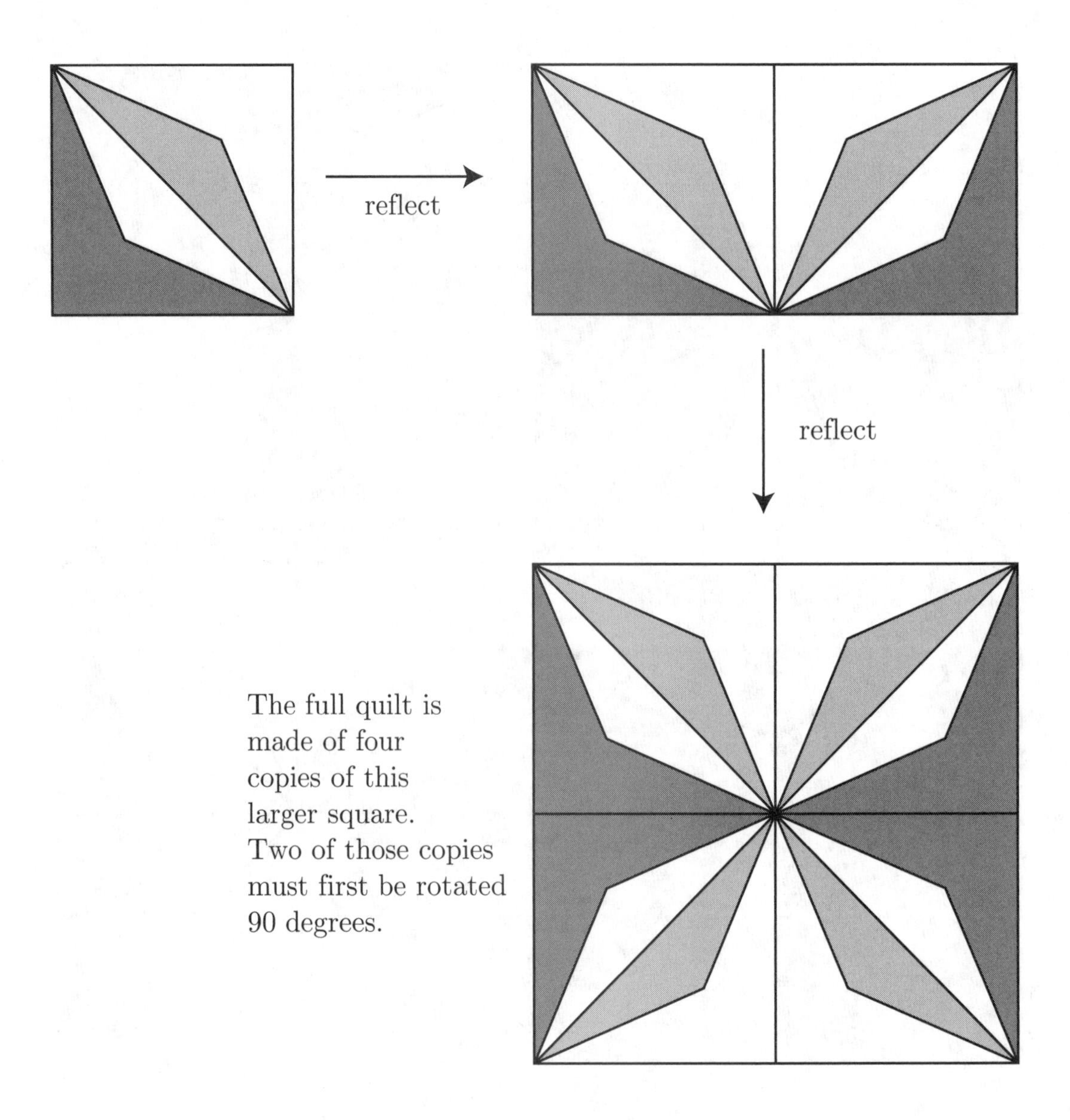

Figure 8.20: Analyzing a Quilt Design

Figure 8.21: Analyzing a Wallpaper Design

ways to do this subdividing—this is only one way. The picture shows that the wallpaper pattern can be created by translating one design over and over so as to fill up a plane (or a wall).

Class Activity 8J: Analyzing Designs

Exercises for Section 8.2 on Symmetry

1. Symmetrical designs are found throughout the world in all different cultures. Figure 8.22 shows a small sample of such designs. Determine what kinds of symmetry these designs have.

2. Practice Class Activities 8G and/or 8H.

3. For both of the patterns in Figure 8.23, find a piece of the pattern so that the pattern can be thought of as made up of repetitions of this smaller design. Show how the pattern is made up of this smaller, repeated design.

Answers to Exercises for Section 8.2

1. The Amish design has 6-fold rotation symmetry in addition to reflection symmetry with respect to 6 different lines: 3 lines that pass through the middles of opposite flower petals, and 3 lines that pass through the middles of opposite hearts.

   The Norwegian knitting design has 2-fold rotation symmetry in addition to reflection symmetry with respect to 2 different lines: one horizontal line and one vertical line. (Notice that the individual "snowflake" designs within the design have 4-fold rotation symmetry and have reflection symmetry with respect to horizontal, vertical, and two diagonal lines, but the entire Norwegian knitting design has less symmetry.)

   The Native American design has the same symmetries as the Norwegian knitting design.

   The design from a Persian rug has 4-fold rotation symmetry in addition to reflection symmetry with respect to 4 different lines: one vertical, one horizontal, and two diagonal lines.

3. See Figure 8.24.

Amish design found in central Pennsylvania

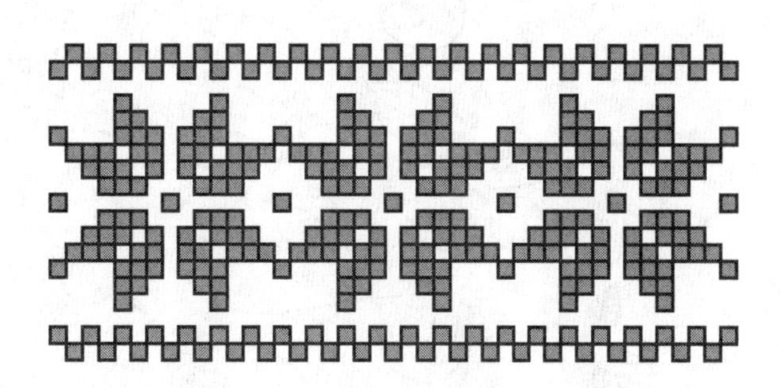

Norwegian knitting design

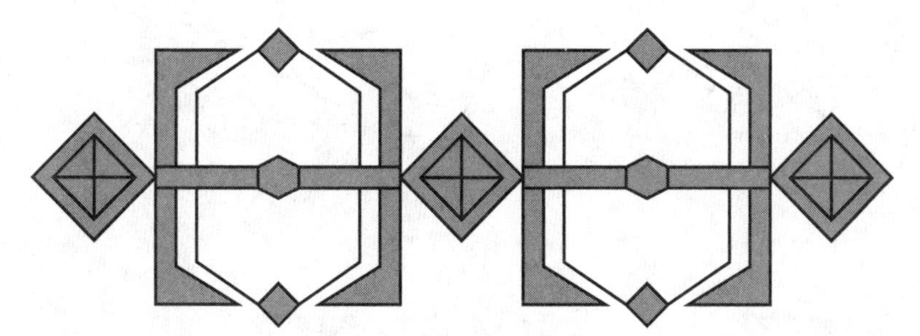

Native American design

design from a Persian rug

Figure 8.22: Some Symmetrical Designs from Around the World

Figure 8.23: Analyze These Patterns

Figure 8.24: Analysis of Patterns

Problems for Section 8.2 on Symmetry

1. Find examples of symmetrical designs that are common to some culture in some part of the world that you are interested in. This can be from a modern-day culture or a historical one. Make copies of the designs (either by photocopying or by drawing them) and determine what kinds of symmetry the designs have.

2. Draw a simple asymmetrical shape or design. Now draw copies of your shape or design so as to create a single new design that has both 2-fold rotation symmetry and translation symmetry *simultaneously.*

3. Draw a simple asymmetrical shape or design. Now draw copies of your shape or design so as to create a single new design that has both 4-fold rotation symmetry and translation symmetry *simultaneously.*

4. Draw a simple asymmetrical shape or design. Now draw copies of your shape or design so as to create a single new design that has both 2-fold rotation symmetry and reflection symmetry *simultaneously.*

5. Draw a simple asymmetrical shape or design. Now draw copies of your shape or design so as to create a single new design that has both 4-fold rotation symmetry and reflection symmetry *simultaneously.*

6. Draw a simple asymmetrical shape or design. Now draw copies of your shape or design so as to create a single new design that has rotation, reflection, and translation symmetry *simultaneously.*

7. Find a symmetrical quilt, rug, or wallpaper pattern, or find a picture of a symmetrical quilt, rug or wallpaper pattern. Determine smaller pieces of your quilt, rug or wallpaper pattern and describe how the quilt, rug, or wallpaper pattern can be thought of as put together from these smaller pieces.

8.3 Congruence

Informally, two shapes (either in a plane or in space) that are the same size and shape are called *congruent.* What information about two shapes will guarantee that they are congruent? Surprisingly, the situation is very different for triangles than it is for other shapes. As we'll see, this is related to the fact that triangles are structurally stable, whereas other polygons are not. This, in turn, explains some standard practices in building construction. So, even though the study of congruence may seem purely abstract and theoretical, it is in fact related to very real, practical issues.

Two shapes or designs in a plane are said to be **congruent** if there is a rotation, a reflection, a translation, or a combination of these transformations, that take one shape or design to the other shape or design. For example, Figure 8.25 shows two congruent shapes. They are congruent because the shape on the right was created by first translating the shape on the left horizontally to the right 2 inches, and then rotating the translated shape 90° about the point shown.

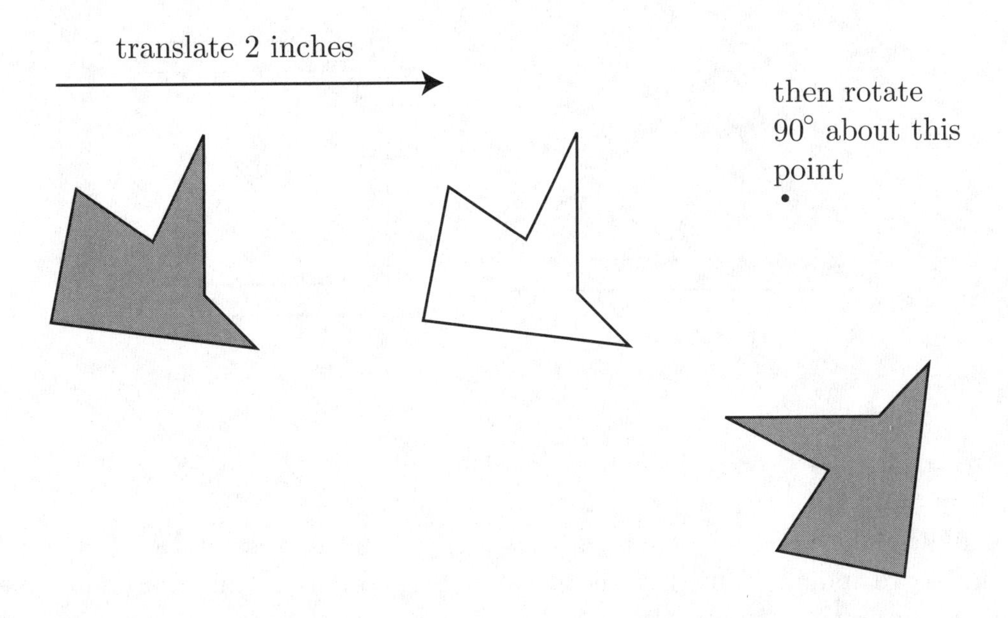

Figure 8.25: Congruent Shapes

In practice, if you have two shapes or designs drawn on two semi-transparent pieces of paper, you can check if they are congruent by "matching them up." This may require sliding the pieces of paper around, rotating them, and perhaps even flipping one piece of paper over (which has the effect of reflecting the shape or design). If it is possible to move the papers around so that the shapes match up, then the shapes are congruent; if it is not possible to match the shapes up, then they are not congruent.

Class Activity 8K: Triangles and Quadrilaterals of Specified Side Lengths

Class Activity 8L: Triangles with an Angle, a Side, and an Angle Specified

Congruence of Triangles

If you did Class Activity 8K, then you probably noticed that the triangle made by threading three pieces of straw together was rigid, whereas the quadrilateral made by threading four pieces of straw together was "floppy" (see Figures 8.26, 8.27, and 8.28). In other words, the triangle's sides

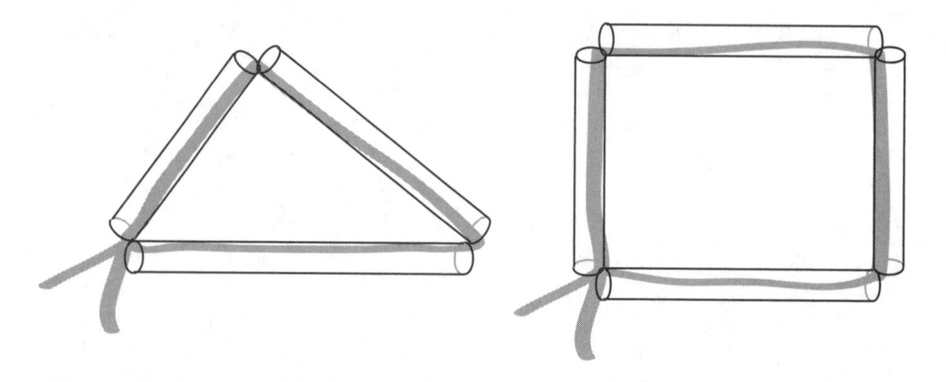

Figure 8.26: A Triangle and a Quadrilateral Made of Straws

could not be moved independently, whereas the quadrilateral's sides *could* be moved independently, so as to form many different quadrilaterals. We can interpret this in terms of congruence: *all triangles with sides of length 3 inches, 4 inches, and 5 inches are congruent*, whereas there are quadrilaterals with sides of length 3 inches, 4 inches, 3 inches, 4 inches (in that order) that are *not* congruent.

What if you had a triangle with sides of lengths other than 3 inches, 4 inches, and 5 inches? Imagine making this different triangle out of straws as well. Don't you think it would also turn out to be rigid like the 3 inch, 4 inch, 5 inch one? In fact, any triangle—no matter what the lengths of the sides are—is structurally rigid. This can be stated in terms of congruence:

Given a triangle that has sides of length a, b, and c units, it is

Figure 8.27: Straws Forming a Rectangle

Figure 8.28: The Same Straws Forming a Parallelogram

congruent to all other triangles that have sides of length a, b, and c units.

This last statement is called **side-side-side congruence** or **SSS congruence** for triangles. This is the mathematical way of expressing the fact that triangles are structurally rigid. This structural rigidity of triangles makes them common in building construction. For example, the frame of house under construction is temporarily braced with additional pieces of wood to keep the walls from falling over (see Figure 8.29). These additional pieces of wood create triangles; since triangles are rigid, the walls are held tightly in place.

Figure 8.29: Extra Pieces of Wood Create Triangles for Stability

Triangles are also used in structures that need to be sturdy but relatively light-weight. For example, that the crane in Figure 8.30 consists of many triangles. Because the triangles are rigid, they make the crane sturdy without the use of solid metal, which would be very heavy.

Another situation where all triangles with the specified properties are congruent occurs when one side of a triangle is specified and the two angles at either end are specified as well. For example, if you did Class Activity 8L, then you probably noticed that you could form two triangles that have a given line segment AB as one side, and have a 40° angle at A and a 60° angle at B (see Figure 8.31). One triangle is "above" AB and the other is "below"

Figure 8.30: A Crane is Made Out of Many Triangles

AB. However, these two triangles are congruent, as can be seen by reflecting across AB.

So all triangles that have AB as one side and that have a 40° angle at A and a 60° angle at B are congruent. The same will be true of other angles as well—as long as the two specified angles add to less than 180° (otherwise a triangle cannot be formed, since all three angles in a triangle must add to 180°). Thus in general:

> if a line segment is specified as a side of a triangle, and if two angles that add to less than 180° are specified at the two ends of the line segment, then all triangles formed from that line segment and those angles are congruent.

This last statement is called **angle-side-angle congruence** or **ASA congruence** for triangles.

Exercises for Section 8.3 on Congruence

1. What is SSS congruence?

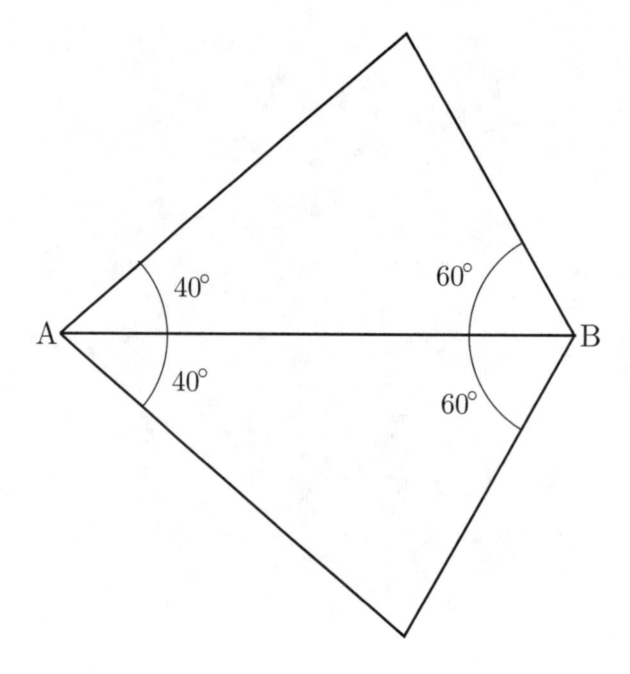

Figure 8.31: The Two Triangles That Have AB as One Side and Have Specified Angles at A and B

2. What is ASA congruence?

3. Suppose someone tells you that they have a garden with four sides, two of which are 10 feet long and opposite each other, and the other two of which are 15 feet long and opposite each other. With this information do you know the exact shape of the garden, or not? Take into account that the person might be eccentric!

Answers to Exercises for Section 8.3

1. See text.

2. See text.

3. No, this information alone does not allow you to determine the shape of the garden. Although most people would probably make their garden rectangular, someone with an artistic flair might make the garden in

the shape of a parallelogram that is not a rectangle. You can simulate this with four strung together straws, as in Class Activity 8L and Figure 8.26.

Problems for Section 8.3 on Congruence

1. Suppose that Ada, Bada and Cada are three cities and that Bada is 20 miles from Ada, Cada is 30 miles from Bada, and Ada is 40 miles from Cada. There are straight line roads between Ada and Bada, Ada and Cada, and Bada and Cada.

   (a) Draw a careful and precise map showing Ada, Bada, and Cada and the roads between them, using a scale of 10 miles = 1 inch. Describe how to use a compass to make a precise drawing.

   (b) If you were to draw another map, or if you were to compare your map to a classmate's, how would they compare? In what ways might the maps differ, in what ways would they be the same? Which criterion for triangle congruence is most relevant to this?

2. This problem continues the investigation of problem 9 in Section7.2 on how big your reflected face appears in a mirror. Figure 8.32 shows a side view of a person looking into a mirror.

   (a) What is the significance of the points F and G in Figure 8.32? Explain why, referring to the principles of reflection.

   (b) Using the theory of similar triangles discussed in this section, explain why triangles ABF and CBF are congruent and explain why triangles CDG and EDG are congruent.

   (c) Use your answer to part (b) to explain why the length of BD is half the length of AE. Therefore explain why the reflection of your face in a mirror is half as long as your face's actual length. (Notice that BD and FG have the same length.)

3. Ann and Kelley are standing at the bank of a river, wondering how wide the river is. Ann is wearing a basball cap, so she comes up with the following idea: she lowers her cap until she sees the tip of the visor just at the opposite bank of the river. She then turns around to face away from the river, being careful not to move her head or cap, and has

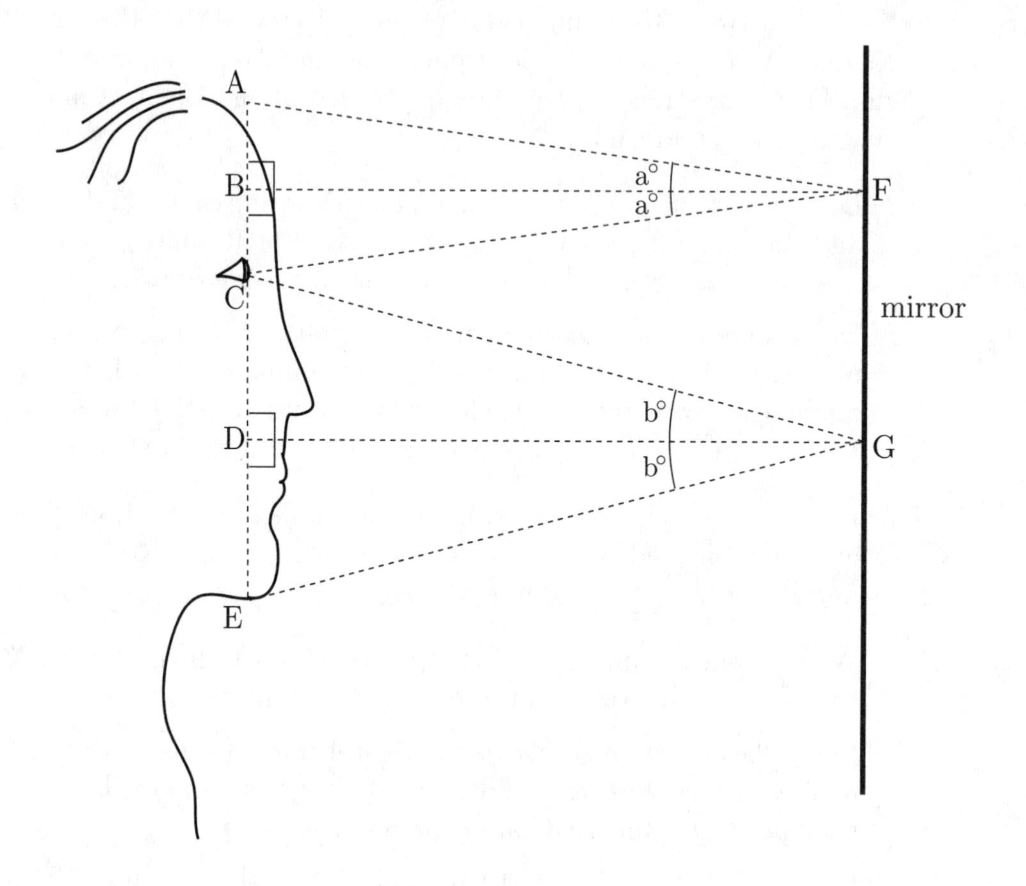

Figure 8.32: A Side View of a Person Looking in a Mirror

Kelly walk to the spot where she can just see Kellys shoes. By pacing off the distance between them, Ann and Kelly figure that Kelly was 50 feet away from Ann. If the ground around the river is level, what, if anything, can Ann and Kelly conclude about how wide the river is? Relate this to triangle congruence!

8.4 Similarity

In the previous section we discussed the notion of congruence, which is the way to say that two geometric objects are "equal": two shapes are congruent if they are the same shape and size. But what if there are two shapes or objects that are identical *except* for their sizes, as in Figure 8.33? These shapes are called *similar*. The notion of similarity is extremely useful. Similarity is used in solving many problems in geometry; it also has a wide range of practical applications, such as in surveying and map making. Representational drawing involves the concept of similarity, whether or not it is used consciously.

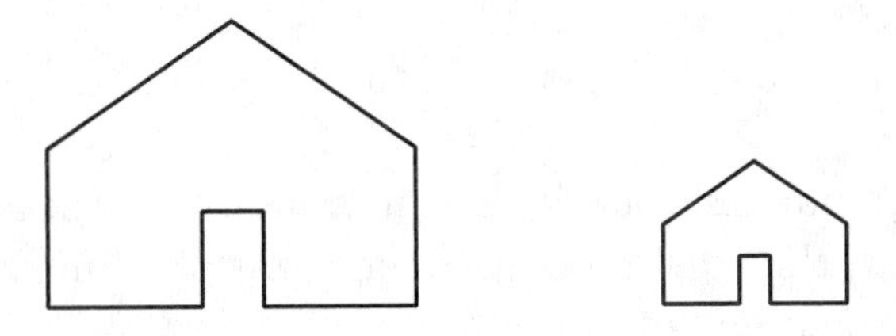

Figure 8.33: Similar Shapes

Informally, we say that two objects that have the same shape, but not necessarily the same size, are **similar**. Another way to say this is: two objects or shapes are similar if one object represents a scaled version of the other (scaled up or down). For example, a scale model of a train is similar (at least on the outside!) to the actual train it is modeled on. The network of streets on a street map is similar to the network of real streets it represents—at least if the streets are on flat ground.

One way to create a shape or design that is similar to another shape or design is by using grids of lines. If you enjoy doing crafts, then you may already be familiar with this technique. Draw a network of equally spaced parallel and perpendicular grid lines over the design you want to scale (or use an overlay of such grid lines), as shown in Figure 8.34. If you want the new design to be, say, twice as wide and twice as long, then make a new network of parallel and perpendicular grid lines that are spaced twice as wide as the original grid lines. Now copy the design onto the new grid lines, square by square. The new design will be similar to the original design.

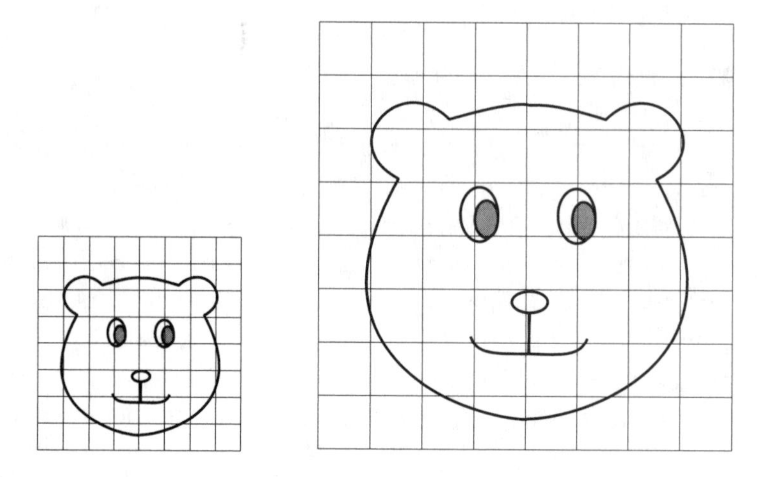

Figure 8.34: Using a Grid to Make Similar Pictures

A key feature of similarity is this: if two objects (in a plane, or even in space) are similar, then distances between corresponding parts of the objects *all scale by the same factor*. More precisely, if two objects are similar, then there is a positive number, k, such that the distance between two points on the second object is k times as long as the distance between the corresponding points on the first object. This number k is called the **scale factor** (or **scaling factor**) from the first object to the second object.

For example, the box of a toy model car might indicate that the toy car is a 24 : 1 scale model of an actual car. This means that the scale factor from the toy car to the actual car is 24, or equivalently, that the scale factor from the actual car to the toy car is $\frac{1}{24}$. In particular, the length, width, and height of the actual car are 24 times the length, width, and height, respectively, of the toy car. Similarly for other distances, such as the width of the windshield, or the length of the hood: each distance on the actual car is 24 times the corresponding distance on the toy car.

Warning: scale factors only apply to lengths, *not* to areas or volumes!

In many practical situations, two shapes or objects are given as being similar. If various lengths on the one object are known, then it's possible to determine the corresponding lengths on the other object, as long as at least one of the corresponding lengths is known. There are three common ways to do this, and all involve the scale factor (even if indirectly). It will be best to illustrate the three ways with a specific example.

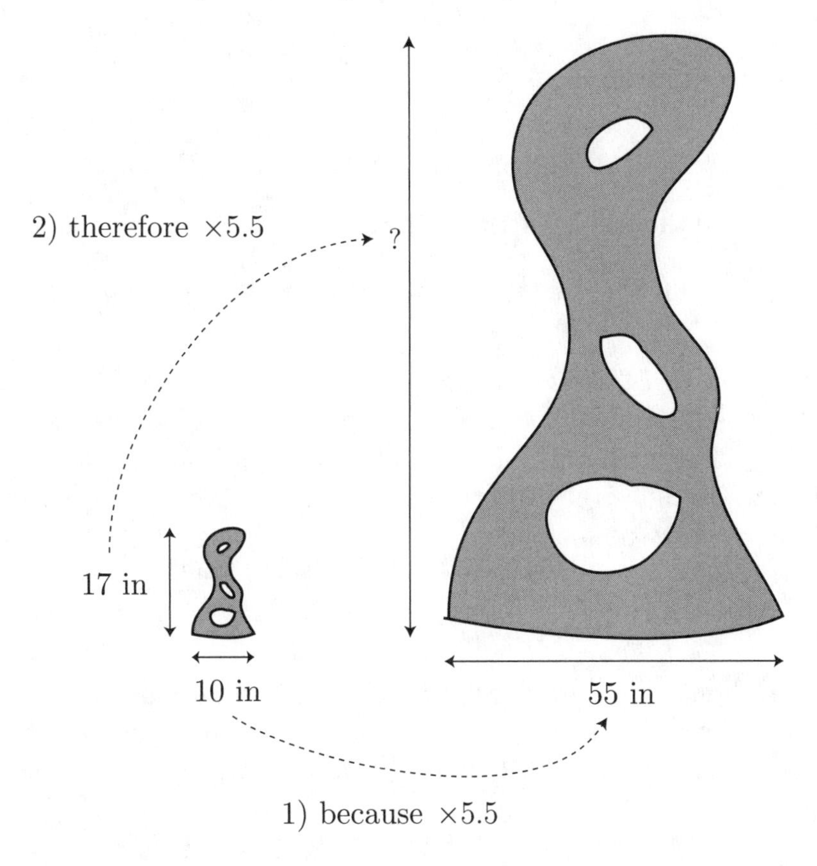

Figure 8.35: Using the "Scale Factor" Method

Problem: Suppose that an artist creates a scale model of sculpture. The scale model is 10 inches wide and 17 inches tall. If the actual sculpture is to be 55 inches wide, then how tall will the actual sculpture be?

Solutions:

1. *"Scale Factor" Method:* (See Figure 8.35.) Since the scale model and the actual sculpture are to be similar, there is a scale factor, k, from

the model to the actual sculpture such that every length on the actual sculpture is k times as long as the corresponding length on the scale model. This means that the width of the actual sculpture is $k \times 10$ inches and the height of the actual sculpture is $k \times 17$ inches. Since the width of the sculpture is given as 55 inches, therefore

$$k \times 10 \text{ inches} = 55 \text{ inches},$$

so

$$k = 55 \div 10 = 5.5.$$

Therefore the height of the sculpture is

$$k \times 17 \text{ inches} = 5.5 \times 17 \text{ inches} = 93.5 \text{ inches}.$$

2. *"Relative Sizes" Method:* (See Figure 8.36.) Since the scale model is 10 inches wide and 17 inches tall, it is $\frac{17}{10} = 1.7$ times as tall as it is wide. The actual sculpture should therefore also be 1.7 times as tall as it is wide. Since the sculpture is to be 55 inches wide, it should be

$$1.7 \times 55 \text{ inches} = 93.5 \text{ inches}$$

tall.

Why should the actual sculpture also be 1.7 times as tall as it is wide? This is because there is a scale factor, k, such that the width and height of the sculpture is $k \times 10$ and $k \times 17$, respectively. Therefore

$$\frac{\text{sculpture height}}{\text{sculpture width}} = \frac{k \times 17}{k \times 10} = \frac{17}{10} = 1.7.$$

3. *"Set up a Proportion" Method:* As before, since the scale model and the actual sculpture are to be similar, there is a scale factor, k, from the model to the actual sculpture. But

$$k = \frac{\text{width of sculpture}}{\text{width of model}},$$

and

$$k = \frac{\text{height of sculpture}}{\text{height of model}},$$

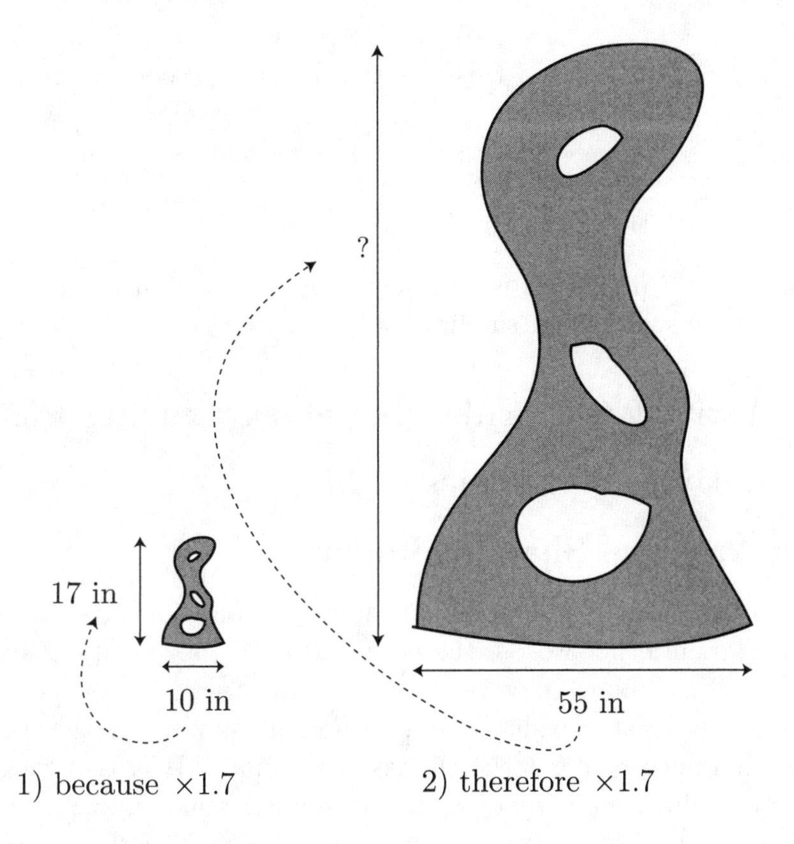

Figure 8.36: Using the "Relative Sizes" Method

therefore

$$\frac{\text{width of sculpture}}{\text{width of model}} = \frac{\text{height of sculpture}}{\text{height of model}}.$$

All the quantities in this last equation are known, except for the height of the sculpture. Let's let h stand for the height of the sculpture in inches. Then, substituting the know quantities, we have

$$\frac{55 \text{ in}}{10 \text{ in}} = \frac{h \text{ in}}{17 \text{ in}}.$$

But we know that we can determine whether two fractions are equal by cross-multiplying. So because the previous equation is true, therefore

$$55 \times 17 = 10 \times h.$$

Dividing both sides of this equation by 10, we see that

$$h = \frac{55 \times 17}{10} = 93.5,$$

so the sculpture is 93.5 inches tall.

Although the third method is more common, notice that the logic behind it is a little more subtle than the first two methods.

Class Activity 8M: Enlarging and Reducing Pictures

Class Activity 8N: Scales of Maps

When Are Two Shapes Similar?

So far, in the situations we have encountered, objects have been *given* as being similar, in other words, the very nature of the situation tells us that the objects are similar. However, there can be cases where it is not entirely obvious or clear that two shapes in question are similar. In those cases, how can we tell whether or not the shapes are similar? It is tempting to think that one can always tell "by eye," but this is not reliable. Figure 8.37 shows two triangles that appear to be similar, but are not. Although we will not study the question of determining similarity in full generality, we will study the special case of triangles, where there is an especially useful criterion for similarity.

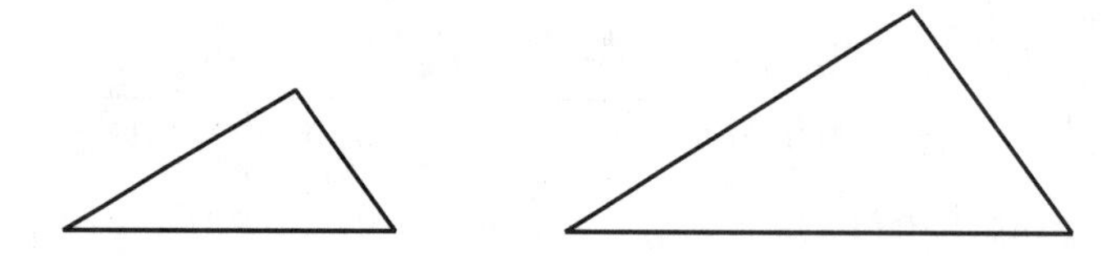

Figure 8.37: Triangles That Appear Similar but Are Not

Here is the criterion for similarity of triangles: two triangles are similar exactly when the two triangles have the same angles. To clarify: two triangles are similar exactly when it is possible to match each angle of the first triangle with an angle of the second triangle in such a way that matched angles are equal. So, for example, we can determine that the triangles in Figure 8.38 are similar because they have the same angles.

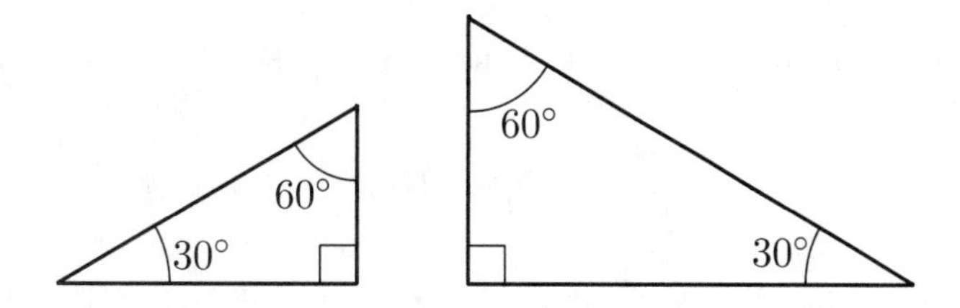

Figure 8.38: Similar Triangles

Actually, notice that in order to check if two triangles are similar, you really only need to check that *two* of their angles are the same. This is because the sum of the angles in a triangle is always 180°, so if two angles are known, the third one is determined as the angle that makes the three add to 180°. Thus if two angles in one triangle are the same as two angles in another triangle, then the third angles will automatically also be equal.

Here's an example of how to use the criterion for triangle similarity. In Figure 8.39, what is the length of side DE? First, notice that the picture

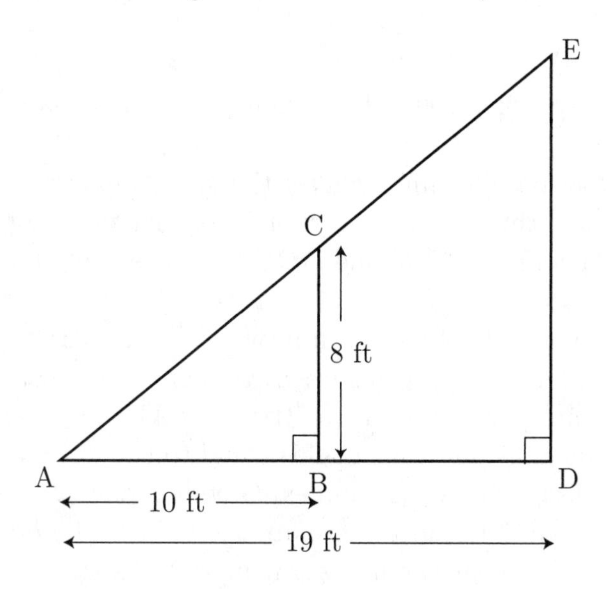

Figure 8.39: What is the Length of DE?

shows that triangles ABC and ADE have two angles in common: the angle at A and a right angle. Therefore these triangles are similar. Now we can apply any of our three methods to determine the length of side DE. Using the "scale factor" method, the scale factor is $\frac{19}{10}$, so the length of DE is

$$\frac{19}{10} \times 8 \text{ ft } = 15.2 \text{ ft.}$$

More generally, as illustrated in Figure 8.40, if two triangles ABC and ADE share an angle at A, and if BC and DE are parallel, then it turns out that the triangles ABC and ADE are similar. This is because of the fact that

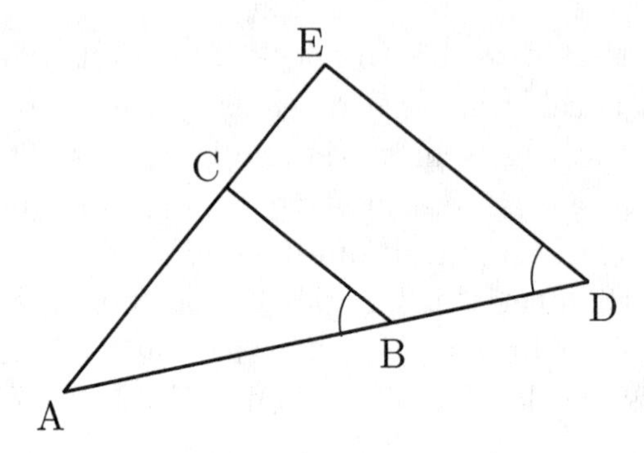

Figure 8.40: Parallel Lines Create Similar Triangles

a line crossing two parallel lines makes the same angle with both parallel lines. We will leave this last fact unexplained, although it probably seems quite plausible. Therefore ABC and ADE have the same angles and so are similar.

Another common type of situation in which similar triangles are created is illustrated in Figure 8.41. In this case, two lines cross at a point A. If the lines BC and DE are parallel, then the triangles ABC and ADE are similar. This is because the angle at A in ABC is equal to the angle at A in ADE, as described in Exercise 5 on page 25 and explained on page 28. Also, the angles at B and at D are equal because these are the points where the parallel line segments BC and DE meet the line segment BD. As stated in the previous paragraph, we simply assume this to be true, and it should seem plausible. Therefore, two out of three angles in ABC and ADE are equal, so all three must be equal, and thus the triangles are similar.

Why does the "three equal angles" criterion for triangle similarity work? This is more advanced, and we won't discuss this thoroughly, but basically, it is because the process of scaling doesn't change angles. If two triangles are

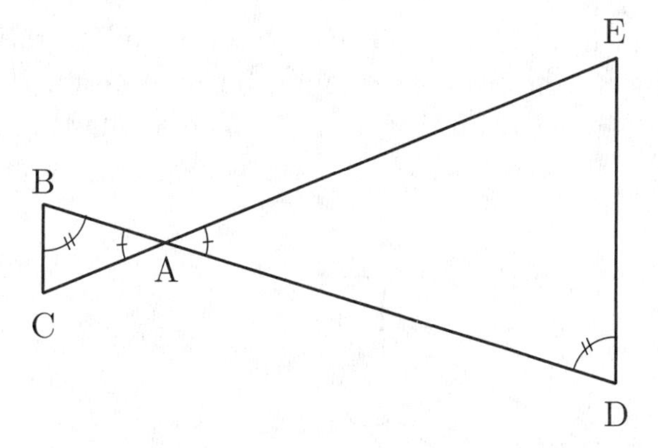

Figure 8.41: Another Way Parallel Lines Create Similar Triangles

similar, then one can be scaled so as to match the other, and since this scaling doesn't change angles, the triangles must have the same angles. Conversely, suppose two triangles have the same angles, say a, b, and c. Then one of the triangles can be scaled so that the sides common to angles a and b now have the same length in both triangles. In the scaling process, the angles don't change, so now if you match up the sides that have the same length, they both have angles a and b at either end, and so these triangles are identical. Therefore the original triangles must have been similar.

Using Similar Triangles to Determine Distances

Similar triangles can be used in practical situations to find an unknown distance or length when several other related distances and lengths are known. Often, the tricky part in applying the theory of similar triangles is figuring out what the similar triangles are and how to draw a sketch of them. You will need to apply your visualization skills to help you make a sketch of the situation, showing the relevant components. In the following examples, think carefully about why the pictures were drawn the way they were. If they had been drawn from a different perspective would you see the relevant similar triangles? (In some cases yes, but in many cases, no.)

Here is an example of a realistic situation where the theory of similar triangles applies. When we look at objects in the distance, they appear to be smaller than they actually are. One way to describe how big an object appears to be is to compare it to the size your thumb by stretching your arm out straight in front of you, closing one eye, and "sighting" from your thumb to the object you are considering. This situation creates a pair of similar triangles, as shown in Figure 8.42. The similar triangles are ABC

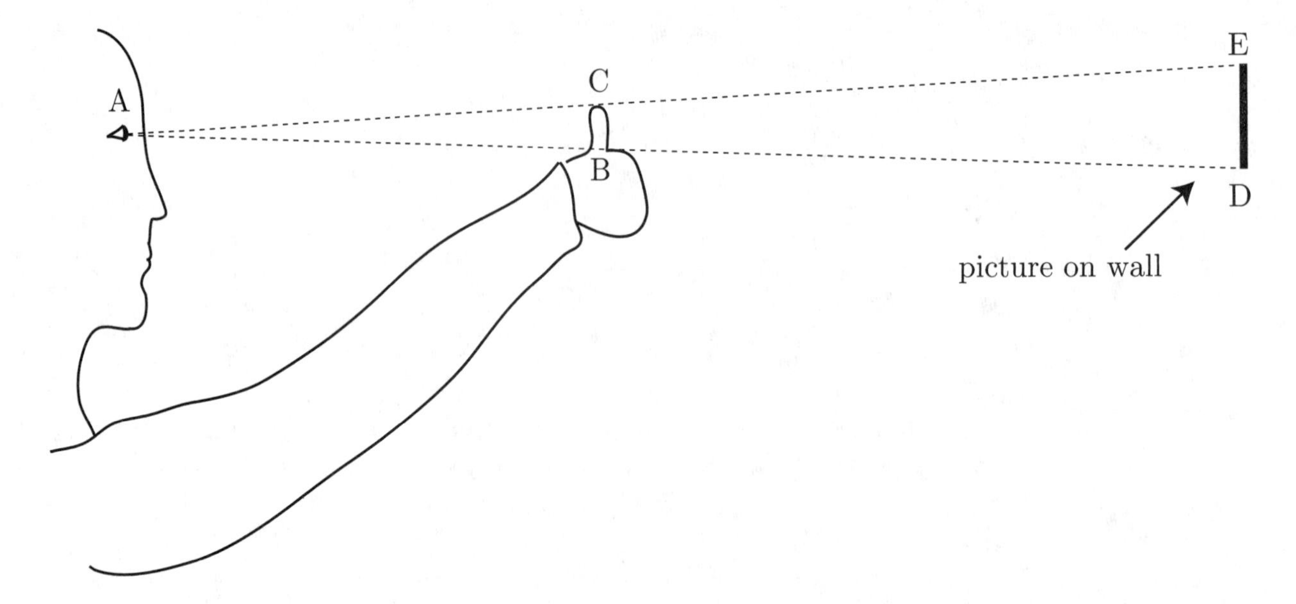

Figure 8.42: "Thumb Sighting" a Picture on a Wall

(eye, base of thumb, top of thumb) and ADE (eye, bottom of picture, top of picture). These triangles are similar because the angles at A are equal and the angles at B and D are equal since the thumb and the picture are parallel (see page 140 and Figure 8.40).

Here's a way to use this "thumb sighting" to find a distance. Let's say that I see a man standing in the distance and let's say that at this distance, he appears to be "1 thumb tall." If the man is actually 6 feet tall, then approximately how far away is he? As explained above, this "thumb sighting" creates similar triangles (see Figure 8.43, which is *not* drawn to scale!). The distance from my sighting eye to the base of my thumb on my outstretched arm is 22 inches. My thumb is 2 inches tall. So the distance from my eye to my thumb is $\frac{22}{2}$ times as long as the length of my thumb. Therefore, according to the "relative sizes" method, the man is

$$\frac{22}{2} \times 6 \text{ feet } = 66 \text{ feet}$$

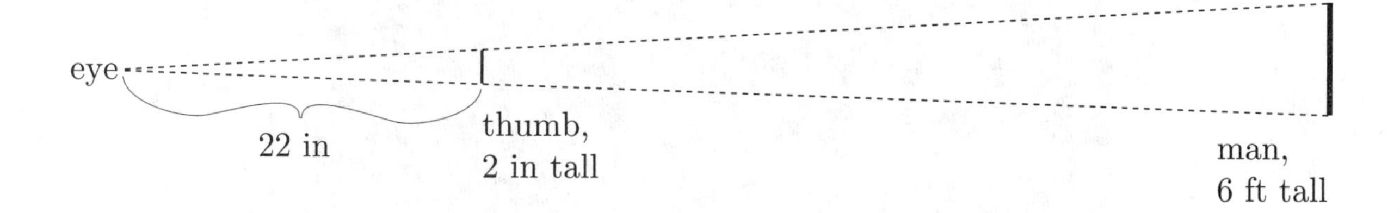

Figure 8.43: "Thumb Sighting" a Man

away. You might object that this tells us the distance from my eye to the man's feet, but, because the man is fairly far away, this is almost identical to the distance along the ground from my feet to his feet. Plus, we are only getting a rough estimate of how far away the man is because all the measurements used are very rough. So this extra bit of inaccuracy is insignificant.

How could you modify the method of "thumb sighting" to make it more useful and more accurate? You could use other objects besides your thumb for sighting—a ruler held vertically would be much more accurate, for example. And what if you didn't just hold the ruler in your hand, but had it attached to a fixed length of pole—that would also create greater accuracy. Making these sorts of improvements leads to some of the surveying equipment used today. When you see people at construction sights looking through a device on a tripod at a pole in the distance (such as in Figure 8.44), these people are surveying.

Some kinds of surveying equipment measure distances based on the theory of similar triangles, using a somewhat more elaborate method than the primitive "thumb sighting". In order to measure the distance between the surveying equipment and a pole of known height in the distance, the person surveying looks through the equipment and "sights" the pole. Instead of a thumb, there is a kind of ruler inside the surveying equipment.

Class Activity 8O: Measuring Distances by "Surveying"

Class Activity 8P: Determining the Height of a Tree

Exercises for Section 8.4 on Similarity

1. Ryan wants to make a large model of the Great Pyramid at Giza in Egypt. The pyramid is about 481 feet tall and each of its four sides is

Figure 8.44: Surveying

about 756 feet long at the base. Ryan wants the sides of his model's pyramid to be 5 feet long at the base. How tall should Ryan's model pyramid be?

Use the three different methods: *scale factor*, *relative sizes*, and *set up a proportion* to solve the problem. In each case, explain the reasoning behind the method.

2. (Continuation of the previous problem.) Now Ryan also wants to make a model of the Sphinx that is near the Great Pyramid at Giza in Egypt. The sphinx is about 240 feet long and 66 feet high. Ryan wants to use the same scale as he did with his model pyramid. How long and how tall should his model sphinx be?

3. Ms. Bullock's class is making a display of the Sun and the planets. Each planet will be depicted as a circle. They want to show the Earth as a circle of diameter 10 cm and they want to show the correct relative sizes of the planets and the Sun. Given the information below, what should the diameters of these scale drawings be? Since the Sun is so big, it wouldn't be practical to make a full model of the Sun; how could the children make a sliver of the Sun of the correct size?

Heavenly Body	Approximate Diameter
Sun	1,392,000 km
Mercury	4,900 km
Venus	12,100 km
Earth	12,700 km
Mars	6,800 km
Jupiter	138,000 km
Saturn	115,000 km
Uranus	52,000 km
Neptune	49,500 km
Pluto	2,300 km

4. A sculptor makes a 15 inch wide, 8 inch deep, and 27 inch high scale model for a sculpture she plans to carve out of marble. She finds a 5 foot wide, 4 foot deep, and 10 foot high block of marble. What will the finished dimensions of the sculpture be if she makes the sculpture as large as possible?

5. The website

`http://www.exploratorium.edu/science_explorer/pringles_pinhole.html`

shows how to make a **camera obscura** from a Pringles® potato chip can. A camera obscura is a fun device that projects images onto a screen through a small hole. It illustrates a fundamental idea of photography: that of projecting an image through a small hole. You can make a camera obscura from any tube by cutting off a piece of the tube (about 2 inches from one end), placing semi-transparent paper or plastic over the place where you cut, and taping the tube back together, with the semi-transparent paper now in the middle of the tube. Cover one end of the tube with aluminum foil and use a pin to poke a small hole in the middle of the foil. Cover the side of the tube with aluminum foil to keep light out. Now put the open end of the tube up to your eye and look at a well lit scene. You should see an upside down projected image of the scene on the semi-transparent paper.

Suppose you make a camera obscura out of a tube of diameter 3 inches, and suppose you put the semi-transparent paper or plastic 2 inches from the hole in the aluminum foil. How far away would you have to stand from a 6 foot tall man in order to see the entire man on the camera obscura's screen?

6. Suppose that you go outside on a clear night when a lot of the Moon is visible and use a ruler to "sight" the Moon (as described in Class Activity 8O). Use the data below to determine how big the Moon will appear to you on your ruler. (You will also need to measure the distance from you eye to a ruler that you are holding with your arm stretched out.) Some night when the Moon is visible, try this out and verify it!

   The distance from the Earth to the Moon is approximately 384,000 km.

   The diameter of the Moon is approximately 3,500 km.

Answers to Exercises for Section 8.4

1. *Scale Factor Method:* The scale factor from the pyramid to the model is a number k such that

$$k \times 756 \text{ ft } = 5 \text{ ft } .$$

   So

$$k = \frac{5}{756}.$$

   Therefore the height of the model should be

$$k \times 481 \text{ ft } = \frac{5}{756} \times 481 \text{ ft } = 3.2 \text{ ft } .$$

   Ryan might want to convert this measurement to inches. He can do this by multiplying by 12, since there are 12 inches in each foot.

Relative Sizes Method: The height of the pyramid is $\frac{481}{756}$ times as long as its width along the base. The same should be true for the scale model. Therefore the height of the scale model is

$$\frac{481}{756} \times 5 \text{ feet} = 3.2 \text{ feet}.$$

Proportion Method: There are two ways of expressing the scale factor k (from the pyramid to the model) as a fraction. One is

$$k = \frac{5}{756}$$

(because the scale factor k is the number you multiply 756 feet by to get 5 feet). The other is

$$k = \frac{(\text{height of model pyramid in ft })}{481}$$

(because the scale factor k is the number you multiply 481 feet by to get the height of the model pyramid in feet). The two fractions that are equal to the scale factor k must also be equal to each other, therefore

$$\frac{5}{756} = \frac{(\text{height of model pyramid in ft})}{481}.$$

But we know that two fractions are equal exactly when their "cross multiples" are equal. Therefore

$$5 \times 481 = 756 \times (\text{height of model pyramid in ft}),$$

and so

$$\begin{aligned} \text{height of model pyramid} &= \frac{5 \times 481}{756} \text{ ft} \\ &= 3.2 \text{ ft.} \end{aligned}$$

2. The *relative sizes method* provides a straightforward way to do this problem. Because the pyramid is 756 feet wide and the Sphinx is 240 feet long, therefore the length of the Sphinx is $\frac{240}{756}$ times as long as the width of the pyramid. Similarly, because the Sphinx is 66 feet high, the height of the Sphinx is $\frac{66}{756}$ times as long as the width of the pyramid.

The same relationships should hold for the model pyramid and model Sphinx. Because the model pyramid is 5 feet wide, therefore the length of the model Sphinx should be

$$\frac{240}{756} \times 5 \text{ feet} = 1.6 \text{ feet},$$

and the height of the model Sphinx should be

$$\frac{66}{756} \times 5 \text{ feet} = .4 \text{ feet}$$

Ryan might want to convert these measurements to inches. He can do this by multiplying by 12, since each foot consists of 12 inches.

3. From the way the problem is stated, we infer that we want the collection of circles representing the planets and Sun to be similar to cross-sections of the actual planets—all with the same scale factor. If you use the *relative sizes* method, then you don't have to worry about converting kilometers to centimeters (or vice versa), and so you won't have to deal with huge numbers.

   Because the Earth's diameter is $12,700$ km and Mercury's diameter is $4,900$ km, therefore Mercury's diameter is $\frac{4,900}{12,700}$ times as long as the Earth's diameter. The same relationship should hold for the models of Mercury and the Earth. Since the model Earth is to have a diameter of 10 cm, therefore the model Mercury should have a diameter of

$$\frac{4,900}{12,700} \times 10 \text{ cm} = 3.9 \text{ cm}.$$

   Similarly for the other heavenly bodies:

Heavenly Body	Approximate Diameter of Circle Representing it
Sun	$\frac{1{,}392{,}000}{12{,}700} \times 10$ cm $= 1096$ cm $= 10.96$ m
Mercury	$\frac{4{,}900}{12{,}700} \times 10$ cm $= 3.9$ cm
Venus	$\frac{12{,}100}{12{,}700} \times 10$ cm $= 9.5$ cm
Earth	given as 10 cm
Mars	$\frac{6{,}800}{12{,}700} \times 10$ cm $= 5.4$ cm
Jupiter	$\frac{138{,}000}{12{,}700} \times 10$ cm $= 108.7$ cm
Saturn	$\frac{115{,}000}{12{,}700} \times 10$ cm $= 90.6$ cm
Uranus	$\frac{52{,}000}{12{,}700} \times 10$ cm $= 40.9$ cm
Neptune	$\frac{49{,}500}{12{,}700} \times 10$ cm $= 39$ cm
Pluto	$\frac{2{,}300}{12{,}700} \times 10$ cm $= 1.8$ cm

The circle representing the Sun should have a diameter of about 11 meters, and therefore a radius of about 5.5 meters. The children could measure a piece of string $5\frac{1}{2}$ meters long. One child could hold one end and stay in a fixed spot, while another child could attach a pencil to the other end and use it to draw a piece of a very large circle representing the Sun.

4. One way to solve this problem is to think about the scale factors you might use. Since each foot is 12 inches, 5 feet is $5 \times 12 = 60$ inches, 4 feet is $4 \times 12 = 48$ inches, and 10 feet is $10 \times 12 = 120$ inches. Thinking only about the width, the scale factor from the model to the sculpture, k, should be such that

$$k \times 15 = 60,$$

so

$$k = \frac{60}{15} = 4.$$

Similarly, thinking only about the depth and the height, the scale factors would be

$$k = \frac{48}{8} = 6,$$

and

$$k = \frac{120}{27} = 4.4,$$

respectively. The artist will need to use the smallest scale factor, otherwise her sculpture would require more marble than she has. So the artist should use $k = 4$, in which case the dimensions of the sculpture are as follows:

$$\begin{aligned} \text{width} &= 4 \times 15 \text{ in} &= 60 \text{ in} , \\ \text{depth} &= 4 \times 8 \text{ in} &= 32 \text{ in} , \\ \text{height} &= 4 \times 27 \text{ in} &= 108 \text{ in} . \end{aligned}$$

5. Figure 8.45 shows how light from an object enters the pinhole (at point A) and projects onto the screen inside a camera obscura. If the object

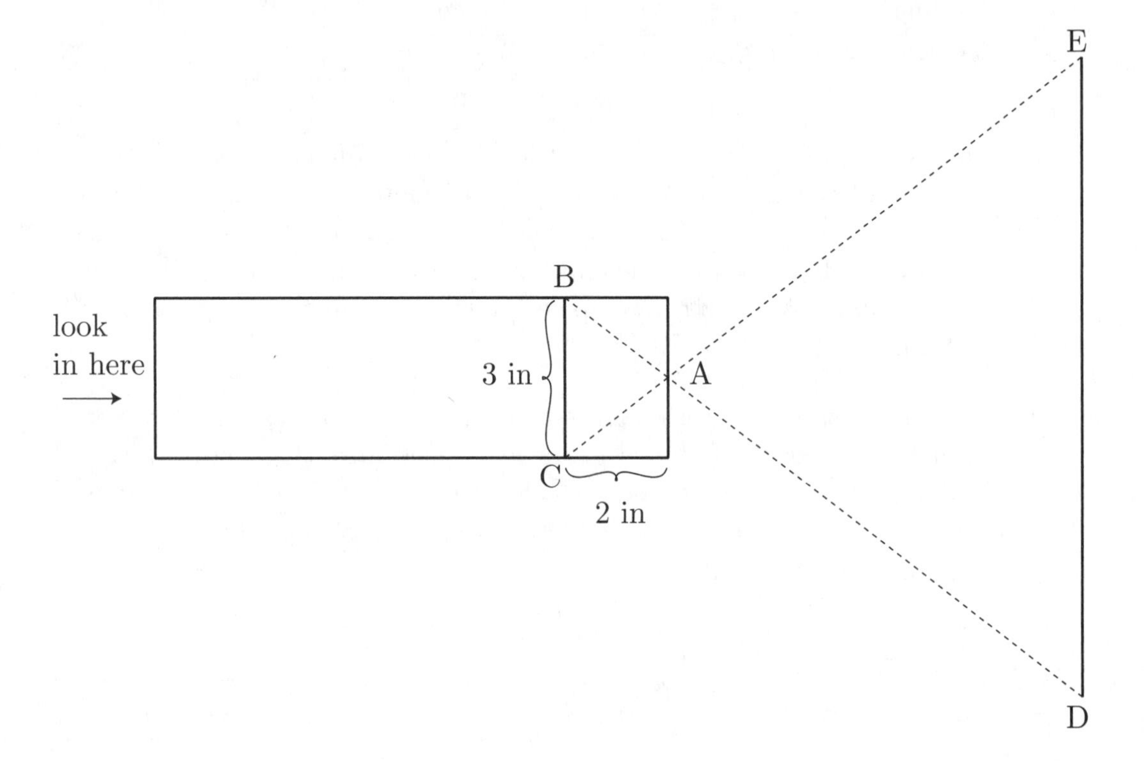

Figure 8.45: A Side View of a Camera Obscura

being looked at and the screen of the camera obscura are both vertical, then BC and DE are both vertical and are therefore parallel. Therefore, as described in the text, ABC and ADE are similar. Since these two triangles are similar, all corresponding distances on the triangle scale by the same scale factor. Since the distance from the screen of the camera obscura to the pinhole is $\frac{2}{3}$ times the length of the screen, therefore, by the *relative sizes* method, the distance from the pinhole to the man should be $\frac{2}{3}$ times the length of the man, which is $\frac{2}{3} \times 6$ feet $= 4$ feet. So you need to stand at least 4 feet from the man to see the full length of him on the camera obscura.

6. When you sight the Moon with a ruler you create similar triangles ABC and ADE as shown in Figure 8.46 (not to scale!). Since the diameter

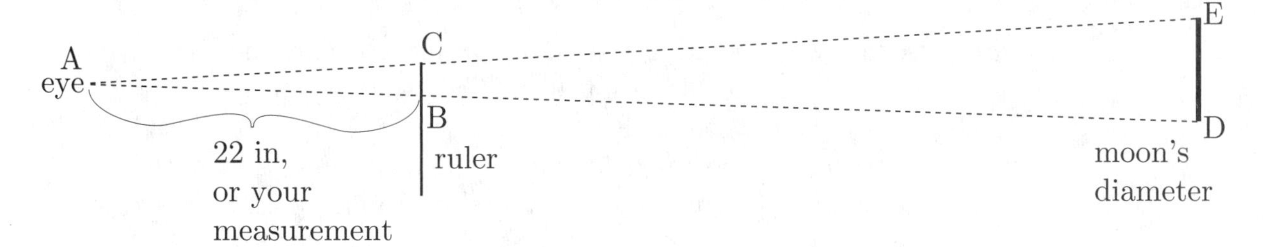

Figure 8.46: Sighting the Moon

of the Moon is $3,500$ km and the distance to the Moon is $384,000$ km, therefore the Moon's diameter (DE) is $\frac{3,500}{384,000}$ times as long as the the distance to the Moon (AD). The same relationship must hold for the apparent size of the Moon on the ruler (BC) and the distance from your eye to the ruler (AB): the apparent size of the Moon on the ruler must be $\frac{3,500}{384,000}$ times as long as the distance from your eye to the ruler (this is the *relative sizes* method). Notice that $\frac{3,500}{384,000}$ is about $\frac{1}{100}$. So if the distance from your eye to the ruler is about 22 inches, then the Moon's diameter will appear to be about .2 inches on your ruler.

Problems for Section 8.4 on Similarity

1. Ms. Winstead's class went outside on a sunny day and measured the lengths of some of their classmates' shadows. The class also measured the length of a shadow of a tree. Inside, the children made the following table:

	Tyler	Jessica	SunJae	Lameisha	tree
shadow length	33 in	34 in	32 in	34 in	22 feet
height	53 in	57 in	52 in	58 in	?
height ÷ shadow length	1.61	1.68	1.63	1.71	

Based on their table, the children estimated that the tree should be about 1.66 times as tall as its shadow was long. So the children figured that the tree is about 1.66×22 feet $= 37$ feet tall.

Explain how the reasoning of the children in Ms. Winstead's class is related to similar triangles (draw a picture to aid your explanation), and name the method that the children used to determine the height of the tree.

2. An art museum owns a paiting that it would like to reproduce in reduced size onto a 24 inch by 36 inch poster. The painting is 85 inches by 140 inches. Give your recommendation for the size of the reproduced painting on the poster—how wide and long do you suggest that it be? Draw a scale picture showing how you would position the reproduced painting on the poster.

3. Most ordinary cameras produce a negative, which is a small picture of the scene that was photographed (with reversed colors). A photograph is produced from a negative by printing onto photographic paper. When a photograph is printed from a negative, one of two things happens: either the printed picture shows the full picture that was captured in the negative, or the printed picture is cropped, showing only a portion of the full picture on the negative. In either case, the rectangle forming the printed picture is similar to the rectangular portion of the negative that it comes from.

   An ordinary 35 mm camera produces a negative in the shape of a $1\frac{7}{16}$ inch by $\frac{15}{16}$ inch rectangle. Some of the most popular sizes for printed pictures are $3\frac{1}{2}$ in $\times$ 5 in, 4 in $\times$ 6 in, and 8 in $\times$ 10 in. Can any of these size photographs be produced without either cropping the picture or leaving blank space around the picture? Why or why not? Explain your reasoning clearly.

4. Let's say that you are standing on top of a mountain, looking down at the valley below, where you can see cars driving on a road. You stretch out your arm, use your thumb to "sight" a car, and find that the car appears to be as long as your thumb is wide. Estimate how far away you are from the car. Explain your method clearly. (You will need to make an assumption to solve this problem. Make a realistic assumption and make it clear what your assumption is.)

5. (a) During a total solar eclipse, the Moon moves in front of the Sun, obscuring the view of the Sun from the Earth (at some locations on the Earth). Surprisingly, the Sun and Moon seem to "match up" during a solar eclipse, appearing to be almost identical in size as seen from the Earth. How does this situation give rise to similar triangles? Draw a sketch (it does not have to be to scale!).

   (b) Use the data below and either the *scale factor* method or the *relative sizes* method to determine the approximate distance of the Earth to the Sun (do not use the *set up a proportion* method). Explain the reasoning behind the method you use.

   The distance from the Earth to the Moon is approximately $384,000$ km.

   The diameter of the Moon is approximately $3,500$ km.

   The diameter of the Sun is approximately $1,392,000$ km.

6. On page 143 there is a description of how the theory of similar triangles can be used to measure distances in land surveying. This problem will help you understand the theory for how the altitude above sea level of a location can be determined by surveying as well.

   Suppose a person stands on the side of a hill at a point A. Using surveying equipment, the person measures that the distance to a point B up the hill is 38 feet. Surveying equipment can also be used to measure angles. So suppose that the surveying equipment reports that the line connecting A and B makes an angle of 15 degrees with a horizontal line, as shown in Figure 8.47. Suppose that by previous surveying, the point A is already known to be 523 feet above sea level.

   (a) The triangle in Figure 8.47 is drawn to scale. Use a ruler to make measurements on this triangle. Then use these measurements, together with mathematical reasoning and the information about points A and B given above to find the altitude of the point B above sea level.

   (b) Even if you measured carefully, your results won't be really accurate, so in practice one needs a better method than measuring a scale drawing. A more accurate method is to use trigonometry.

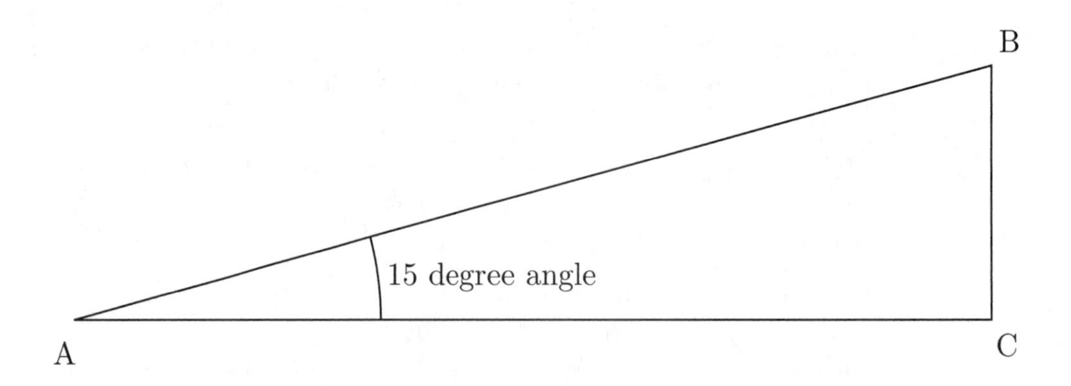

Figure 8.47: Surveying on a Hill

Here is what trigonometry tells us in this situation: if we declare the the distance from A to B to be 1 unit, then the distance from B to C is $\sin(15°)$ units $= .258819$ units and the distance from A to C is $\cos(15°)$ units $= .965926$ units.

Use the more accurate information from the paragraph above to find the altitude of point B above sea level *to the nearest inch.* Assume that point A is actually 523 feet and 3 inches above sea level and that the distance from point A to point B is actually 38 feet and 2 inches. Give your answer in feet and inches (for example: 682 feet and 10 inches).

7. Which TV screen appears bigger: a 50 inch screen viewed from 8 feet away or a 25 inch screen viewed from 3 feet away? Explain carefully. (TV screens are typically measured on the diagonal, so to say that a TV has a "50 inch screen" means that the diagonal of the screen is 50 inches long.)

8. A pinhole camera is a very simple camera made from a closed box with a small hole on one side. Film is put inside the camera, on the side opposite the small hole. The hole is kept covered until one wants to take a picture, when light is allowed to enter the small hole, producing an image on the film. Unlike an ordinary camera, a pinhole camera does not have a view finder, so with a pinhole camera, you can't see what the picture you are taking will look like.

(a) Suppose that you have a pinhole camera for which the distance from the pinhole to the opposite side (where the film is) is 3 inches. Suppose that the piece of film to be exposed (opposite the pinhole) is about 1 inch tall and $1\frac{1}{2}$ inches wide. Let's say you want to take a picture of a bowl of fruit that is about 15 inches wide and piled 6 inches high with fruit, and let's say that you want the bowl of fruit to fill up most of the picture. Approximately how far away from the fruit bowl should you locate the camera? Explain, using similar triangles.

(b) Explain why the image produced on film by a pinhole camera is upside down.

9. Imagine that the Moon were to remain the same size as it is now, but were to orbit the Earth at *half* the distance from the Earth as it currently does. If you went outside at night and "sighted" the Moon with a ruler (as described in Class Activity 8O and Exercise 6), how would its new sighted diameter compare to its current sighted diameter? Use pictures (which do not need to be to scale!) to help you explain your answer clearly. (The distance from the Earth to the Moon is approximately 384,000 km and the diameter of the Moon is approximately 3,500 km.)

10. Sue has a rectangular garden. If Sue makes her garden twice as wide and twice as long as it is now, will the area of her garden be twice as big as its current area? Examine this *carefully* by working out some examples and drawing pictures. Explain your conclusion. If the garden is not twice as big (area-wise), how big is it compared to the original?

Chapter 9

Measurement and Geometry

In this chapter we will examine some fundamental aspects of measurement. We will focus especially on measurement in geometry, namely on length, area, and volume. In particular, we will look at some of the familiar area and volume formulas and see why they are valid and where they come from.

9.1 The Concept of Measurement

A variety of items that we encounter in daily life and that are used in scientific applications can be measured. For example: the passing of time, the amount of gasoline pumped into a car, the amount of electricity used by a household, the area of land owned by a person, the distance between to cities, are all examples of quantities that can be measured. What does it mean to measure a quantity? The idea of measurement is simple, but fundamental. In order to measure a quantity, one first needs a fixed "reference amount" of this quantity. This "reference amount" is called a **unit**. For example, a mile is a unit of length, an acre is a unit of area, a gallon is a unit of capacity (or volume), and a kilowatthour is a unit of electical power. To **measure** a given quantity means to compare the quantity with a unit of the quantity. Usually, this means determining how many units of the quantity make up the given quantity.

How do we measure quantities? There are often a variety of ways to measure quantities, and we often have special devices for measuring them. But the simplest and most direct way to measure a quantity is to make a count of how many of the unit are in the quantity to be measured. For example, to measure the amount of rice in a small bag, you can simply scoop it out, cup by cup, and count how many cups you scooped. But to measure the amount of water in your bathtub, it would be quite tedious to scoop it out cup by cup or gallon by gallon. So you would probably want to think up an indirect method for measuring this amount of water.

The simplest measuring device that all of us are familiar with is a ruler. A ruler whose unit is an inch displays a number of inch long lengths. In order to use the ruler to measure how long an object is in terms of inches, we simply read off the number of inches on the ruler, rather than repeatedly laying down an inch-long object and counting how many it took to cover the length of the object. Similarly, other measuring devices such as scales, calipers, and speedometers allow us to measure quantities indirectly, rather than making direct counts of a number of units.

What quantities can be used as units? *Any fixed amount* of a measurable quantity can be a unit, but for purposes of clear communication it usually makes sense to use commonly accepted standard units. For example, in order to encourage a child to develop her concept of measurement, you might have her measure the length of a piece of ribbon by laying pens end to end. She might report that the ribbon is $5\frac{1}{2}$ pens long. In this case, the unit of length is "a pen," which is not a standard unit of length. On the other hand, if you

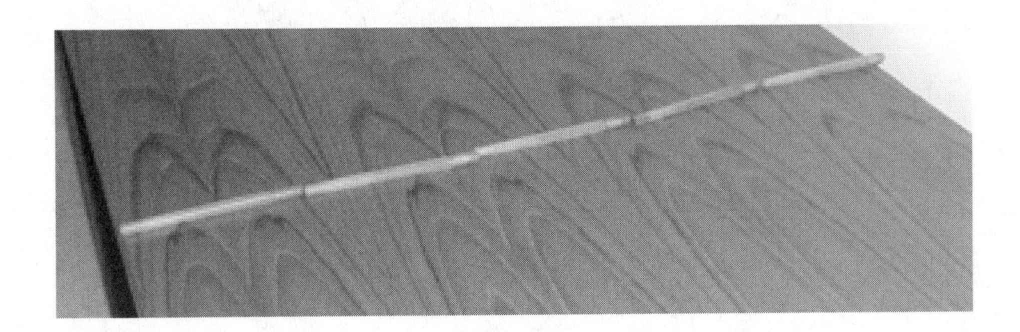

Figure 9.1: Using a Non-Standard Unit to Measure: This Desk is 5 Pens Wide

want to buy ribbon from a spool at a store, you will probably use yards, feet, inches, or some combination of these to say how long a piece you want. That is, if the store is in the U.S.! If you were in France, you would use meters and centimeters to describe the length of ribbon.

Here is an example where it is natural to use a non-standard unit. Look at the design in Figure 9.2. What is its area? The easiest way to describe this area in terms of one of the pieces that makes up the design. A single piece making up the figure is shown in Figure 9.3. We can decide to call the area of this single piece *one squiggle*, and to use *one squiggle* as our unit for the purpose of measuring the area of the whole design. The design in Figure 9.2 is made up of 36 congruent pieces (4 rows, with 5 + 4 pieces in each row), each of which has area *one squiggle*, so the total area of the design is 36 *squiggles*.

What is considered a standard unit varies not only from place to place in the world, but has also varied throughout history.

Figure 9.2: What is the Area of This Design?

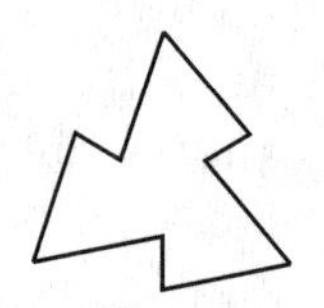

Figure 9.3: One Squiggle

Systems of Measurement

A system of measurement is a collection of standard units. In the United States today, there are two systems of measurement that are in common use: the U.S. customary system of measurement and the metric system (also known as the International System of Units, or the SI system).

The U.S. customary system

The following are some standard units of measurement that are commonly used in the U.S. customary system of measurement.

Units of length		
unit	abbreviation	some relationships
inch	in	
foot	ft	1 ft = 12 in
yard	yd	1 yd = 3 ft
mile	mi	1 mi = 1760 yd = 5280 ft

For any unit of length, there are corresponding units of area and volume: a square unit, and a cubic unit, respectively. A square unit is the area of a square that is one unit wide and one unit long. A cubic unit is the volume of a cube that is one unit wide, one unit deep, and one unit high. For example, a square inch is the area of a square that is one inch wide and one inch long. A cubic inch is the volume of a cube that is one inch wide, one inch deep, and one inch high, as depicted in Figure 9.4.

Units of area		
unit	abbreviation	some relationships
square inch	in^2	
square foot	ft^2	$1 \text{ ft}^2 = 12^2 \text{ in}^2 = 144 \text{ in}^2$
square yard	yd^2	$1 \text{ yd}^2 = 3^2 \text{ ft}^2 = 9 \text{ ft}^2$
square mile	mi^2	
acre		$1 \text{ acre} = 43{,}560 \text{ ft}^2$

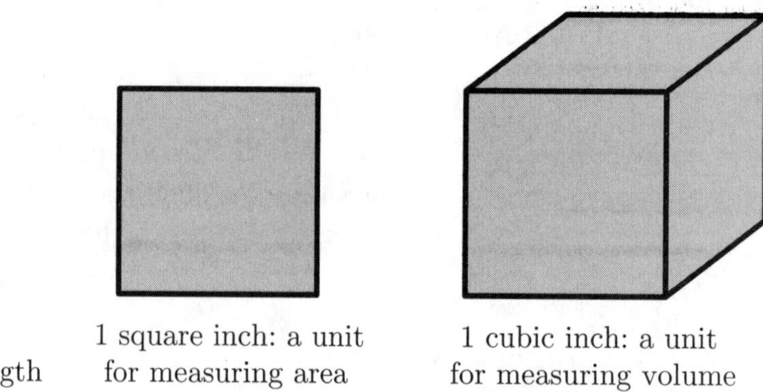

Figure 9.4: Inch, Square Inch, And Cubic Inch

Units of volume		
unit	abbreviation	some relationships
cubic inch	in^3	
cubic foot	ft^3	
cubic yard	yd^3	$1\ yd^3 = 3^3\ ft^3 = 27\ ft^3$

Units of capacity (volume)		
unit	abbreviation	some relationships
teaspoon	tsp.	
tablespoon	T. or tbs. or tbsp.	1 T. = 3 tsp.
fluid ounce (or liquid ounce)	fl. oz.	1 fl. oz. = 2 T.
cup	c.	1 c. = 8 fl. oz.
pint	pt.	1 pt. = 2 c. = 16 fl. oz.
quart	qt.	1 qt. = 2 pt. = 32 fl. oz.
gallon	gal.	1 gal. = 4 qt. = 128 fl. oz.

Units of weight (avoirdupois)		
unit	abbreviation	some relationships
ounce	oz	
pound	lb.	1 pound = 16 ounces
ton	t.	1 ton = 2000 pounds

Notice the distinction between an *ounce*, which is a unit of weight, and a *fluid ounce*, which is a unit of capacity or volume.

Unit of temperature		
unit	abbreviation	comments
degree Fahrenheit	° F	water freezes at 32° F water boils at 212° F

It is interesting to research the history of units because their definitions have changed over time. For example, the inch was originally defined as the width of a thumb or as the length of three barleycorns. Since widths of thumbs and lengths of barleycorns can vary, these definitions are obviously not very precise. The inch is currently defined to be *exactly* 2.54 centimeters (see [2]).

The metric system

The metric system came into being in France around the time of the French revolution (1790). Around that period of time, which is often called the *Age of Enlightenment*, there was a focus on rational thought, and scientists sought to organize their subjects in a rational way. The metric system is a natural outcome of this desire since it is an efficient, organized system that is designed to be compatible with the decimal system for writing numbers.

The metric system gives names to units in a uniform way. For each kind of quantity to be measured, there is a base unit (such as: meter, gram, liter). Each related unit is labeled with a prefix that indicates the unit's relationship to the base unit. For example, the prefix *kilo* means *thousand*, so a *kilometer* is a thousand meters, and a *kilogram* is a thousand grams. Many of the metric system prefixes are only used in scientific contexts, and not in everyday situations. Some of the metric system prefixes are shown in the table below. For more information on the metric system, see the websites

`http://lamar.ColoState.edu/~hillger/`
`http://www.hlalapansi.demon.co.uk/Metric/`

The website

`http://lamar.ColoState.edu/~hillger/week.htm`

has some ideas for teaching the metric system.

Some metric system prefixes		
prefix	meaning	
nano-	$10^{-9} = \frac{1}{1000000000}$	billionth
micro-	$10^{-6} = \frac{1}{1000000}$	millionth
milli-	$10^{-3} = \frac{1}{1000}$	thousandandth
centi-	$10^{-2} = \frac{1}{100}$	hundredth
deci-	$10^{-1} = \frac{1}{10}$	tenth
deca-	10	ten
hecto-	$10^2 = 100$	hundred
kilo-	$10^3 = 1,000$	thousand
mega-	$10^6 = 1,000,000$	million
giga-	$10^9 = 1,000,000,000$	billion

The following are some of the most commonly used units of the metric system.

Units of length		
unit	abbreviation	some relationships
millimeter	mm	1 mm $= \frac{1}{1000}$ m
centimeter	cm	1 cm $= \frac{1}{100}$ m, 1 cm = 10 mm
meter	m	1 m = 100 cm
kilometer	km	1 km = 1000 m

Figure 9.5 shows lines of length 1 millimeter, 1 centimeter, and 1 inch. The inch and the centimeter are related by the fact that 1 inch = 2.54 cm.

Figure 9.6 indicates the lengths of 1 meter and 1 kilometer. A meter is a little over a yard (about a yard and 3 inches). A kilometer is about .6 miles, so a bit more than half a mile.

As with any unit of length, there are the associated units of area and volume. Figure 9.7 shows a square centimeter and a cubic centimeter.

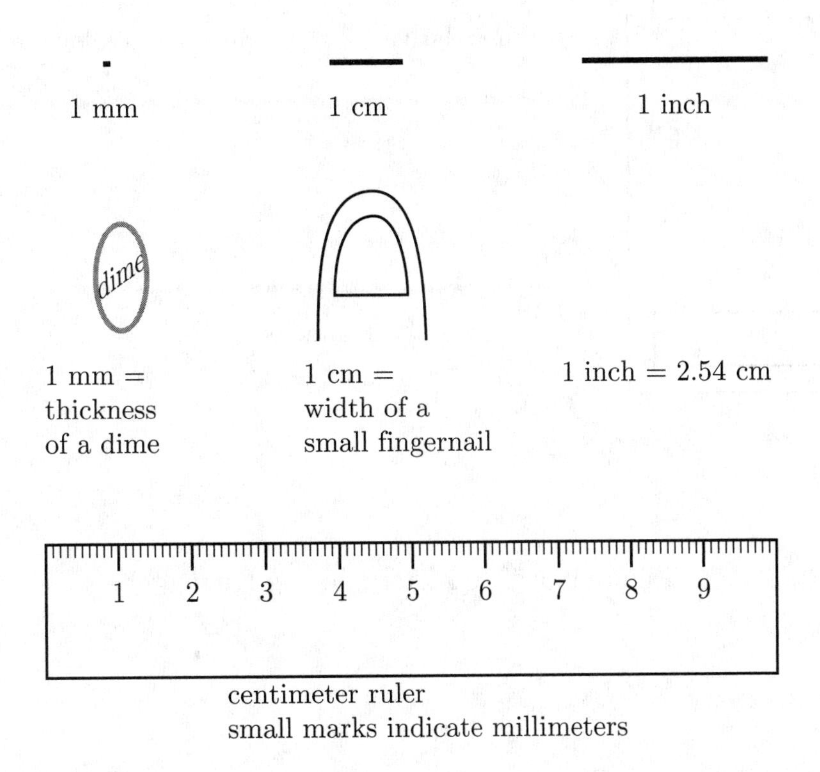

Figure 9.5: Comparing Millimeters, Centimeters, and Inches

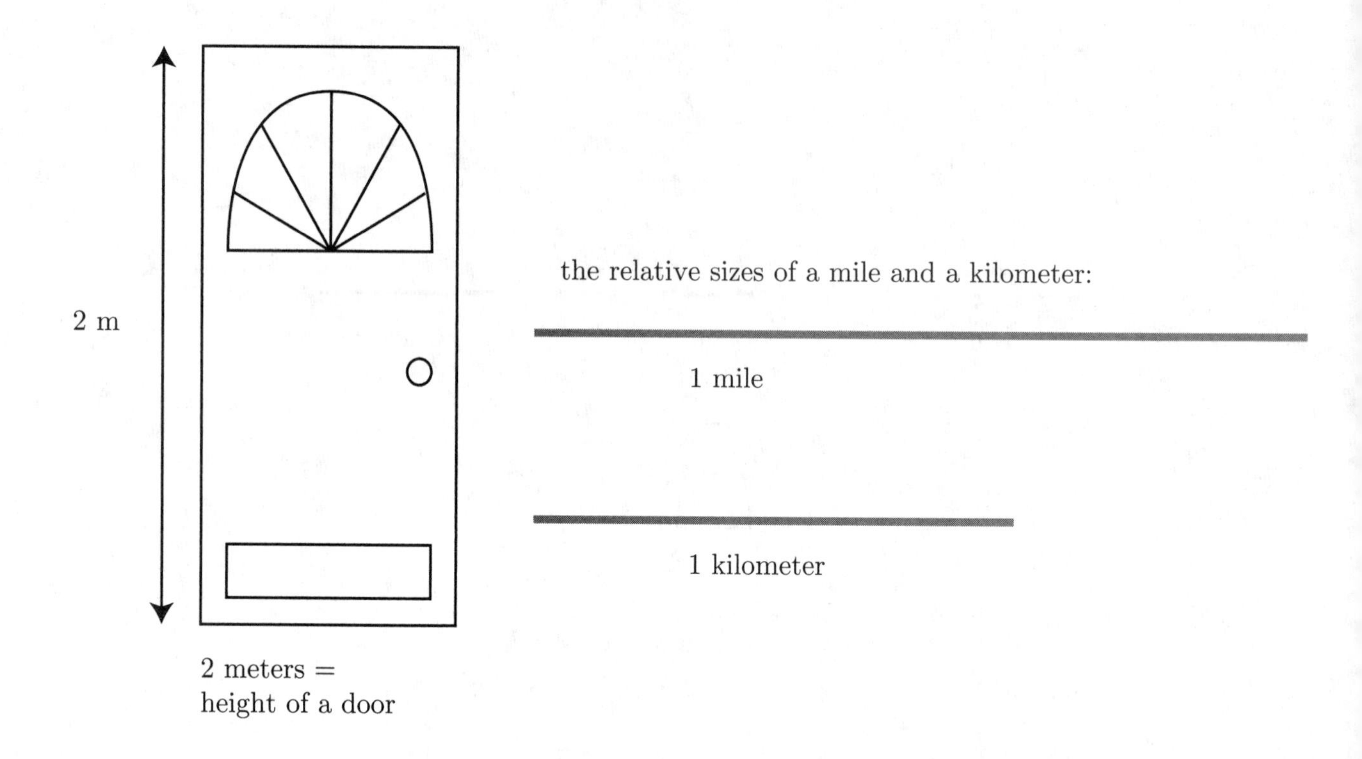

Figure 9.6: The Meter and the Kilometer

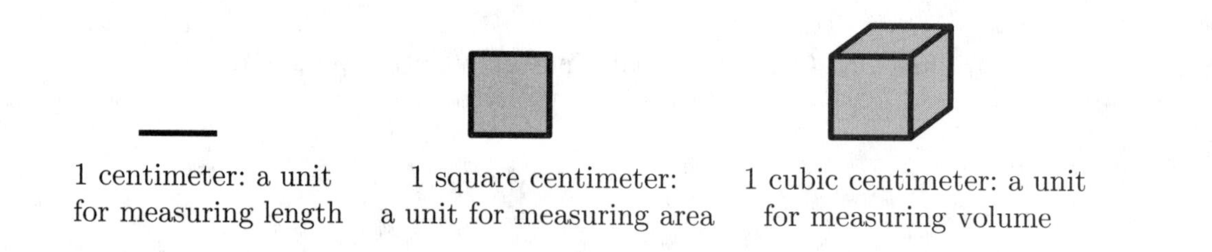

Figure 9.7: Centimeter, Square Centimeter, and Cubic Centimeter

Units of area		
unit	abbreviation	some relationships
square millimeter	mm^2	
square centimeter	cm^2	$1\ cm^2 = 10^2\ mm^2 = 100\ mm^2$
square meter	m^2	$1\ m^2 = 100^2\ cm^2 = 10{,}000\ cm^2$
square kilometer	km^2	$1\ km^2 = 1000^2\ m^2 = 1{,}000{,}000\ m^2$

Units of volume		
unit	abbreviation	some relationships
cubic millimeter	mm^3	
cubic centimeter	cm^3 or cc	$1\ cm^3 = 10^3\ mm^3 = 1000\ mm^3$
cubic meter	m^3	$1\ m^3 = 100^3\ cm^3 = 1{,}000{,}000\ cm^3$
cubic kilometer	km^3	$1\ km^3 = 1000^3\ m^3 = 1{,}000{,}000{,}000\ m^3$

Units of capacity		
unit	abbreviation	some relationships
milliliter	ml	$1\ ml = \frac{1}{1000}\ l$
liter	l	$1\ l = 1000\ ml$

Popular sodas often come in 1 liter bottles. One liter is a little more than a quart. Doses of liquid medicines are often measured in milliliters. In this case, a milliliter is also often called a **cc**, for cubic centimeter.

Units of mass (weight)		
unit	abbreviation	some relationships
gram	g	
kilogram	kg	$1\ kg = 1000\ g$

Nutrition labels often refer to grams. For example, food packages usually give serving sizes in both ounces and grams. One ounce is about 28 grams. 1 liter of water weighs 1 kilogram, which is about 2.2 pounds.

Also, although 1000 kilograms should be called a megagram, it is usually called a **metric ton** or **tonne**.

Relationships between cm, ml, and g
1 ml of water weighs one g
1 ml of water has a volume of 1 cm^3

As seen in the table above, in the metric system, the units of length, capacity, and mass (weight) are related in a simple and logical way: 1 milliliter of water weighs one gram, and 1 milliliter of water fills a cube that is 1 cm wide, 1 cm deep, and 1 cm high.

Unit of temperature		
unit	abbreviation	comments
degree Celsius	° C	water freezes at 0° C water boils at 100° C

Some basic relationships between the metric and U.S. customary systems of measurement are shown in the table below.

Basic metric/U.S. customary relationships	
length	1 in=2.54 cm (exact)
capacity	1 gal=3.79 l (good approximation)
weight (mass)	1 kg=2.2 lb (good approximation)

Even though the metric system was designed to be more logical and organized than the old customary systems, the definitions of units in the metric system have changed over time too. As technology and scientific understanding advance, it is possible to give more and more precise definitions of units. For example, the meter was originally defined as one ten-millionth of the distance from the equator to the North Pole, whereas it is currently defined as the length of a path traveled by light in a vacuum in

$$\frac{1}{299,792,458}$$

of a second (see [2]).

Class Activity 9A: Becoming Familiar with Centimeters and Meters

Class Activity 9B: Becoming Familiar with Units of Area

Class Activity 9C: Becoming Familiar with Units of Volume

Reporting and Interpreting Measurements

When it comes to measuring actual physical quantities, a certain amount of error and uncertainty is unavoidable. Any reported measurement of an actual physical quantity is necessarily only approximate—you can never say that a measurement of a real object is *exact*. For example, if you use a ruler to measure the width of a piece of paper, you may report that the paper is $8\frac{1}{2}$ inches wide. But the paper isn't *exactly* $8\frac{1}{2}$ inches wide, it is just that the mark on your ruler that is closest to the edge of the paper is at $8\frac{1}{2}$ inches.

The way that a measurement is reported should reflect its accuracy. For example, suppose that the distance between two cities is reported as 1200 miles. Then, since there are zeros in the tens and ones places, we assume that this measurement is *rounded to the nearest hundred.* In other words, the actual distance between the two cities is between 1100 miles and 1300 miles, but closer to 1200 miles than to either 1100 miles or 1300 miles. Thus the actual distance between the two cities could be anywhere between 1150 and 1250 miles, as indicated in Figure 9.8.

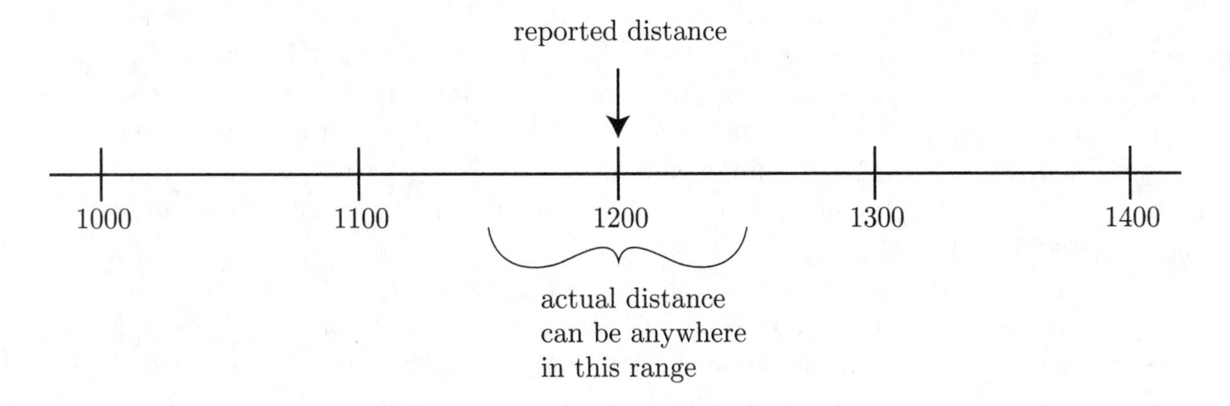

Figure 9.8: Interpreting a Reported Distance

On the other hand, if the distance between two cities is reported as 1230 miles, then we assume that this measurement is *rounded to the nearest ten.* In other words, the actual distance between the cities is between 1220 miles and 1240 miles, but closer to 1230 miles than to either 1220 miles or 1240 miles.

In this case, the actual distance between the two cities could be anywhere between 1225 and 1235 miles, as indicated in Figure 9.9. Notice that in this case, we are getting a much smaller range of numbers that the actual distance could be than in the previous example: here it's a range of 10 miles (1225 miles to 1235 miles), whereas in the previous case it's a range of 100 miles (1150 miles to 1250 miles).

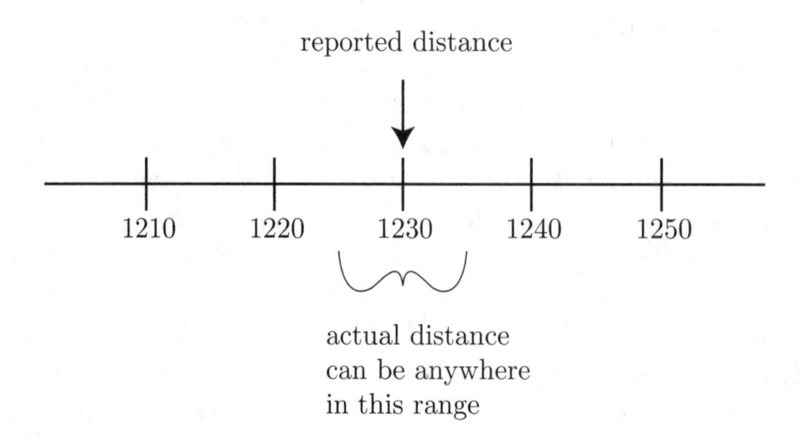

Figure 9.9: Interpreting Another Reported Distance

Similarly, if you step on a digital scale and your weight is reported as 130.4 pounds, then (assuming the scale is reasonably accurate) this is your actual weight, *rounded to the nearest tenth.* In other words, your actual weight is between 130.3 pounds and 130.5, but closer to 130.4 pounds than to either 130.3 pounds or 130.5 pounds. In this case, your actual weight is between 130.35 and 130.45 pounds.

Notice that there is a big difference in reporting that the weight of an object is 130.0 pounds and reporting the weight as 130 pounds. When the weight is reported as 130.0 pounds, we assume that this weight is *rounded to the nearest tenth.* In this case, the actual weight is between 129.95 and 130.05 pounds. On the other hand, when the weight of an object is reported as 130 pounds, we assume this weight is *rounded to the nearest ten.* In this case, we only know that the actual weight is between 125 and 135 pounds, which allows for a much wider range of actual weights than the range of 129.95 to 130.05 pounds (a 10 pound range versus a .1 pound range).

Exercises for Section 9.1 on the Concept of Measurement

1. What does it mean to measure a quantity?

2. What is the simplest, most basic way to measure?

3. What is a unit?

4. What are the two main systems of measurement used in the U.S. today? Describe their units.

5. How many feet are in one mile? How many ounces are in one pound? How many pounds are in one ton?

6. What is the difference between an ounce and a fluid ounce?

7. What is special about the way units in the metric system are named?

8. How is one milliliter related to one gram and to one centimeter?

9. What is the difference between reporting that something weighs 130 pounds and that it weighs 130.0 pounds?

10. If the distance between two cities is reported as 2500 miles, does that mean that the distance is exactly 2500 miles? If not, what can you say about the exact distance?

11. Use square centimeter tiles or centimeter graph paper to show or to draw three different shapes that have an area of 8 cm^2.

12. Use cubic inch blocks (or cubic centimeter blocks) to make a variety of solid shapes that have a volume of 24 in^3 (or 24 cm^3).

Answers to Exercises for Section 9.1 on the Concept of Measurement

1. – 9. See text.

10. If the distance between two cities is reported as 2500 miles, then we assume that this number is the actual distance, rounded to the nearest hundred. Therefore, the actual distance is between 2400 miles and 2600 miles, but closer to 2500 miles than to either 2400 miles or 2600 miles. Therefore the actual distance between the two cities could be anywhere between 2450 miles and 2550 miles.

11. Arrange 8 square centimeter tiles in any way to form a shape, or draw a shape that encloses 8 squares, as in Figure 9.10.

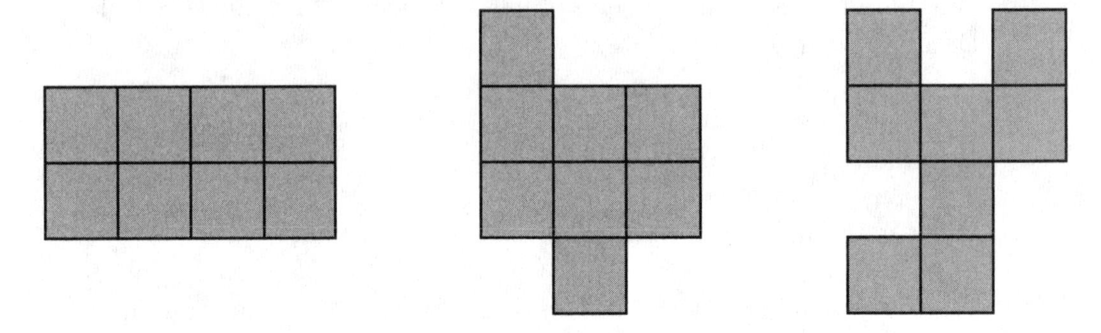

Figure 9.10: Three Ways to Show 8 cm^2

12. Arrange 24 cubic inch blocks (or cubic centimeter blocks) in any way to make a solid shape.

9.2 Measurable Attributes

Most objects have several different attributes, aspects, or parts that could be measured. For example, suppose you buy a spool of wire at a hardware store. What measurable quantities are associated with this spool of wire? There is the length of wire that is wound onto the spool. There is the diameter of the wire (the cross-section of the wire is a small circle). There is the diameter of the spool and the length of the spool. There is the weight of the wire, and the weight of the wire and spool. Any one object can have many measurable aspects. What we choose to measure should depend on what we want to know. In this section, we will consider some of the different measurable attributes objects can have. In particular, this will lead us to a consideration of dimension.

Class Activity 9D: The Biggest Tree in the World

Length, Area, Volume, and Dimension

Since the units for length, area, and volume are related, it is especially easy to get confused about which unit to use when. Most real objects have one or more aspect or part that should be measured by a length, another aspect or part that should be measured by an area, and yet another aspect or part that should be measured by a volume.

When do we use a unit of *length* (such as inches, centimeters, miles, kilometers etc.) to measure or to describe the size of something? When we want to answer one of the following kinds of questions:

How far?

How long?

How wide?

For example, "How much moulding (i.e., *how long* a piece of moulding) will I need to go around this ceiling?". The answer, in feet, is the number of foot long segments can be put end to end to go all around the ceiling (see Figure 9.11).

How many of these
1 foot lengths ——
does it take to go around
the ceiling of this room
(i.e., around the outside)?

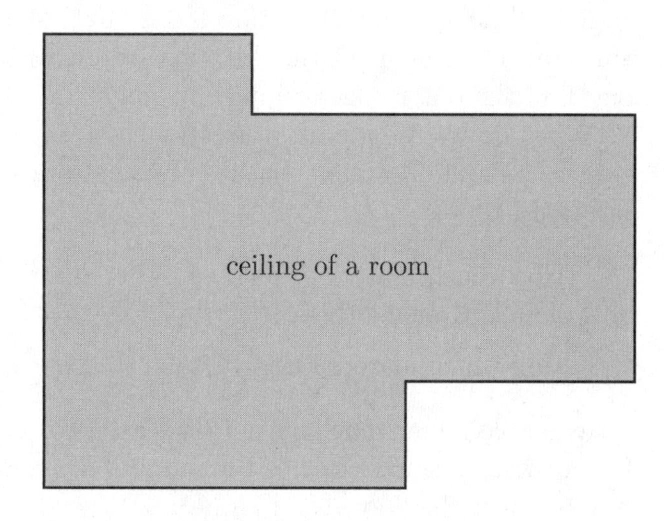

Figure 9.11: A Length Problem

A **length** describes the size of something (or a part of something) that **length** is one-dimensional; the length of that one-dimensional thing is how many of a chosen unit of length (such as inches, cm, etc.) it takes to cover the thing without overlapping. Roughly speaking, a thing is **one-dimensional** if at each location, there is only one independent direction along which to move within the thing. An imaginary creature living in a one-dimensional world could only move forwards and backwards in that world. Living in a one-dimensional world would be like being stuck in a tunnel. Here are some examples of one-dimensional things:

> A line segment,
>
> a circle (only the outer part, not the inside),
>
> the four line segments making a square (not the inside of the square),
>
> a curved line,
>
> the equator of the earth,
>
> the edge where two walls in a room meet,
>
> an imaginary line drawn from one end of a student's desk to the other end.

In general, the "outer edge" around a (flat) shape is one-dimensional. The **perimeter** of a shape is the distance around a shape; it is therefore the total length of the outer edge around the shape.

When do we use a unit of *area* (such as square feet or square kilometers) to measure or to describe the size of something? When we want to answer a question like:

> How much material (such as paper, fabric, or sheet metal) does it take to make this?
>
> How much material does it take to cover this?

For example, "how much carpet do I need to cover the floor of this room?". The answer, in square feet, is the number of 1 foot by 1 foot squares it would take to cover the floor (see Figure 9.12).

An **area** describes the size of something (or a part of something) that is two-dimensional; the area of that two-dimensional thing is how many of a chosen unit of area (such as square inches, square centimeters, etc.) it takes to cover the thing without overlapping. Roughly speaking, a thing is

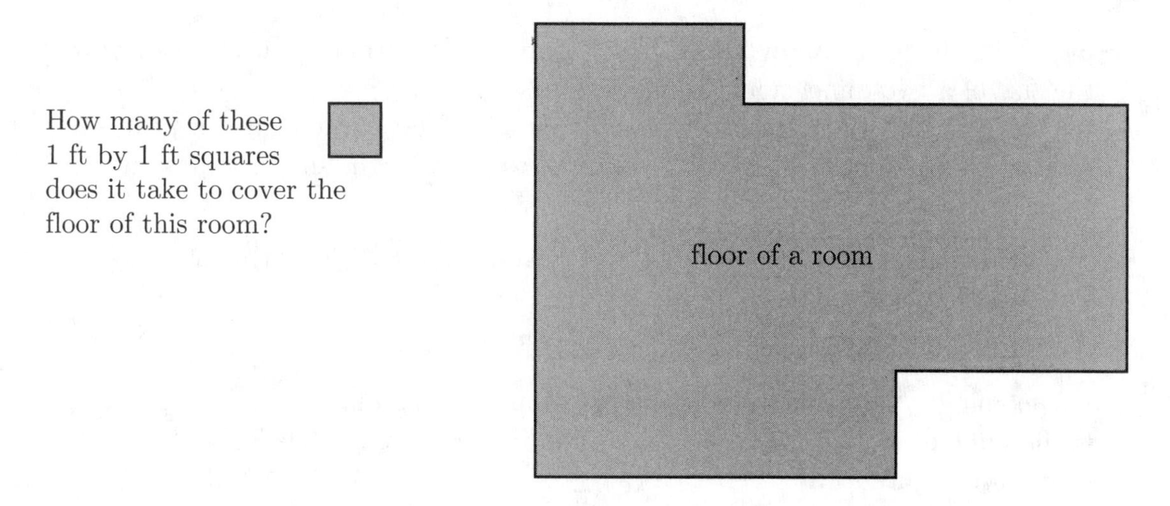

Figure 9.12: An Area Problem

two-dimensional if at each location, there are two independent directions along which to move within the thing. An imaginary creature living in a two-dimensional world could only move in the forwards/backwards and right/left directions, as well as in "in between" combinations of these directions (such as diagonally), but *not* up/down. What would it be like to live in a two dimensional world? The book *Flatland* by A. Abbott is just such an account. Here are some examples of two-dimensional things:

A plane,

the *inside* of a circle (this is also called a disk),

the *inside* of a square,

a piece of paper (only if you think of it as having no thickness!),

the *surface* of a balloon (not counting the inside!),

the *surface* of a box,

some farm land (only counting the surface, not the soil below),

the surface of the earth.

Notice that most of the units that are used to measure area, such as

cm^2, m^2, km^2, in^2, ft^2, mi^2, have a "2" in their abbreviation. This "2" reminds us that we are measuring the size of a two-dimensional thing.

When do we use a unit of *volume* (such as cubic yards or cubic centimeters, or even gallons or liters) to measure or to describe the size of something? When we want to answer questions like:

> How much substance (such as air, water, or wood) is in this object?
>
> How much substance does it take to fill this object?

For example, "how much air is in this room?". The answer, in cubic feet, is the number of 1 foot by 1 foot by 1 foot cubes that are needed to fill the room (see Figure 9.13).

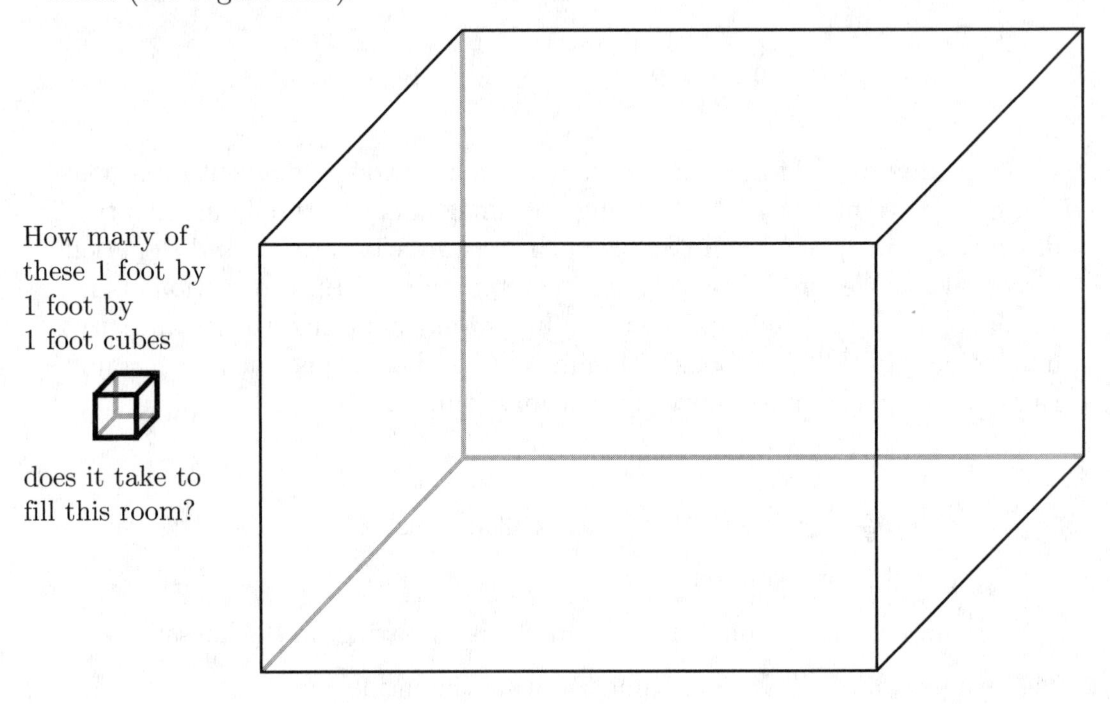

Figure 9.13: A Volume Problem

A **volume** describes the size of something (or a part of something) that is three-dimensional; the volume of that three-dimensional thing is how many of a chosen unit of volume (such as cubic inches, cubic centimeters, etc.) it takes to fill the thing without overlapping. Roughly speaking, a thing is **three-dimensional** if at each location, there are three independent directions along which to move within the thing. Our world is three-dimensional because we can move in the three independent directions: forwards/backwards, right/left, and up/down, as well as all "in between" combinations of these directions. Here are some examples of three-dimensional things:

the air around us,

the inside of a balloon,

the inside of a box,

the water in a cup,

the compost in wheelbarrow,

the inside of the earth.

Notice that many of the units that are used to measure volume, such as

cm^3, m^3, in^3, ft^3

have a "3" in their abbreviation. This "3" reminds us that we are measuring the size of a three-dimensional thing.

One of the difficulties with understanding whether to measure a length, an area, or a volume is that many objects have parts or aspects of different dimensions. For example, consider a balloon. Its outside *surface* is two-dimensional, whereas the air filling it is three-dimensional. Then there is the circumference of the balloon, which is the length of an imaginary one-dimensional circle drawn around the surface of the balloon. The size of each of these parts should be described in a different way: the air in a balloon can be described by its volume, the surface of the balloon is described by its area, and the circumference of the balloon is described by its length.

Here is another example where different parts have different dimensions: the earth. The *inside* of the earth is three-dimensional, its *surface* is two-dimensional, and the equator is a one-dimensional imaginary circle drawn around the surface of the earth. Once again, the sizes of these different parts of the earth should be described in different ways: by volume, area, and length, respectively.

Another example is farm land. The size of farm land is often measured in acres, which is a measure of area—this only takes into account how big the two-dimensional top surface of the farm land is. This is a practical way to talk about how big farm land is, even though it doesn't take into account the most important part of the land—the three-dimensional soil below the surface!

For a fascinating look at various dimensions, see the website:

`http://mam2000.mathforum.com/765/index.html`

Class Activity 9E: Dimension and Size

Class Activity 9F: Explaining Why We Add to Calculate Perimeters of Polygons

Class Activity 9G: Explaining Why We Multiply to Determine Areas of Rectangles and Volumes of Boxes

Exercises for Section 9.2 on Measurable Attributes

1. Review the section on Multiplication and Areas of Rectangles and the section on Multiplication and Volumes of Boxes in Chapter 4. Be able to explain why it makes sense to calculate the area of a rectangle by multplying its length times its width and be able to explain why it makes sense to calculate the volume of a box (rectangular prism) by multiplying its width times its depth times its height.

2. (a) Use centimeter or inch graph paper to make a pattern for a closed box (the box should have six sides, and when you fold up the pattern, there should be no overlapping pieces of paper).

   (b) How much paper is your box made of? (Be sure to use an appropriate unit in your answer.)

   (c) Describe one-dimensional, two-dimensional, and three-dimensional parts or aspects of your box. In each case, give the size of the part or aspect of the box using an appropriate unit.

3. Describe one-dimensional, two-dimensional, and three-dimensional parts or aspects of a water tower. In each case, name an appropriate U.S. customary unit and an appropriate metric unit for measuring or describing the size of that part or aspect of the water tower. What are practical reasons for wanting to know the sizes of these parts or aspects of the water tower?

4. Describe one-dimensional, two-dimensional, and three-dimensional parts or aspects of a store. In each case, name an appropriate U.S. customary unit and an appropriate metric unit for measuring or describing the size of that part or aspect of the store. What are practical reasons for wanting to know the sizes of these parts or aspects of the store?

Answers to Exercises for Section 9.2 on Measurable Attributes

1. See text.

2. (a) Figure 9.14 shows a *scaled* picture of a pattern for a box. Each small square represents a 1 inch by 1 inch square.

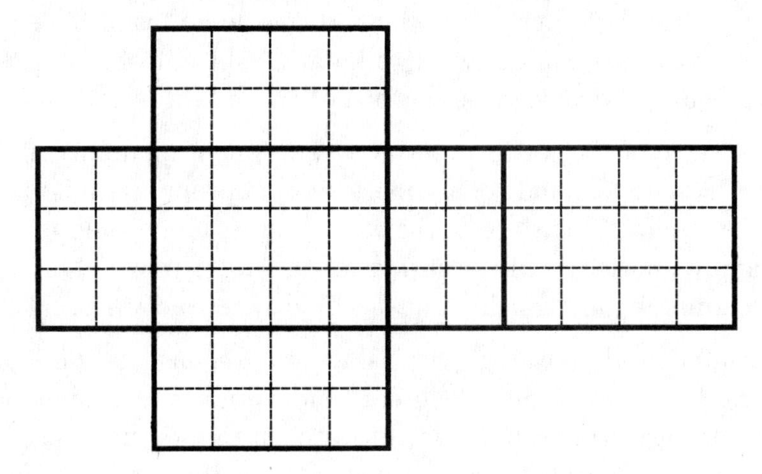

Figure 9.14: A Pattern for a Box

   (b) The box formed by the pattern in Figure 9.14 (in its actual size) is made of 52 square inches of paper because the pattern that was used to create the box consists of 52 squares that each have area 1 square inch.

   (c) The edges of the box are one-dimensional aspects of the box. Depending on how the box is oriented, the lengths of these edges are the height, width, and depth of the box. These measure 2 inches, 3 inches, and 4 inches. Another one-dimensional aspect of the box is its *girth*, i.e., the distance around the box. Depending on where you measure around, the girth is either 10 inches, 12 inches, or 14 inches.

The surface of the box is a two-dimensional part of the box. Its area is 52 square inches, as discussed in part b.

The air inside the box is a three-dimensional aspect of the box. If you filled the box with 1 inch by 1 inch by 1 inch cubes there would be 2 layers with 3×4 cubes in each layer. Therefore, according to the meaning of multiplication, there would be $2 \times 3 \times 4 = 24$ cubes in the box. Therefore the volume of the box is 24 cubic inches.

3. The height of the water tower (or an imaginary line running straight up through the middle of the water tower) is a one-dimensional aspect of the water tower. Feet and meters are appropriate units for describing the height of the water tower. Some towns do not want high structures, therefore the town might want to know the height of a proposed water tower before it decides whether or not to build it.

   The surface of the water tower is a two-dimensional part of the water tower. Square feet and square meters are appropriate units for describing the size of the surface of the water tower. One would need to know the approximate size of the surface of the water tower in order to know approximately how much paint it would take to cover it.

   The water inside a water tower is a three-dimensional part of a water tower. Cubic feet, gallons, liters, and cubic meters are appropriate units for measuring the volume of water in the water tower. Clearly, it makes sense for the owner of a water tower to know how much water it can hold.

4. The edge where the store meets the sidewalk is a one-dimensional aspect of the store. Its length is appropriately measured in feet, yards, or meters. This length is the width of the store front, which is one way to measure how much exposure the store has to the public eye.

   The surface of the store facing the sidewalk is a two-dimensional part of the store. Its area is appropriately described in square feet, square yards, or square meters. The area of the store front is another way to measure how much exposure the store has to the public eye (a store front that is narrow but tall might attract as much attention as a wider, lower store front).

Another important two-dimensional part of a store is its floor. The area of the floor is appropriately measured in square feet, square yards, or square meters, or possibly even in acres if the store is very large. The area of the floor indicates how much space there is on which to place items to sell to customers.

The air inside a store is a three-dimensional aspect of the store. Its volume is appropriately measured in cubic feet, cubic yards, cubic meters, or possibly even in liters or gallons. One might want to know the volume of air in the store in order to figure how many air conditioning units are needed.

Problems for Section 9.2 on Measurable Attributes

1. Describe one-dimensional, two-dimensional, and three-dimensional parts or aspects of a soda bottle. In each case, name an appropriate U.S. customary unit and an appropriate metric unit for measuring or describing the size of that part or aspect of the soda bottle. What are practical reasons for wanting to know the sizes of these parts or aspects of the soda bottle?

2. Describe one-dimensional, two-dimensional, and three-dimensional parts or aspects of a car. In each case, name an appropriate U.S. customary unit and an appropriate metric unit for measuring or describing the size of that part or aspect of the car. What are practical reasons for wanting to know the sizes of these parts or aspects of the car?

3. Describe one-dimensional, two-dimensional, and three-dimensional parts or aspects of the blocks in Figure 9.15. In each case, compare the sizes of the three blocks, using an appropriate unit. Use this to show that each block can be considered "biggest" of all three.

4. The Lazy Daze Pool Club and the Slumber-N-Sunshine Pool Club have a friendly rivalry going. Each club claims that they have the bigger swimming pool. Make up realistic sizes for the clubs' swimming pools, so that both clubs have a legitimate basis for saying that they have the larger pool. Explain clearly why each club can say their pool is biggest. Be sure to use appropriate units. Considering the function of a pool, what is a good way to compare sizes of pools? Explain.

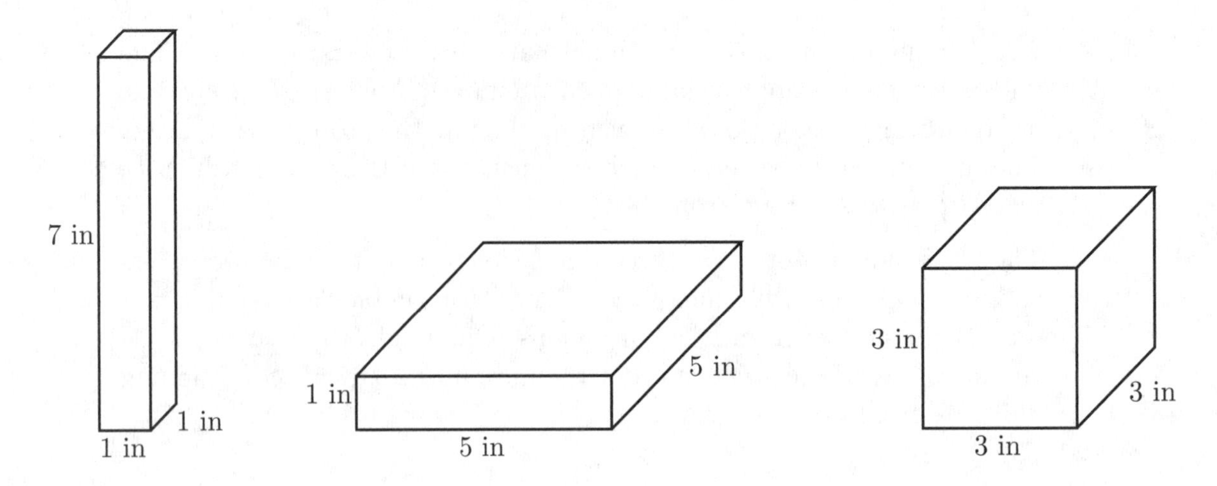

Figure 9.15: Which Block is Biggest?

5. Tim needs a sturdy cardboard box that is 3 ft tall by 2 ft wide by 1 ft deep. Tim wants to make the box out of a large piece of cardboard that he will cut, fold and tape. The box must close up completely, so it needs a top and a bottom. Show Tim how to make such a box out of one rectangular piece of cardboard: tell Tim what size cardboard he'll need to get (how wide, how long) and explain or show how he should cut, fold and tape the cardboard to make the box. Be sure to specify exact lengths of any cuts Tim will need to make. Include pictures where appropriate. Your instructions and box should be practical—one that Tim can actually make and use. You might want to make a scale model out of paper to test it! Note: many boxes have flaps around all four sides that one can fold down and interlock to make the top and bottom of the box. You can make this kind of top and bottom, or something else if you prefer, but make sure it will make a sturdy box that closes completely.

9.3 Converting From one Unit of Measurement to Another

As we've seen, different units can be used to measure the same quantity. For example, a length can be measured in miles, yards, feet, inches, kilometers,

meters, centimeters, or millimeters. If you know a length in terms of one unit, how can you describe it in terms of another unit? This kind of problem is a conversion problem, and this is what we will study in this section.

To convert a measurement in one unit to another unit, you first need to know how the two units are related. For example, if you want to convert the weight of a bag of oranges given in kilograms into pounds, then you need to know how kilograms and pounds are related. Refering back to the tables in Section 9.1, we see that 1 kg $=$ 2.2 lb.

Once you know how the units are related, to convert from one unit to another will require either multiplication or division. How do you know which one to use? The safest way to figure it out is to think back to the *meanings of multiplication and division.* For example, let's say a bag of oranges weighs 2.5 kg. What is the weight of the bag of oranges in pounds? If one kilogram is 2.2 pounds, then 2.5 kilograms weighs 2.5 times as much, i.e., $2.5 \times 2.2 = 5.5$ pounds. We can think of the 2.5 kilograms as 2.5 groups, with 2.2 pounds in each group, so this goes straight back to the meaning of multiplication.

Here's another example. If a container holds 25 liters, then how many gallons does it hold? Refering back to the tables in Section 9.1, we see that 1 gal $=$ 3.79 l. So this means that every group of 3.79 liters is equal to 1 gallon. Therefore, to figure out how many gallons are in 25 liters, we need to find out how many groups of 3.79 liters are in 25 liters. This is a division problem, using the "how many groups?" interpretation of division! So the answer is that there are $25 \div 3.79 = 6.6$ gallons in 25 liters.

In some cases, the tables in Section 9.1 do not tell you directly how two units are related. For example, none of the tables relate kilometers and miles (although the text does say that 1 km is about .6 miles). This was done purposely so that you would not be confronted with an overwhelming amount of data. But even so, the tables of data in Section 9.1 do provide you with enough information to do any conversion—although some conversions may require several steps.

For example, let's say you ran 3 miles and your friend ran 5 kilometers. Who ran farther? We can compare the two by converting 3 miles to kilometers. According to Section 9.1, 1 inch is exactly 2.54 cm, so this provides a link between the U.S. customary and metric units of length. To convert 3 miles to kilometers, you can go through the following chain of conversions:

$$\text{miles} \longrightarrow \text{feet} \longrightarrow \text{inches} \longrightarrow \text{centimeters} \longrightarrow \text{meters} \longrightarrow \text{kilometers}\ .$$

I think of this as "ratcheting down," "going across," and then "ratcheting back up."

So here's how you can convert 3 miles to kilometers. First, miles to feet: since 1 mile is 5280 feet, 3 miles is 3 times as many feet (3 groups of 5280 feet), so

$$3 \text{ miles } = 3 \times 5280 \text{ feet } = 15840 \text{ feet.}$$

Next, feet to inches: since 1 foot is 12 inches, 15840 feet are 15840 times as many inches (15840 groups of 12), so

$$15840 \text{ feet} = 15840 \times 12 \text{ inches } = 190080 \text{ inches.}$$

Now inches to centimeters: since 1 inch is 2.54 centimeters, 190080 inches are 190080 times as many centimeters (190080 groups of 2.54 centimeters), so

$$190080 \text{ inches } = 190080 \times 2.54 \text{ cm } = 482803.2 \text{ cm.}$$

Next, centimeters to meters: since 100 centimeters make 1 meter, we need to figure out how many groups of 100 centimeters are in 482803.2 centimeters. According to the "how many groups?" interpretation of division, this is answered by dividing:

$$482803.2 \text{ cm } = 482803.2 \div 100 \text{ m } = 4828.032 \text{ m.}$$

Finally, meters to kilometers: since 1000 meters make 1 kilometers, we need to figure out many groups of 1000 meters are in 4828.032 meters. As before, according to the "how many groups?" interpretation of division, this problem is answered by dividing:

$$4828.032 \text{ m } = 4828.032 \div 1000 \text{ km } = 4.828032 \text{ km.}$$

Therefore 3 miles is exactly 4.828032 kilometers, which is just a little less than 5 kilometers. So your friend ran farther.

Notice that every step above is perfectly logical and makes complete sense as long as you are thinking about the meanings of multiplication and division!

Notice also that it's very easy to carry these calculations out on a calculator: after each step, simply *leave the result of your calculation in your calculator*, and use this number for the next calculation. For example, when you get the 15840 feet from multiplying 3 times 5280 feet, just leave the 15840 in your calculator for the next calculation (multiplying by 12). That way you don't risk making an error by punching the 15840 back in. Here is a "flow chart" of the calculator work for the problem of converting 3 miles to kilometers. The arrows point to results calculated by the calculator:

punch in 3,

times 5280 $\rightarrow$ 15840,

times 12 $\rightarrow$ 190080,

times 2.54 $\rightarrow$ 482803.2,

divided by 100 $\rightarrow$ 4828.032,

divided by 1000 $\rightarrow$ 4.828032.

Dimensional Analysis

Many people like to use the method called **dimensional analysis** in order to convert a measurement from one unit to another. This is a method whereby one multplies by fractions that incorporate the relationships between units. As we'll see, it boils down to exactly the same calculations you would do if you used the logical method that relies on the meanings of multiplication and division that was described above.

For example, suppose that a German car is 5.1 meters long. How long is it in feet? To solve this, we will use the information provided in the tables of Section 9.1 in order to carry out the following chain of conversions:

$$\text{meters} \rightarrow \text{centimeters} \rightarrow \text{inches} \rightarrow \text{feet.}$$

$$5.1 \text{ m} = 5.1 \text{ m} \times \frac{100 \text{ cm}}{1 \text{ m}} \times \frac{1 \text{ in}}{2.54 \text{ cm}} \times \frac{1 \text{ ft}}{12 \text{ in}} = 16.7 \text{ ft}$$

So the car is 16.7 feet long.

How are the fractions that one multiplies by chosen? They are chosen so that the units will "cancel" and so that all that's left is feet—because we want an answer in feet:

$$5.1 \text{ m} = 5.1 \cancel{\text{m}} \times \frac{100 \cancel{\text{cm}}}{1 \cancel{\text{m}}} \times \frac{1 \cancel{\text{in}}}{2.54 \cancel{\text{cm}}} \times \frac{1 \text{ ft}}{12 \cancel{\text{in}}} = 16.7 \text{ ft}$$

Also, each fraction is chosen so that it is equal to 1:

$$\frac{100 \text{ cm}}{1 \text{ m}} = 1, \quad \frac{1 \text{ in}}{2.54 \text{ cm}} = 1, \quad \frac{1 \text{ ft}}{12 \text{ in}} = 1.$$

We can use this to explain why dimensional analysis is a valid procedure: dimensional analysis is really just repeatedly multiplying by one. Therefore the quantity remains unchanged, but is only expressed in a different unit.

Area and Volume Conversions

Once you know how to convert from one unit of length to another, you can use this information to convert between the related units of area and volume. However, in order to calculate area and volume conversions correctly, it's important to *remember the meanings* of the units for area and volume (such as square feet, cubic inches, square meters, and so on). If you don't, then it's very easy to make the following kind of mistake.

> A common mistake: "Since 1 yard is 3 feet, therefore 1 square yard is 3 square feet."

Why is this not correct? The reason is that one square yard is the area of a square that is 1 yard wide and 1 yard long. Such a square is 3 feet wide and 3 feet long. Figure 9.16 shows that such a square is made up of $3 \times 3 = 9$ square feet, *not* 3 square feet!

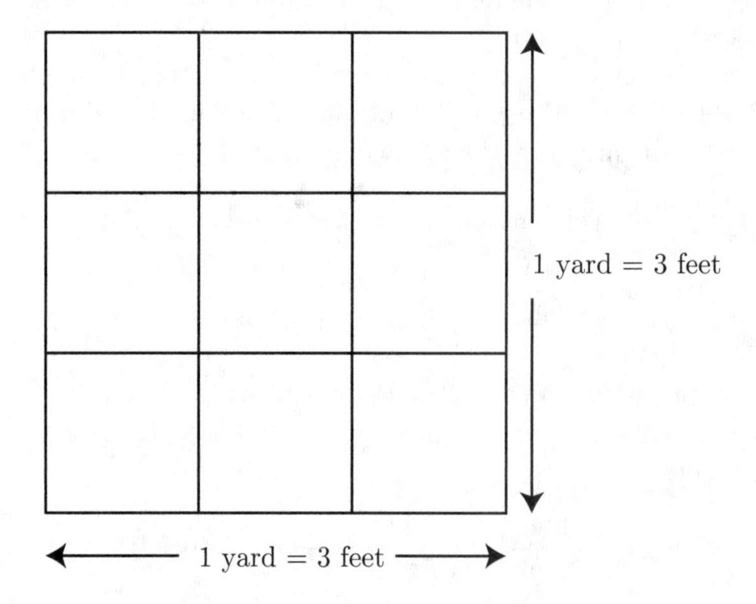

Figure 9.16: One Square Yard is 9 Square Feet, Not 3 Square Feet

Let's say you are living in an apartment that has a floor area of 800 square feet, and you want to tell your Italian pen pal how big that is in square meters. There are several ways you might approach such an area conversion problem. One good way is this:

first find what one (*linear*) foot is in terms of meters and then use that to find what one *square* foot is in terms of *square* meters.

Here's how we can find 1 foot in terms of meters:

1 ft = 12 in.

12 in = 12×2.54 cm = 30.48 cm because each inch is 2.54 cm and there are 12 inches.

30.48 cm = $30.48 \div 100$ m = .3048 m because 1 m = 100 cm, so we must find how many 100s are in 30.48 to find how many meters are in 30.48 cm.

So 1 ft = .3048 m.

Now remember that one square foot is the area of a square that is one foot wide and one foot long. According to the calculation above, such a square is also .3048 meters wide and .3048 meters long. Therefore, by the length times width formula for areas of rectangles,

$$1 \text{ ft}^2 = .3048 \times .3048 \text{ m}^2 = .0929 \text{ m}^2.$$

This means that 800 square feet is $800 \times .0929 = 74.32$ square meters, so the apartment is about 74 square meters.

Volume conversions work similarly. Say we want to find 13 cubic feet in terms of cubic meters, for example. Above, we already calculated that

$$1 \text{ ft} = .3048 \text{ m}.$$

Now 1 cubic foot is the volume of a cube that is 1 foot high, 1 foot deep, and 1 foot wide. And such a cube is also .3048 meters high, .3048 meters deep, and .3048 meters wide. So by the height times width times depth formula for volumes of boxes,

$$1 \text{ ft}^3 = .3048 \times .3048 \times .3048 \text{ m}^3 = .0283 \text{ m}^3.$$

Therefore 13 cubic feet is 13 times as much as .0283 m^3:

$$13 \text{ ft}^3 = 13 \times .0283 \text{ m}^3 = .37 \text{ m}^3.$$

Approximate Conversions and Checking Your Work

Above, we focused on exact (or fairly exact) conversions. However, in many practical situations, you really don't need an exact answer—a ballpark estimate will do. Being able to make a quick, ballpark estimate also gives you a way to check your work for an exact conversion: if your answer comes out way off from your ballpark estimate, you've probably made a mistake somewhere. If it's not too far off, there's a good chance your work is correct.

The following "ballpark" relationships will come in handy for the purpose of making "ballpark" conversions:

> An inch is about $2\frac{1}{2}$ centimeters, so 2 inches is about 5 centimeters.
>
> A meter is a little more than a yard.
>
> A kilometer is about $\frac{6}{10}$ of a mile (so a little over half a mile).
>
> A liter is a little more than a quart.
>
> A kilogram is a bit more than 2 pounds.

So, for example, you can calculate quickly, and mentally:

> that a 5 kilometer run is about 3 miles;
>
> that gas that costs $1.00 per liter in Europe costs about $1.00 per quart, in other words, $4.00 per gallon;
>
> that a 5 pound bag of oranges is about $2\frac{1}{2}$ kilograms (a little less).

Class Activity 9H: Using the Meanings of Multiplication and Divsion to Convert Measurements

Class Activity 9I: Conversions: When do We Multiply? When do We Divide?

Class Activity 9J: Area and Volume Conversions

Class Activity 9K: Size Differences in Area and Volume Conversions

Exercises For Section 9.3 on Conversion

1. Do the following conversions, using the basic fact 1 inch = 2.54 cm in each case.

   (a) A track is 100 meters long. How long is it in feet?

(b) If the speed limit is 70 miles per hour, what is it in kilometers per hour?

(c) Some farmland covers 2.4 square kilometers. How many square miles it is?

(d) Convert the volume of a compost pile, one cubic yard, to cubic meters.

(e) A man is 1.88 meters tall. How tall is he in feet?

(f) How many miles is a 10 kilometer race?

2. According to a research study, the average American consumes 63 doughnuts per year. Let's say that each doughnut is about one inch thick. If all the doughnuts consumed in the U.S. in one year were placed in one tall stack, would this stack reach to the moon? The moon is about 240 thousand miles away from the earth and the population of the U.S. is about 275 million.

3. One liter is the capacity (how much it can hold) of a cube that is 10 cm wide, 10 cm deep, and 10 cm tall. One gallon is .134 cubic feet. One quart is one quarter of a gallon. Which is more: one quart or one liter? Use the basic fact 1 inch $=$ 2.54 cm to figure this out.

4. One acre is 43,560 square feet. The area of land is often measured in acres.

   (a) What is a square mile in acres?

   (b) What is a square kilometer in acres?

5. Suppose you want to cover a football field with artificial turf. You'll need to cover an area that's bigger than the actual field—let's say you'll cover a rectangle that's 60 yards wide and 130 yards long. How many square feet (not square yards!) of artificial turf will it take? The artificial turf comes off a roll that is 12 feet wide. Let's say the turf costs $8 per linear foot cut from the roll (so if you cut off 10 feet from the roll it would cost $80 and you'd have a rectangular piece of artificial turf that's 10 feet long and 12 feet wide). How much will the turf you need cost? (I've been told that preparing the field costs even more than the artificial turf itself.)

Answers To Exercises For Section 9.3 on Conversion

1. (a) 1 m = 100 cm = $100 \div 2.54$ in = 39.37 in = $39.37 \div 12$ ft = 3.28 feet. So 100 m = 328 feet.

   (b) 1 mile = 5280 feet = 63360 inches = 160934.4 cm = 1609.344 m = 1.609 km. So 70 miles per hour is 70×1.609 kilometers per hour, which is 113 kilometers per hour.

   (c) 1 km = 1000 m = 100,000 cm = 39370.08 in = 3280.84 ft = .621 miles. Therefore 1 square kilometer is $.621 \times .621 = .39$ square miles. So 2.4 square kilometers are $2.4 \times .39$ square miles, which is .9 square miles.

   (d) 1 yd = 3 ft = 36 in = 91.44 cm = .9144 m. So 1 cubic yard is $.9144 \times .9144 \times .9144$ cubic meters = .76 cubic meters.

   (e) 1 m = 100 cm = 39.37 in = 3.28 feet. So 1.88 meters = 6.17 feet, which is 6 feet, 2 inches.

   Why is .17 feet equal to 2 inches? This is *not* explained by rounding the .17 to 2! Instead, since 1 foot = 12 inches, therefore .17 feet = $.17 \times 12$ inches = 2.04 inches, which rounds to 2 inches. Remember that you want to keep the accuracy of your answer in line with the accuracy of the initial data, so it would be misleading to report the man's height as 6 feet, 2.04 inches.

   (f) 1 km = 1000 m = 100,000 cm = $100,000 \div 2.54$ in = 39379 in = $39379 \div 12$ ft = 3280.8 ft = $3280.8 \div 5280$ miles = .62 miles.

2. The stack of doughnuts would be 273,438 miles tall which is more than the distance to the moon.

3. 1 liter = 1000 cubic centimeters. Convert 1 cm to feet: 1 cm = .3937 in = .0328 feet. So one cubic centimeter is .000035314 cubic feet. Therefore 1 liter = .035314 cubic feet. Since each gallon is .134 cubic feet, 1 liter = $.035314 \div .134$ gallons = .2635 gallons which is a little more than $\frac{1}{4}$ gallon. So a liter is a little more than a quart.

4. (a) 1 mile = 5280 feet, so 1 square mile = 5280×5280 square feet = $27,878,400$ square feet. Since each acre is $43,560$ square feet, therefore we want to know how many $43,560$ are in $27,878,400$. This is a division problem. $27,878,400 \div 43,560 = 640$. So there are 640 acres in a square mile.

(b) 1 km = 1000 m = 100,000 cm = $100,000 \div 2.54$ in = 39379 in = $39379 \div 12$ ft = 3280.8399 ft. So 1 km^2 = 3280.8399^2 ft^2 = $10,763,910.42$ ft^2 = $10,763,910.42 \div 43,560$ acres = 247 acres.

5. You'll need $60 \times 130 = 7,800$ square yards of turf. Each square yard is $3 \times 3 = 9$ square feet, so you'll need $7,800 \times 9 = 70,200$ square feet of turf. (Or, notice that 60 yards = 180 feet and 130 yards = 390 feet, so you'll need $180 \times 390 = 70,200$ square feet.)

   \$8 buys you 12 square feet of turf, so it will cost at least

$$8 \times (70,200 \div 12) = \$46,800$$

   for the turf. But if you lay the turf in strips across the width of the field then you will need thirty-three strips that are 60 yards = 180 feet long, which will cost \$47,520. (The 33 strips comes from the fact that each strip is 4 yards wide and $130 \div 4 = 32.5$.)

Problems For Section 9.3 on Conversion

1. A car is 16 feet, 3 inches long. How long is it in meters? Give a complete and thorough explanation of the reasoning behind your calculations. Do not use dimensional analysis.

2. A car is 415 centimeters long. How long is it in feet and inches? Give a complete and thorough explanation of the reasoning behind your calculations. Do not use dimensional analysis.

3. The distance between two cities is described as 260 kilometers. What is this distance in miles? Calculate your answer in two ways: 1) using dimensional analysis and 2) using logical thinking and the meanings of multiplication and division. Remember to round your answer appropriately: the way you write your answer should reflect its accuracy. Your answer should not be reported more accurately than your starting data.

4. A house is described as having a floor area of 800 square meters. What is the floor area of this house in square feet? Use two different methods to calculate your answer.

9.4 The *Moving* and *Combining* Principles About Area

In this section we will study the two most fundamental principles that are used in determining the area of a shape or 2-dimensional object. These principles agree entirely with our common sense. In fact, you have probably used these principles without being consciously aware of them.

The *Moving* and *Combining* Principles About Area

1. If you move a shape rigidly without stretching it, then its area does not change.

2. If you combine (a finite number of) shapes *without overlapping* them, then the area of the resulting shape is the sum of the areas of the individual shapes.

These two principles are extremely useful and enable us to determine the areas of a wide variety of shapes.

A common way to use the *combining* principle is as follows: to find the area of some shape, subdivide the shape into pieces whose areas are easy to figure out, and then add up the areas of these pieces. The resulting sum is the area of the original shape because we can think of the original shape as the combination of the pieces.

For example, what is the surface area of the swimming pool pictured in Figure 9.17 (a bird's eye view)? To answer this, imagine dividing the surface of the pool into two rectangular parts, such as shown in Figure 9.18. We can then think of the surface of the whole pool as the combination of these two rectangular pieces. The rectangle on the left is 28 feet wide and 16 feet long, so its area is

$$28 \times 16 \text{ ft}^2 = 448 \text{ ft}^2.$$

The rectangle on the right is 20 feet wide and 30 feet long, so its area is

$$20 \times 30 \text{ ft}^2 = 600\text{ft}^2.$$

Therefore, according to the *combining* principle, the area of the pool is

$$448 + 600\text{ft}^2 = 1048\text{ft}^2.$$

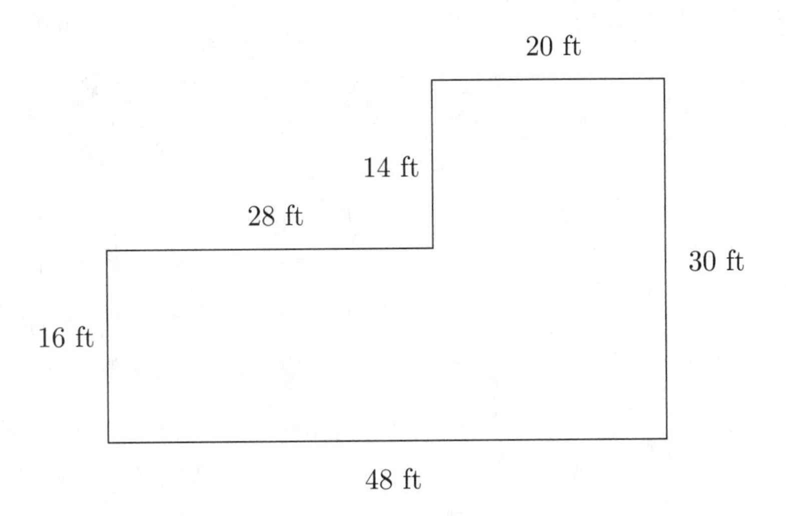

Figure 9.17: A Pool

Another way to use the *moving* and *combining* principles to determine the area of a shape is to subdivide the shape into pieces, then move and recombine those pieces (without overlapping!) to make a new shape whose area is easy to determine.

For example, what is the area of the patio pictured in Figure 9.19? To answer this, imagine slicing off the "bump" and sliding it down to fill in the corresponding "dent", as indicated in Figure 9.20. Since the "bump" was moved rigidly without stretching, and since the "bump" fills the "dent" perfectly, without any overlaps, therefore by the *moving* and *combining* principles, the area of the resulting rectangle is the same as the area of the original patio. Since the rectangle has area

$$24 \times 60 \text{ ft}^2 = 1440 \text{ ft}^2,$$

therefore the patio also has an area of 1440 square feet.

Class Activity 9L: Using the *Moving* and *Combining* Principles

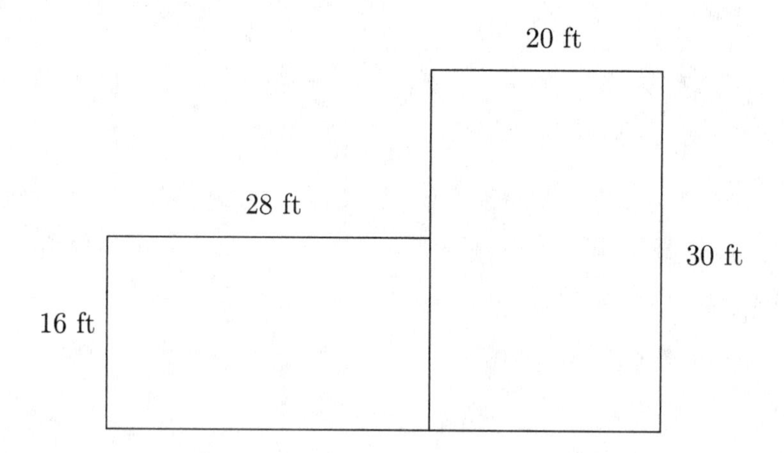

Figure 9.18: Subdividing the Pool

Class Activity 9M: Using the *Moving* and *Combining* Principles to Determine Surface Area

Exercises for Section 9.4 on the *Moving* and *Combining* Principles About Area

1. Figure 9.22 shows the floor plan of a loft that will be getting a new wood floor. How many square feet of flooring will be needed?

2. Figure 9.23 shows the floor plan of an apartment. How many square feet is the apartment? The notation $8'$ stands for 8 feet.

3. What is the surface area of a closed box that is 4 feet wide, 3 feet deep, and 5 feet tall? Draw a scaled down pattern for the box in order to help you determine its surface area.

4. What is the area of the shape in Figure 9.24? (This is not a perspective drawing, it is a flat shape.)

5. Use the *moving* and *combining* principles about area to determine the area of the triangle in Figure 9.25 in *two different ways.* In both cases, *do not* use a formula for areas of triangles. The grid lines in Figure 9.25 are 1 cm apart.

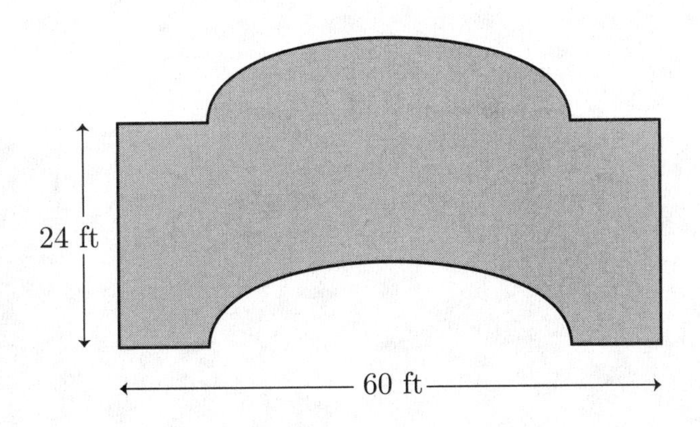

Figure 9.19: A Patio

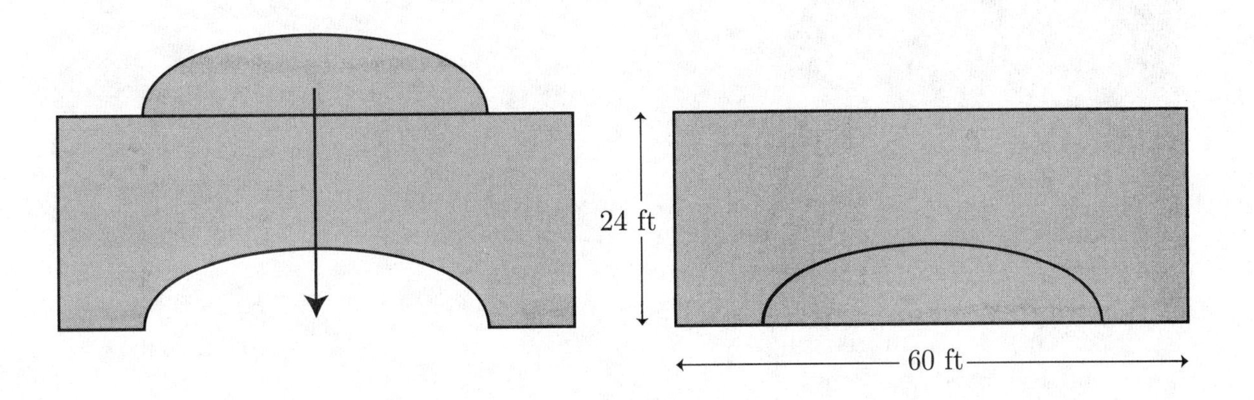

Figure 9.20: Determining the Area of the Patio

Figure 9.21: Children Show 35 Square Inches in Several Ways by Moving And Recombining Parts

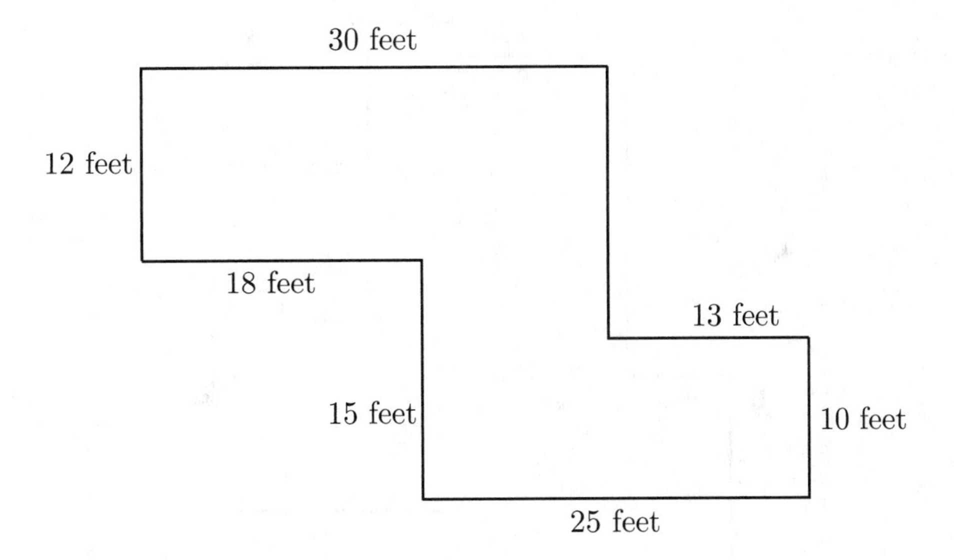

Figure 9.22: Floor Plan of a Loft

6. Use the fundamental principles about area to determine the area of the octagon in Figure 9.26.

Answers to Exercises for Section 9.4 on the *Moving* and *Combining* Principles About Area

1. First, what is the length of the unlabeled side? To solve this, notice that the "vertical" distance across the floor is equal to $12 + 15$ feet (considering the left side of the room), as well as *unlabeled length* $+10$ feet (considering the right side of the room). Therefore

   $$12 + 15 = \text{ unlabeled length } + 10,$$

   so the unlabeled length must be 17 feet.

   Next, notice that we can subdivide the floor into three rectangular pieces: one 30 ft by 12 ft, one 12 ft by 5 ft, and one 25 ft by 10 ft. Since we can think of the floor as the combination of these rectangular pieces, therefore, according to the *combining* principle about area, the number of square feet of flooring needed is $30 \times 12 + 12 \times 5 + 25 \times 10 = 670$.

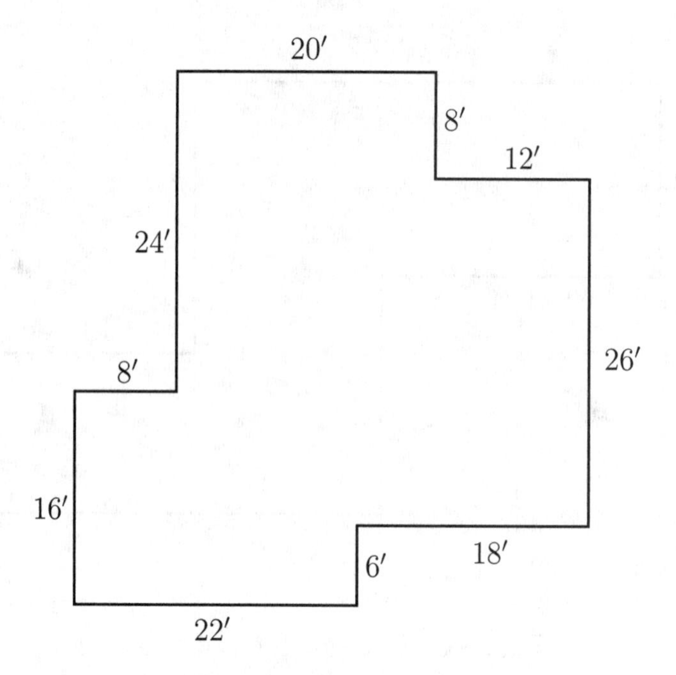

Figure 9.23: A Floor Plan for an Apartment

2. Figure 9.27 shows a way to subdivide the floor of the apartment into four rectangles. According to the *combining* principle about area, the area of the apartment is the sum of the areas of the rectangles. Therefore the area of the apartment is

$$20 \times 8 + 32 \times 26 + 8 \times 16 + 14 \times 6 \text{ ft}^2 = 1204 \text{ ft}^2.$$

3. Figure 9.28 shows a scaled down pattern for the desired box. By subdividing the pattern into rectangles and using the *combining* principle about areas, we see that the surface area of the box is

$$2 \times 12 + 2 \times 20 + 2 \times 15 \text{ ft}^2 = 94 \text{ ft}^2.$$

4. The "wiggly shape" of Figure 9.24 can be subdivided into pieces which can be recombined as shown in Figure 9.29 to make a rectangle that

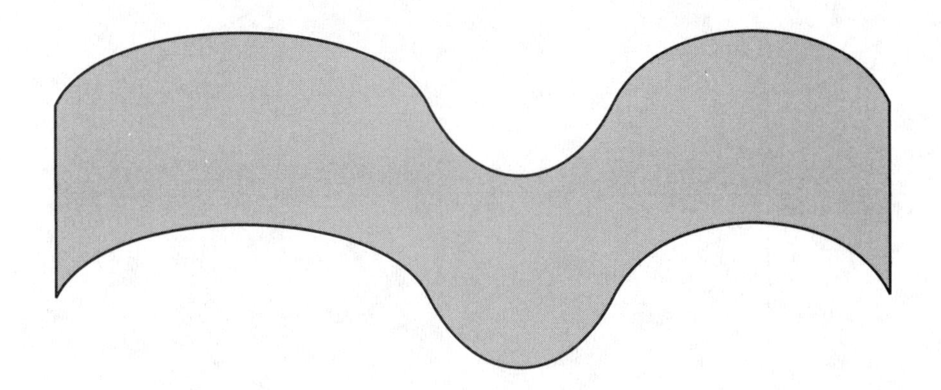

Figure 9.24: A Wiggly Shape

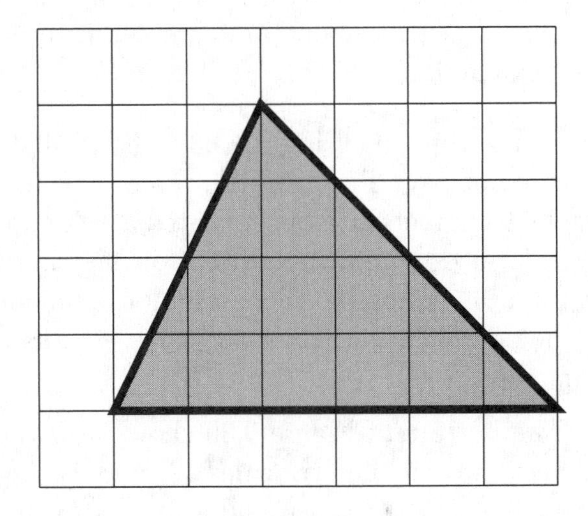

Figure 9.25: Find the Area of the Triangle

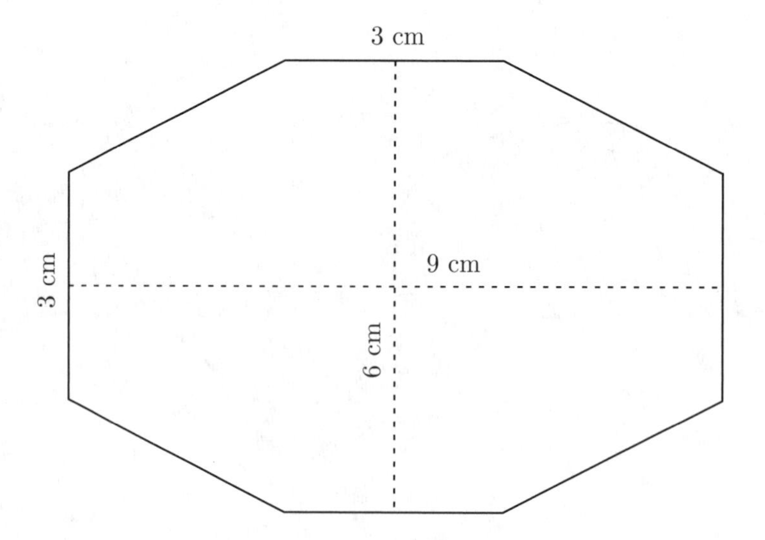

Figure 9.26: An Octagon

is 1 inch wide and $4\frac{1}{2}$ inches long (slice off the two "bumps" on top and the bump on the bottom, and move them to fill the corresponding "dents"). The rectangle has area $4\frac{1}{2}$ square inches, so by the *moving* and *combining* principles about area, the "wiggly shape" must also have area $4\frac{1}{2}$ square inches.

5. Method 1: At the top left of Figure 9.30, the triangle is shown subdivided into three pieces. Two of those pieces can be rotated down to form a 2 cm by 6 cm rectangle (as indicated at the top right). Since the rectangle was formed by moving portions of the triangle and recombining them without overlapping, therefore by the *moving* and *combining* principles, the original triangle has the same area as the 2 cm by 6 cm rectangle, namely 12 cm^2.

   Method 2: The bottom of Figure 9.30 shows how the triangle can be subdivided into two smaller triangles, A and B, and that copies of triangles A and B can be rotated and attached to the original triangle so as to form a 4 cm by 6 cm rectangle. In terms of areas,

$$2\text{A} + 2\text{B} = 24 \text{ cm}^2,$$

   according to the *moving* and *combining* principles. Therefore the area

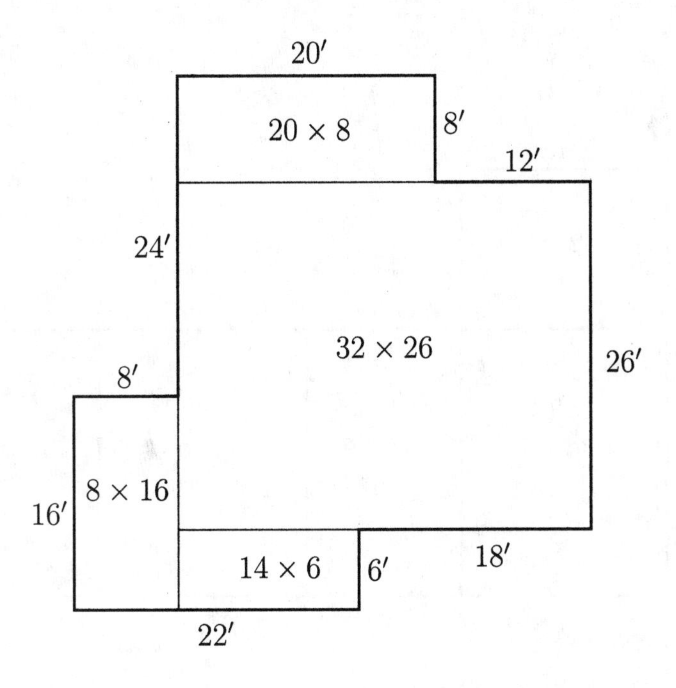

Figure 9.27: Subdividing the Floor into Rectangles

of the original triangle is

$$A + B = 12 \text{ cm}^2,$$

6. The area of the octagon is 45cm^2. One way to determine the area is to think of the octagon as a 6 cm by 9 cm rectangle with four triangles cut off, as indicated in Figure 9.31. Then, according to the *combining* principle about areas,

$$\text{area of octagon} + \text{area of four triangles} = \text{area of rectangle}$$

But the four triangles can be combined, without overlapping, to form two small 1.5 cm by 3 cm rectangles. So, by the *moving* and *combining* principles, we can say that

$$\text{area of octagon } + 2 \times 1.5 \times 3 \text{ cm}^2 = 6 \times 9 \text{ cm}^2.$$

Therefore

$$\text{area of octagon } = 54 - 9 \text{ cm}^2 = 45 \text{ cm}^2.$$

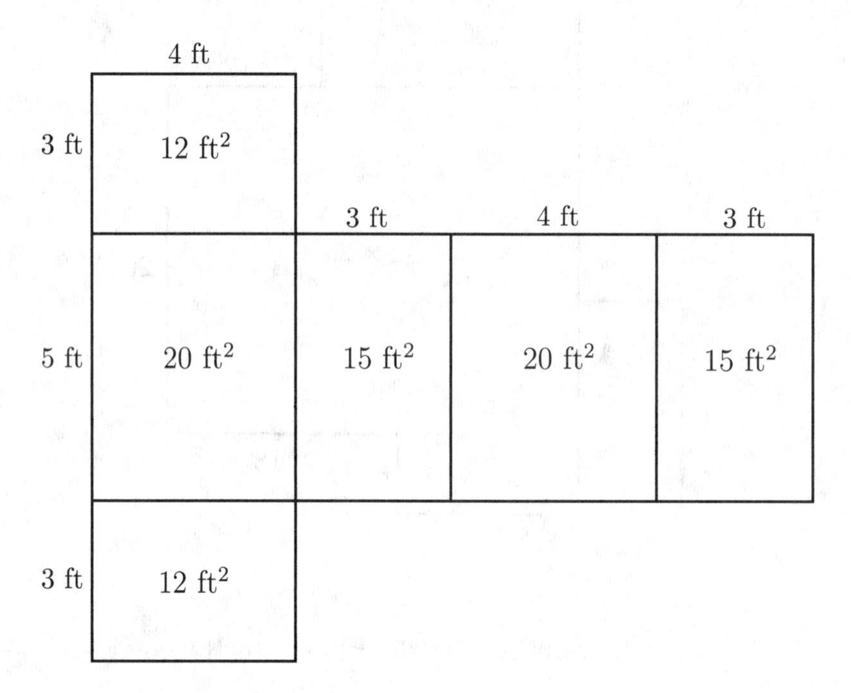

Figure 9.28: A Pattern for a Box

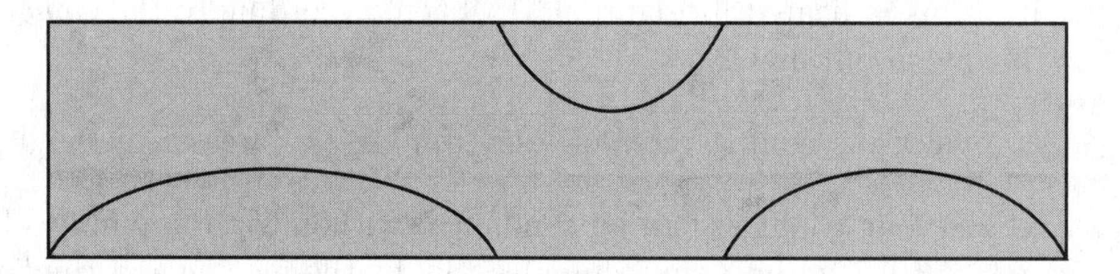

Figure 9.29: The Wiggly Shape Becomes a Rectangle After Subdividing and Recombining

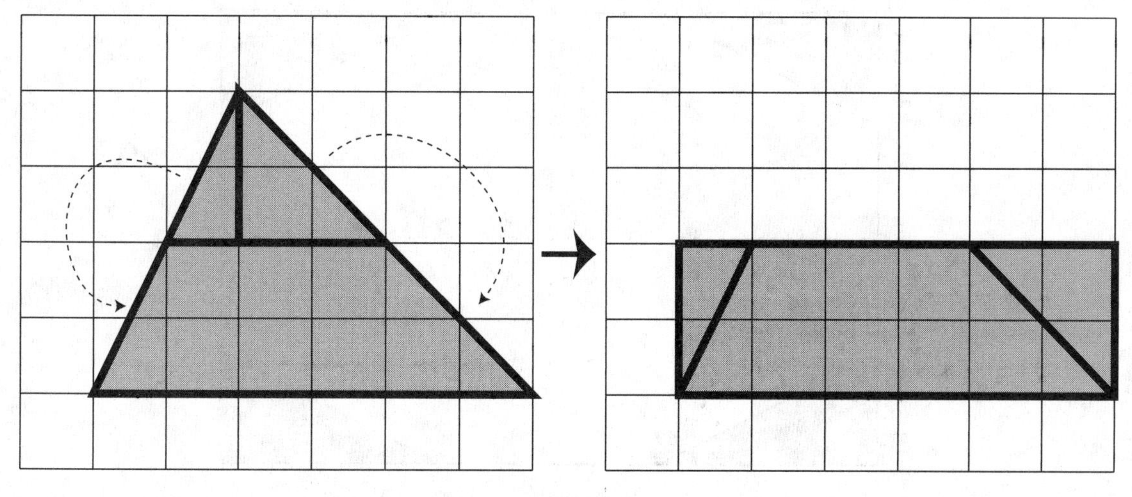

method 1

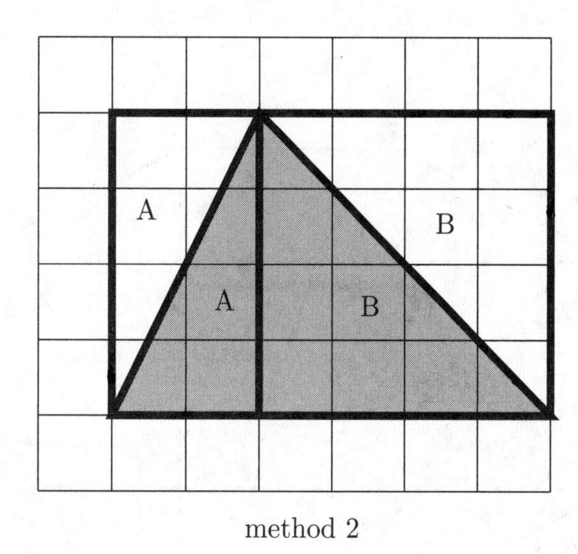

method 2

Figure 9.30: Two Ways to Determine the Area of the Triangle

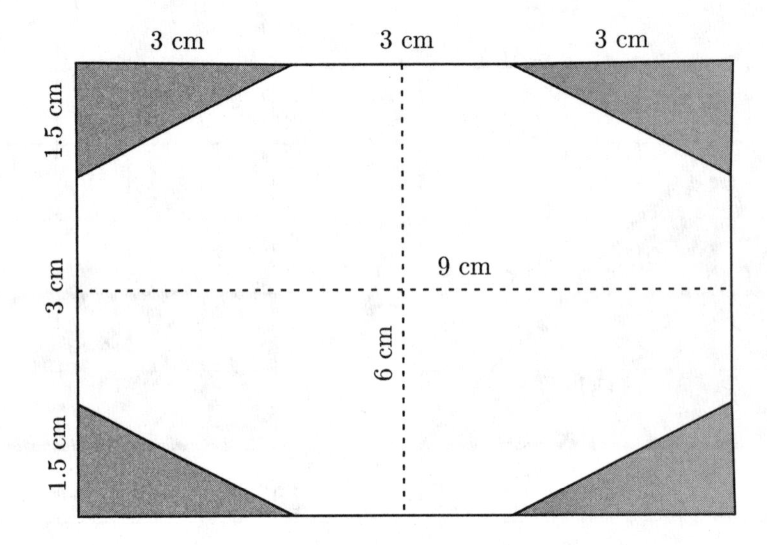

Figure 9.31: The Octagon is a Rectangle With Four Triangles Removed

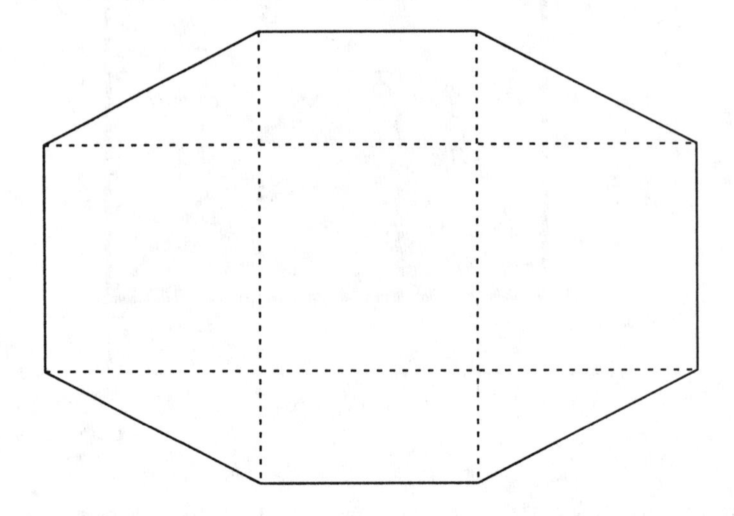

Figure 9.32: An Octagon

Another way to determine the area of the octagon is to subdivide it as shown in Figure 9.32. Once again, the four triangles can be combined to make two rectangles.

Problems for Section 9.4 on the *Moving* and *Combining* Principles About Area

1. Make a shape that has area exactly 25 in^2 but that has *no* (or almost no) straight edges. Explain how you know that your shape has area 25 in^2.

2. If a cardboard box with a top, a bottom and four sides is W feet wide, D feet deep, and H feet tall, then what is the surface area of this box? Find a formula for the surface area in terms of W, D, and H. Explain clearly why your formula is valid.

3. Figure 9.33 shows the floor plan for a one-story house. Calculate the area of the floor of the house. Explain your method clearly.

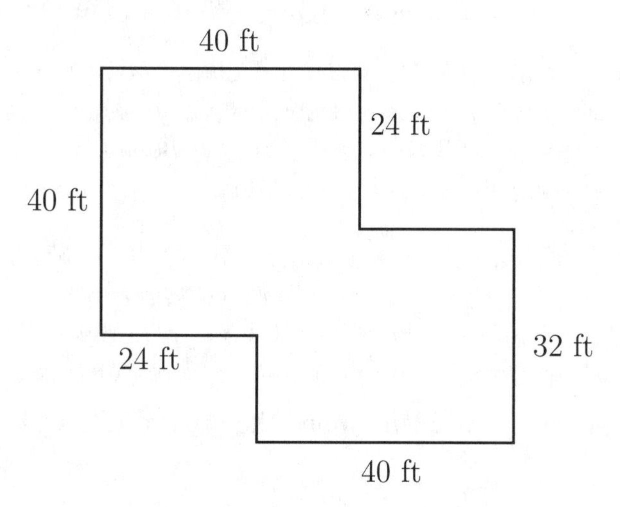

Figure 9.33: The Floor Plan of a House

4. Figure 9.34 shows the floor plan for a one-story house. Calculate the area of the floor of the house. Explain your method clearly.

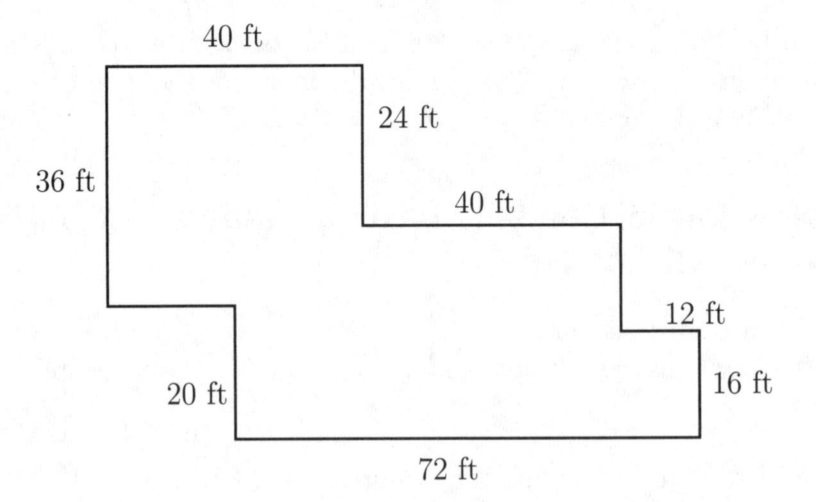

Figure 9.34: The Floor Plan of a House

5. Use the *moving* and *combining* principles about area to determine the exact area of the triangle in Figure 9.35. *Do not* use a formula for areas of triangles. The grid lines in Figure 9.35 are 1 cm apart.

6. An area problem: The Johnsons are planning to build a 5 foot wide brick walkway around their rectangular garden that is 20 feet wide and 30 feet long. What will the area of this walkway be? Before you solve the problem yourself, use Kaitlyn's idea, which follows.

   (a) Kaitlyn's idea is to "take away the area of the garden". Explain how to solve the problem about the area of the walkway *using Kaitlyn's idea.* Explain clearly how to apply one or both of the *moving* and *combining* principles on area in this case.

   (b) Now solve the problem about the area of the walkway in another way.

7. Figure 9.36 shows the floor plan for a modern, one-story house. Bob calculates the area of the floor of the house this way:

$$36 \times 72 - 18 \times 18 = 2268 \text{ ft}^2.$$

What must Bob have in mind? Explain why *Bob's method* is a legitimate way to calculate the floor area of the house and explain clearly

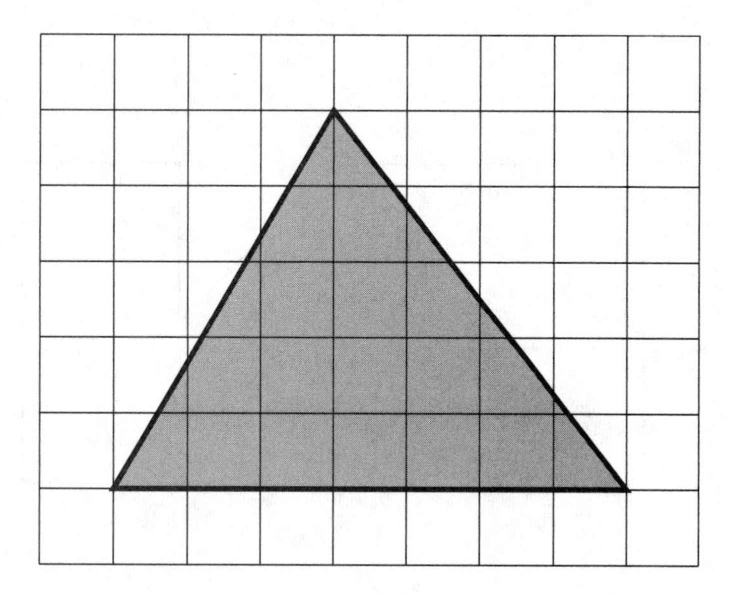

Figure 9.35: Determine the Area

how one or both of the *moving* and *combining* principles on area apply in this case.

8. Use the *moving* and *combining* principles about area to determine the area, in square inches, of the shaded design in Figure 9.37. The shape is a 2 inch by 2 inch square, with a square, placed diagonally inside, removed from the middle. In determining the area of the shape, use no *area* formulas other than the one for areas of rectangles. Explain your method clearly. (Although it is not necessary, you may use another kind of formula—not for calculating areas—that we have studied in a previous section.)

9. Use the *moving* and *combining* principles about area to determine the area, in square inches, of the shaded flower design in Figure 9.38. In doing this, use no *area* formulas other than the one for areas of rectangles. Explain your method clearly. (Although it is not necessary, you may use another kind of formula—not for calculating areas—that we have studied in a previous section.)

10. Use the *moving* and *combining* principles about area to determine the exact area of the triangle in Figure 9.39. *Do not* use a formula for areas

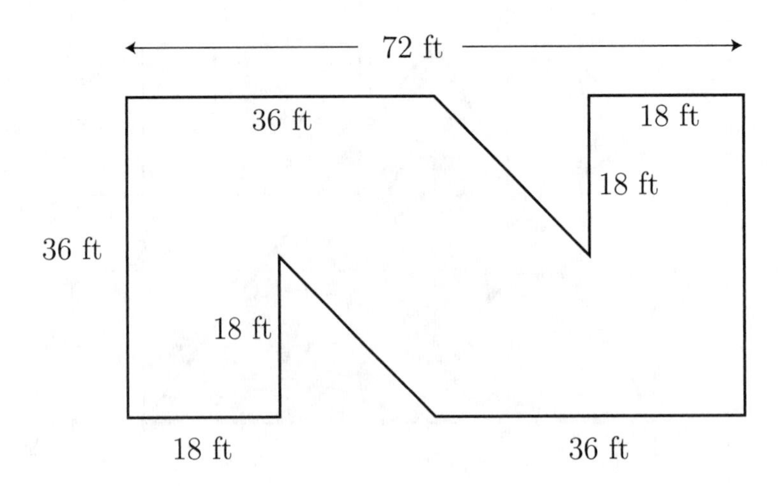

Figure 9.36: A Floor Plan for a Modern House

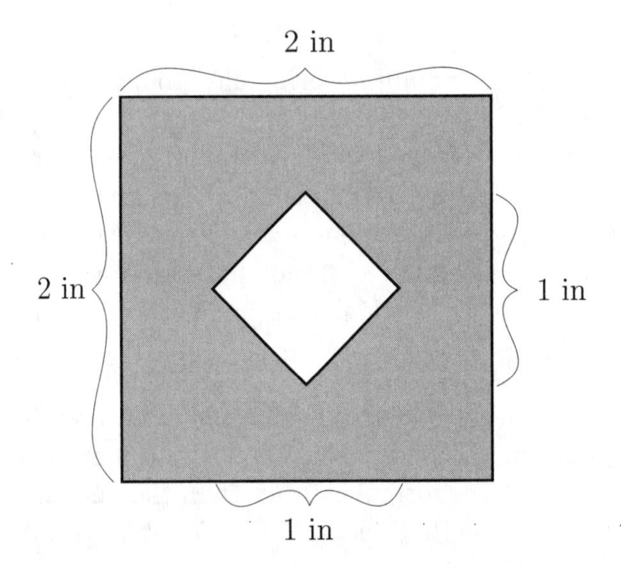

Figure 9.37: A Shape with a Hole

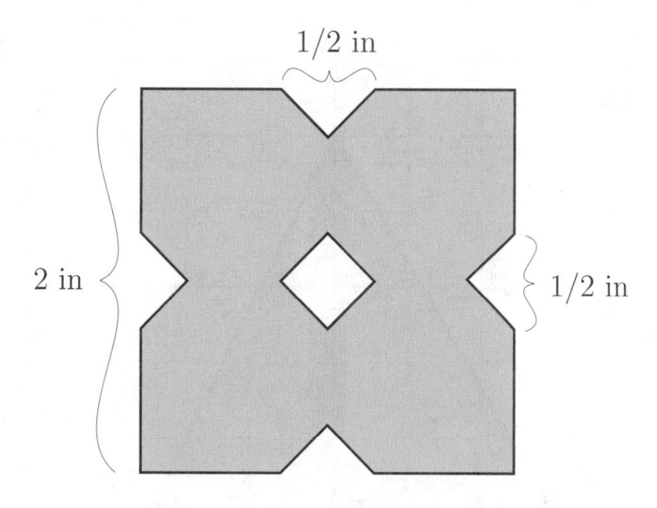

Figure 9.38: A Flower Design

of triangles. Explain your method clearly. The grid lines in Figure 9.39 are 1 cm apart.

11. An area puzzle:

    (a) Trace the 8 unit by 8 unit square in Figure 9.40.

    (b) Now cut out the pieces and reassemble them to form the 5 unit by 13 unit rectangle in Figure 9.41.

    (c) Calculate the area of the square and the area of the rectangle. Is there a problem? Can the pieces from your 8 unit by 8 unit square really be fitting together perfectly, without overlaps or gaps, to form a 5 unit by 13 unit rectangle?

    (d) Explain the problem! *Hints:* Can the two pieces shown in Figure 9.42, that you cut out of the square, really fit together as shown in the rectangle to form an actual triangle? If so, wouldn't there be some similar triangles? Look at the lengths of corresponding sides of the supposedly similar triangles. Do these numbers work they way they should for similar triangles?

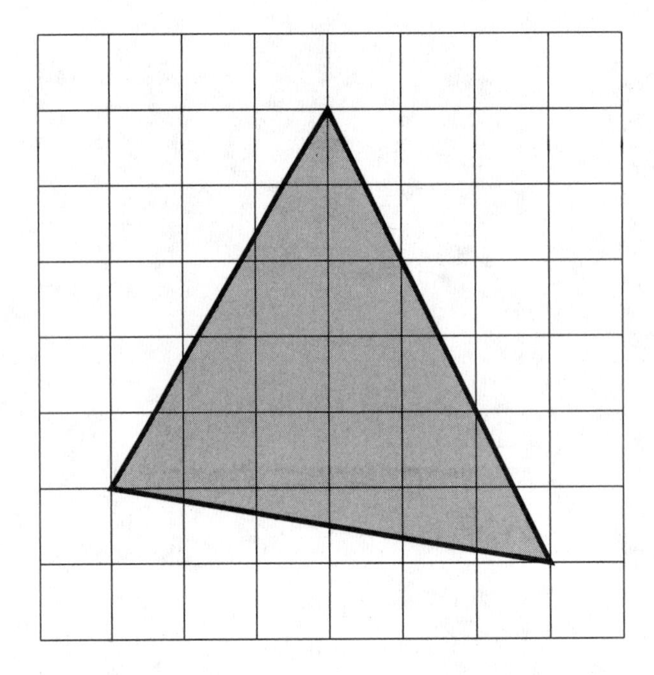

Figure 9.39: Determine the Area of the Triangle

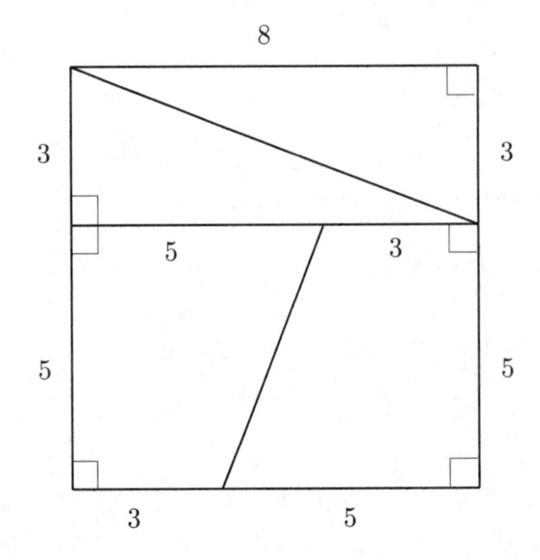

Figure 9.40: A Subdivided 8 by 8 Square

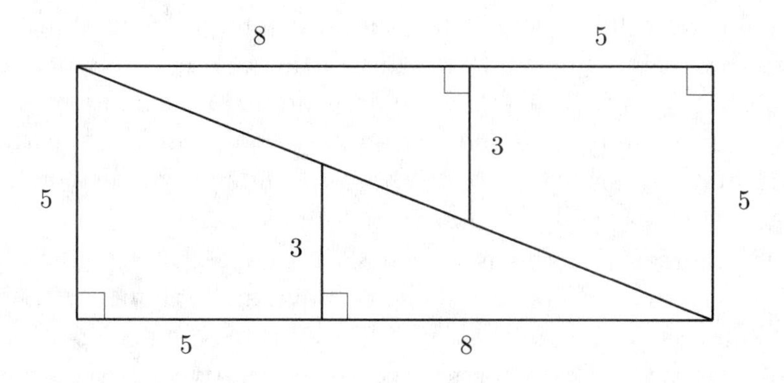

Figure 9.41: A Subdivided 5 by 13 Rectangle

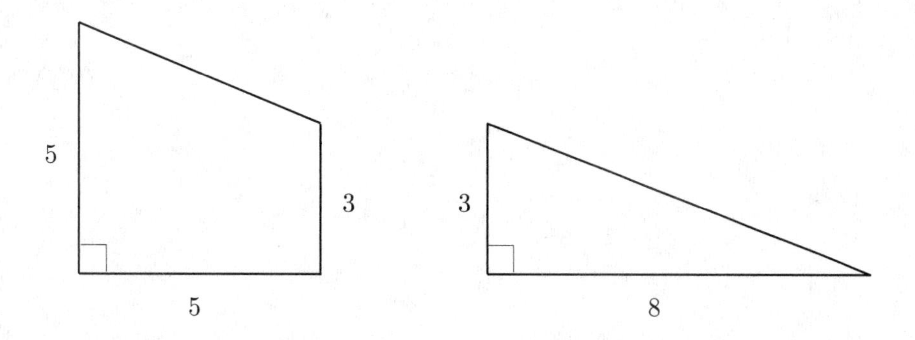

Figure 9.42: Two Pieces Put Together

9.5 Using the Moving and Combining Principles to Prove the Pythagorean Theorem

In this section we will see how to use the *Moving and Combining Principles* about area to explain why the Pythagorean theorem is true. Although most children do not study the Pythagorean theorem in elementary school, the *Moving and Combining Principles*, which can and should be learned informally in elementary school, pave the way to understanding more advanced mathematics such as the Pythagorean theorem.

The Pythagorean Theorem is named after the Greek mathematician Pythagoras, who lived around 500 B.C. There is evidence, though, that this theorem was known long before that, perhaps even by the ancient Babylonians in around 2000 B.C.. Pythagoras founded a secretive school, whose motto was "All is number." The members of this school, the Pythagoreans, tried to explain all of science, philosopy and religion in terms of numbers.

The Pythagorean Theorem (or Pythagoras's theorem) is a theorem about right triangles. Recall that in a right triangle, the side opposite the right angle is called the hypotenuse. The Pythagorean theorem says this:

> In a right triangle, the square of the length of the hypotenuse is equal to the sum of the squares of the lengths of the other two sides. In other words, if c is the length of the hypotenuse of a right triangle, and if a and b are the lengths of the other two sides, then
>
> $$a^2 + b^2 = c^2. \tag{9.1}$$

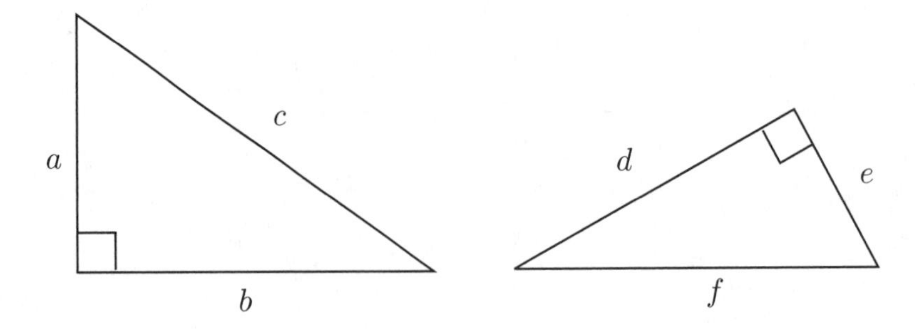

Figure 9.43: Right Triangles

For example, in Figure 9.43, the right triangle on the left has sides of lengths a, b, and c, and the hypotenuse has length c, so we can say:

$$a^2 + b^2 = c^2.$$

The right triangle on the right of Figure 9.43 has sides of lengths d, e, and f, and the hypotenuse has length f, so in this case:

$$d^2 + e^2 = f^2.$$

In Equation 9.1 it is understood that *all sides of the triangle are measured in the same units*—for example, all in centimeters, or all in feet, etc. If you have a right triangle where one side is measured in feet and the other in inches, for example, then you will need to convert both sides to inches or to feet (or to some other unit) in order to use Equation 9.1.

If you have a particular right triangle in front of you, such as one of the two shown in Figure 9.43, you can measure the lengths of its sides and check that the Pythagorean Theorem really is true in that case. For example, in the triangle on the left of Figure 9.43,

$$a = 3 \text{ cm}, \quad b = 4 \text{ cm}, \quad \text{and } c = 5 \text{ cm},$$

so

$$a^2 + b^2 = 25, \quad \text{and} \quad c^2 = 25.$$

So yes, $a^2 + b^2$ really is equal to c^2. For the triangle above on the right, $d =$ 2 cm, $e =$ 4 cm, and f is approximately 4.4 cm, so $d^2 + e^2$ is approximately 19, and f^2 is approximately 19 too. We could continue to check lots of right triangles triangles to see if the Pythagorean theorem really is true in those cases. If we did this many times it would be compelling evidence that the Pythagorean Theorem is always true. But one of the cornerstones of mathematics, an idea that goes all the way back to the time of the mathematicians of ancient Greece, is that lots of evidence is not enough—*proof* is required in order to know for sure that a statement is really true.

Proofs are one of the important aspects of this book too, even if we don't usually call our explanations proofs. A **proof** is a thorough, precise, logical explanation for *why* something is true, based on assumptions or facts that are already known or assumed to be true. So a proof is what establishes that a theorem is true.

Class Activity 9N: Using the Pythagorean Theorem

There are *hundreds* of known proofs of the Pythagorean Theorem. The next class activity will help you work through and "discover" one of these proofs.

Class Activity 9O: A Proof of the Pythagorean Theorem

Exercises for Section 9.5 on the Pythagorean Theorem

1. A garden gate that is 3 feet wide and 4 feet tall needs a diagonal brace to make it stable. How long a piece of wood will be needed for this diagonal brace? See Figure 9.44.

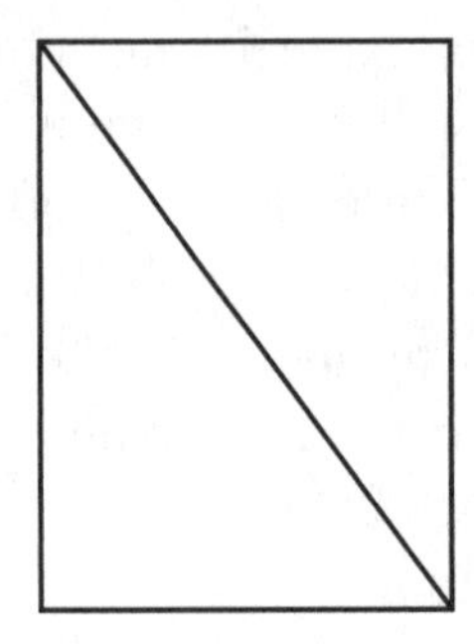

Figure 9.44: Garden Gate With Diagonal Brace

2. An elevator car is 8 feet wide, 6 feet deep and 9 feet tall. What is the longest pole you could fit in the elevator?

3. Imagine a pyramid with a square base. Suppose that the sides of the square base are all 200 yards long and that the distance from one corner of the base to the very top of the pyramid (along an outer edge) is 245 yards. How tall is the pyramid?

Answers to Exercises for Section 9.5 on the Pythagorean Theorem

1. 5 feet (by the Pythagorean theorem).

2. The length of the longest pole that will fit is the distance between point A and point C in Figure 9.45, which is 13.5 feet. To determine

the distance from A to C we will use the Pythagorean theorem twice. First with the right triangle on the floor of the elevator in order to determine the length of AB, and then with the triangle ABC in order to determine the length of AC. First, by the Pythagorean theorem, $8^2 + 6^2 = AB^2$, so $AB = \sqrt{100} = 10$ feet. Notice that ABC is also a right triangle, with the right angle at B. Therefore, by the Pythagorean theorem, $AB^2 + BC^2 = AC^2$, so $AC = \sqrt{181} = 13.5$ feet. Thus the longest pole that can fit in the elevator is 13.5 feet.

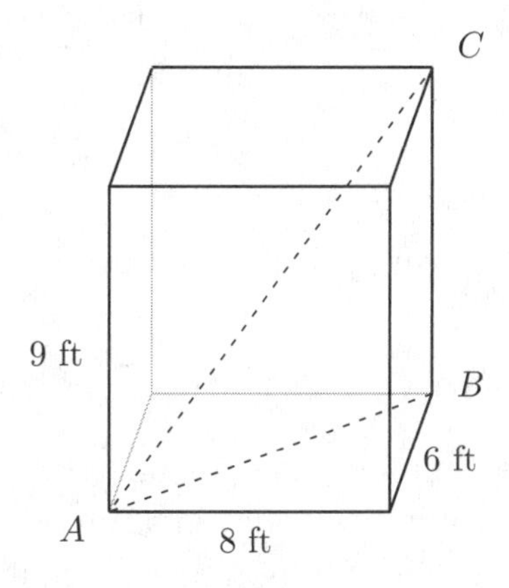

Figure 9.45: A Pole in an Elevator

3. The pyramid is about 200 yards tall. Here's why. Notice that the height of the pyramid is the length of BC in Figure 9.46. We will determine the length of BC by using the Pythagorean theorem twice, first with the triangle ABD, in order to determine the length of AB, and then with the triangle ABC, in order to determine the length of BC. First, by the Pythagorean theorem, $AD^2 + BD^2 = AB^2$, so $AB = \sqrt{20000} = 141.4$ yards (actually, as you'll see in the next step, we really only need AB^2, not AB, so we don't even have to calculate the square root). The triangle ABC is a right triangle, with right angle at B. So by the Pythagorean theorem, $141.4^2 + BC^2 = 245^2$, and therefore $BC = \sqrt{40025} = 200$ yards. So the pyramid is about 200 yards tall.

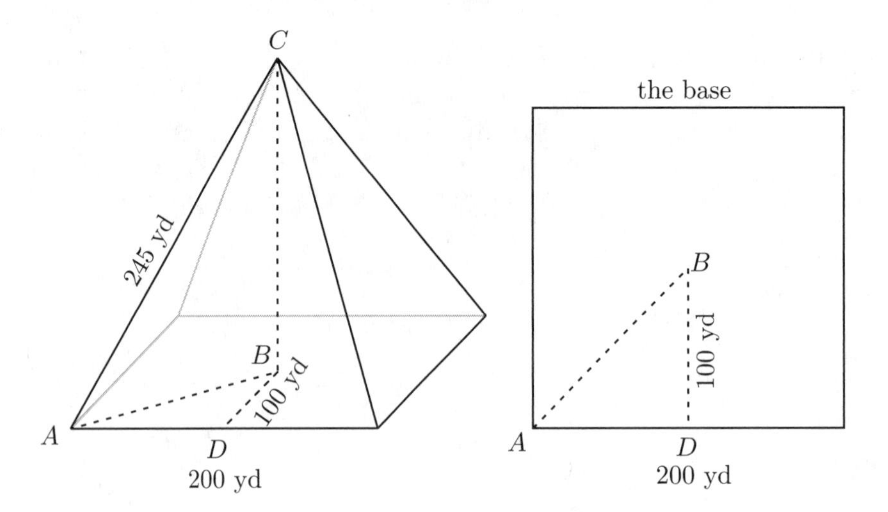

Figure 9.46: A Pyramid

Problems for Section 9.5 on the Pythagorean Theorem

1. In your own words, explain how to use the *Moving and Combining Principles* to explain why it is that given any right triangle with short sides of lengths a and b and hypotenuse of length c,

$$a^2 + b^2 = c^2.$$

2. A tent will be constructed to have a square, 15 foot $\times$ 15 foot base, four vertical walls, each 10 feet high, and a pyramid-shaped top made out of four triangular pieces of cloth. At its tallest point in the center of the tent, the tent will reach a height of 15 feet. See Figure 9.47.

   Draw a pattern for one of the four triangular pieces of cloth that will make the pyramid-shaped top. Label the pattern with enough information so that someone making the tent would know exactly how to measure and cut the material for the top of the tent.

3. Assuming that the earth is a perfectly round, smooth ball of radius 4000 miles, and that 1 mile = 5280 feet, how far away does the horizon appear to be to a 5 foot tall person on a clear day? Explain your reasoning. To solve this problem, start by drawing a picture that shows the cross-section of the earth, a person standing on the surface of the

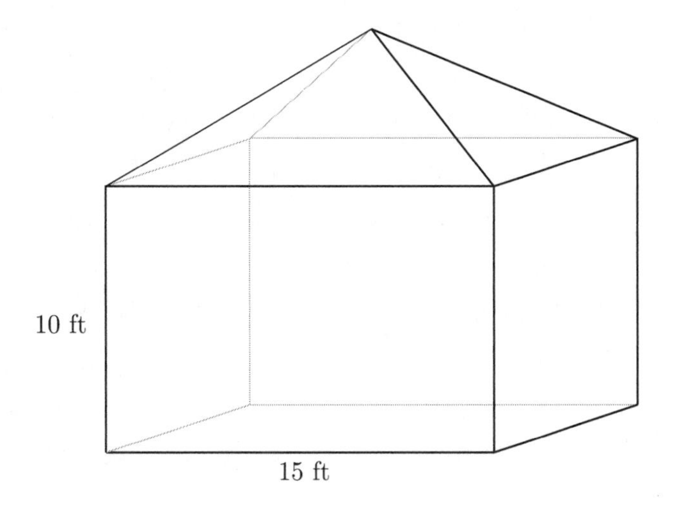

Figure 9.47: A Tent

earth, and the straight line of the person's gaze reaching to the horizon (obviously, you won't want to draw this to scale). You will need to use the following geometric fact: if a line is *tangent* to a circle at a point P (meaning it just "grazes" the circle at the point P; it meets the circle only at that one point) then the line is *perpendicular* to the line connecting P and the center of the circle, as illustrated in Figure 9.48.

9.6 More Ways to Determine Areas

In this section we will examine several methods for determining areas besides the methods using the *moving* and *combining* principles discussed in the previous section. The methods of the previous section work wonderfully for determining areas—when they works. Unfortunately, they don't always work. For example, how could we determine the area of the bean-shaped region in Figure 9.49? There isn't any way to subdivide and recombine the bean-shaped region into regions whose areas are easy to figure out.

Even so, there are a number of ways to determine the area (if only approximately). The next class activity will help you think about how you might determine the area of an irregular shape.

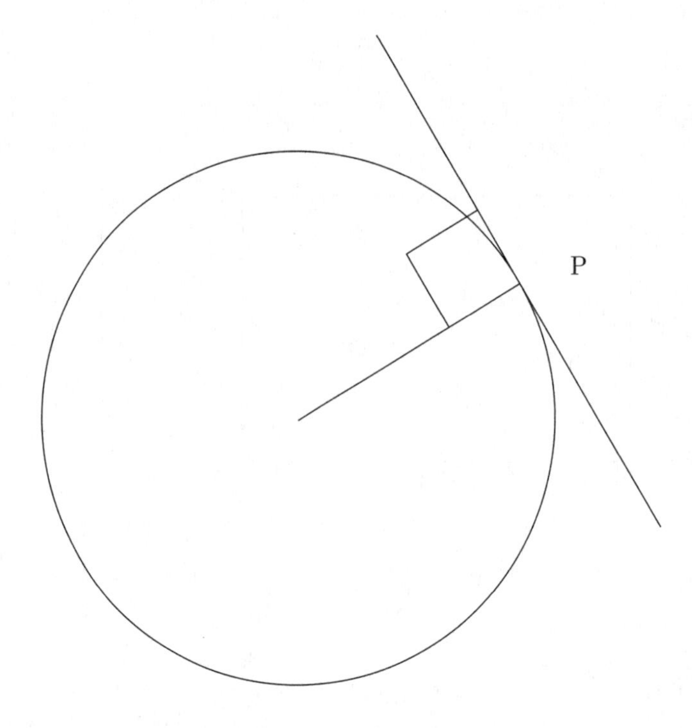

Figure 9.48: A Line Tangent to a Circle

Class Activity 9P: Determining the Area of an Irregular Shape

Practical and Approximate Methods for Determining Areas of Irregular Regions

One way you could find the area of the bean-shaped region in Figure 9.49 would be to cut it out, weigh it, and compare its weight to the weight of a full piece of paper. Then use proportional reasoning to find the area of the bean-shaped region. For example, if the bean-shaped region weighs $\frac{3}{8}$ as much as a whole piece of paper, then its area is also $\frac{3}{8}$ as much as the area of the whole piece of paper.

A good way to determine the area *approximately* is to work with graph paper, as shown in Figure 9.50. The lines on this graph paper are spaced $\frac{1}{2}$ cm apart, with heavier lines spaced 1 cm apart (so that 4 small squares make 1 square centimeter).

By counting the number of 1 cm by 1 cm squares (each consisting of 4

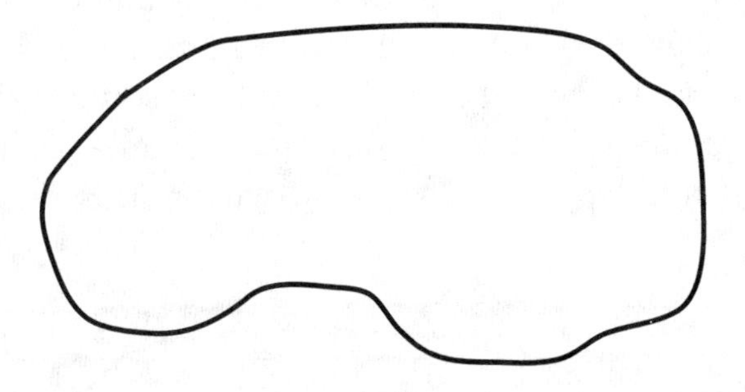

Figure 9.49: A Bean-Shaped Region

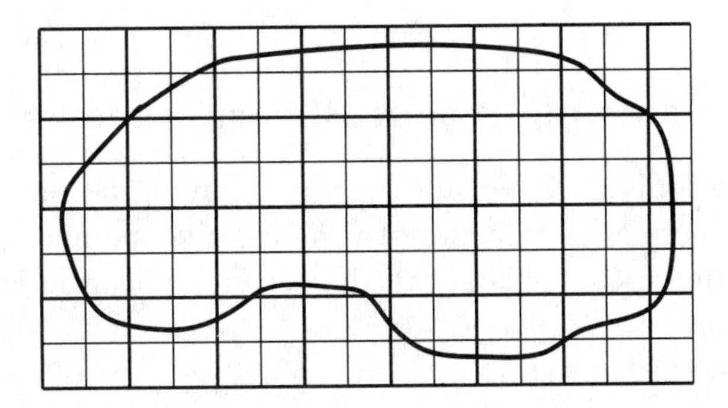

Figure 9.50: Determining the Area of a Bean-Shaped Region

small squares) inside the bean-shaped region, and by mentally combining the remaining portions of the region that are near the boundary, we can determine that the bean-shaped region has an area of about 19 square centimeters.

We can also use the graph paper to find under- and over-estimates for the area of the bean-shaped region. As shown in Figure 9.51, by considering the squares that lie entirely inside the region, and the squares that contain some portion of the region, we can say for sure that the area of the region must be between $14\frac{1}{2}$ cm^2 and $24\frac{1}{4}$ cm^2. We could narrow this range by using finer graph paper.

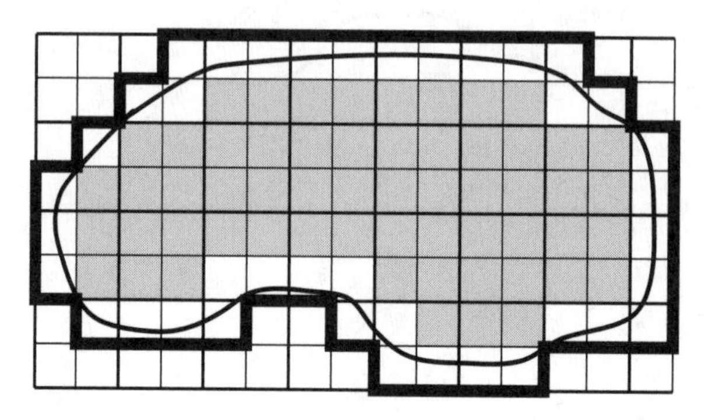

Figure 9.51: Determining Over- and Under-estimates for the Area of the Bean-Shaped Region

Cavalieri's Principle About Shearing and Area

In addition to subdividing a shape and recombining its parts without overlapping them, there is another way to change a shape and get a new shape that has the same area. This is called *shearing*. Shearing is a process that fits with its name.

To illustrate shearing, start with a polygon, pick one of its sides, and then imagine slicing the polygon into extremely thin (really, infinitely thin) strips that are parallel to the chosen side. Now imagine giving those thin strips a push from the side, so that the chosen side remains in place, but so that the thin strips slide over, remaining parallel to the chosen side, and remaining the same distance from the chosen side throughout the sliding process. Then you will have a new polygon, as indicated in Figures 9.52 and 9.53. This process of "sliding infinitely thin strips" is called **shearing**.

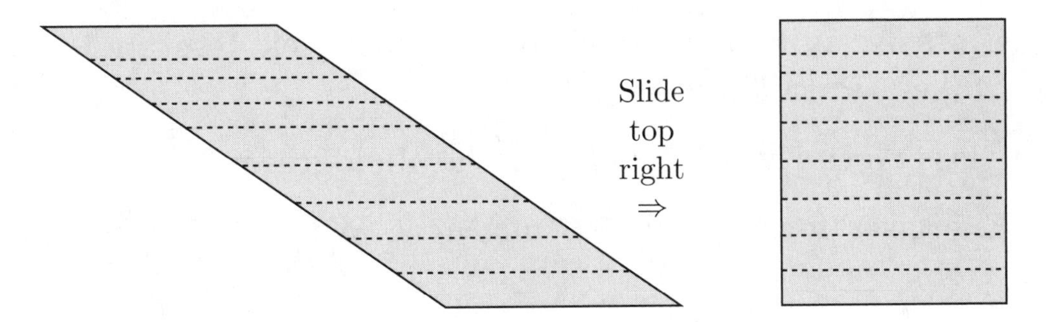

Figure 9.52: Shearing a Parallelogram

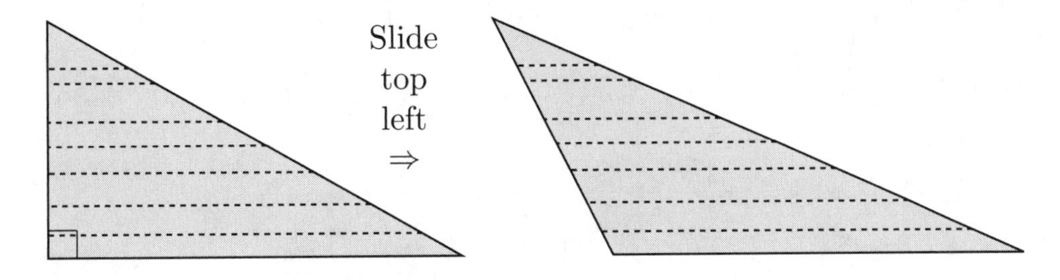

Figure 9.53: Shearing a Triangle

In order to understand shearing, it may help you to think of the thin strips as being like toothpicks. If you give a stack of toothpicks a push from the side, they will slide over, as in shearing (see Figure 9.54).

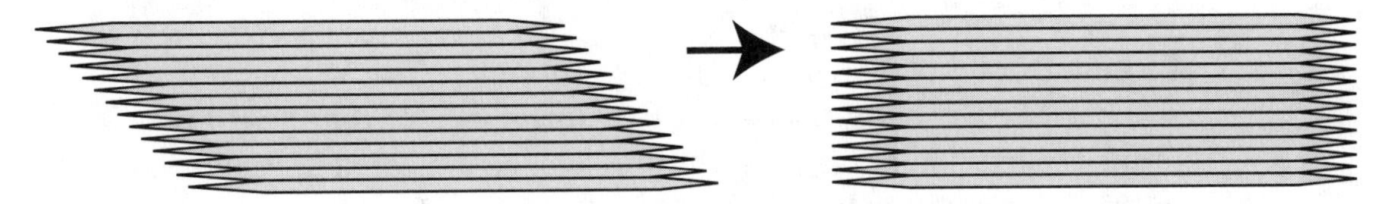

Figure 9.54: Shearing a Toothpick Parallelogram

It's important to note that in the shearing process, the thin strips *remain the same width and length.* They are just slid over, they are not compressed either in width or in length.

Cavalieri's principle for areas says that when you shear a shape as described above, the area of the original and sheared shapes are equal.

Class Activity 9Q: Shearing a Toothpick Rectangle to Make a Parallelogram

Class Activity 9R: Shearing Parallelograms

Exercises for Section 9.6 on More Ways to Determine Areas

1. In Figure 9.55, the small squares in the grid are 1 cm by 1 cm. Find an underestimate and an overestimate for the area of the shaded region. Do this by considering the squares that lie entirely within the region, and the squares that contain a portion of the region (as indicated in Figure 9.51).

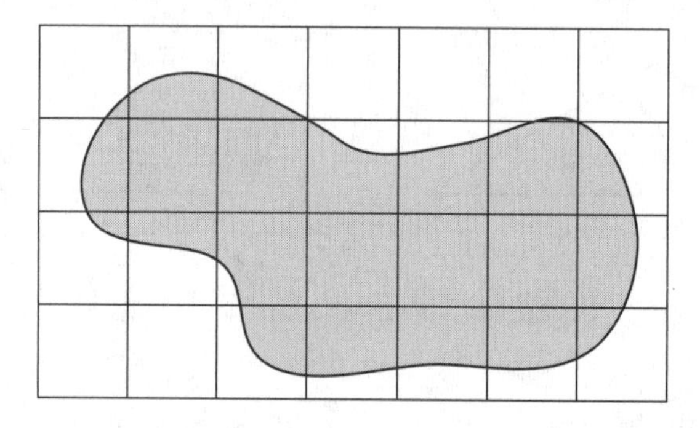

Figure 9.55: Find Under- and Over-estimates of the Area

2. Continuing the previous exercise, use the finer grid in Figure 9.56 to give better over- and under-estimates of the area of the shaded region, (using the method of the previous problem, as indicated in Figure 9.51). In this finer graph paper, each small square is $\frac{1}{2}$ cm by $\frac{1}{2}$ cm.

3. Suppose you don't know the formula for the area of a circle, but you do know about areas of squares. What can you deduce about the area of a circle of radius r units from Figure 9.57?

4. What is Cavalieri's Principle?

5. Figure 9.58 shows a parallelogram on a pegboard. (You can think of the parallelogram as made out of a rubber band which is hooked around

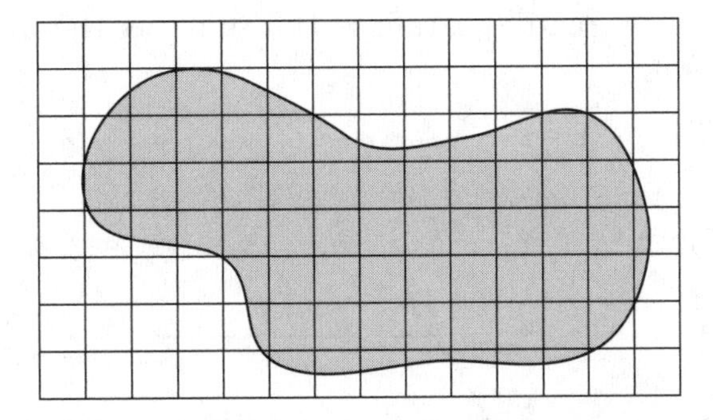

Figure 9.56: Find New Under- and Over-estimates of the Area

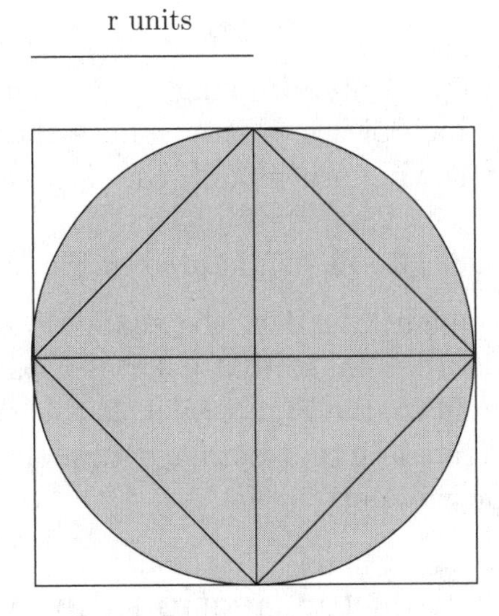

Figure 9.57: Estimating the Area of the Circle

four pegs.) Show two ways to move points C and D of the parallelogram to other pegs, keeping points A and B fixed, in such a way that the area of the new parallelogram is the same as the area of the original parallelogram.

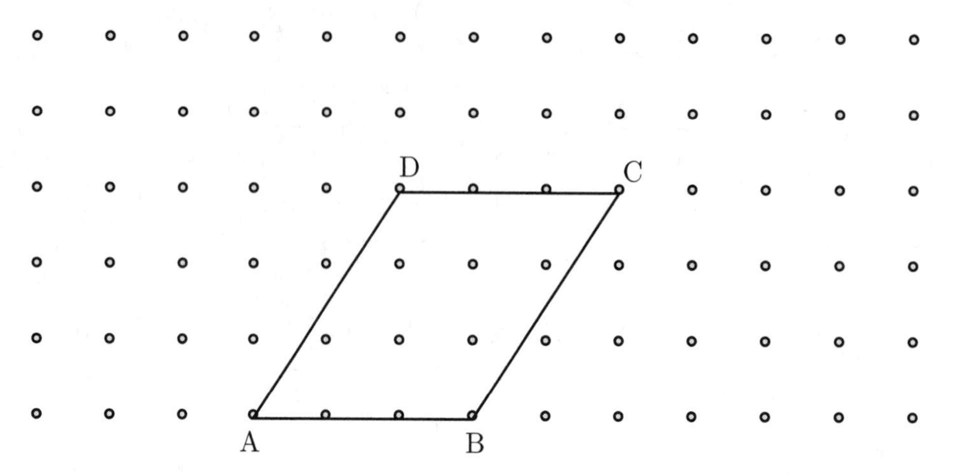

Figure 9.58: A Parallelogram on a Pegboard

6. Using two ordinary plastic drinking straws, cut two 4 inch pieces of straw and two 3 inch pieces of straw. Lace these pieces of straw onto a string in the following order: a 3 inch piece, a 4 inch piece, a 3 inch piece, a 4 inch piece. Tie a knot in the string so that the four pieces of straw form a quadrilateral, as pictured in Figure 9.59.

   Put your straw quadrilateral in the shape of a rectangle. Gradually "squash" the quadrilateral so that it forms a parallelogram that is not a rectangle, as indicated in Figure 9.59. Is this "squashing" process for changing the rectangle into a parallelogram the same as the shearing process? Why or why not?

Answers to Exercises for Section 9.6 on More Ways to Determine Areas

1. Underestimate: 5 square centimeters (counting the squares inside the region) Overestimate: 22 square centimeters (counting the squares that contain portions of the region).

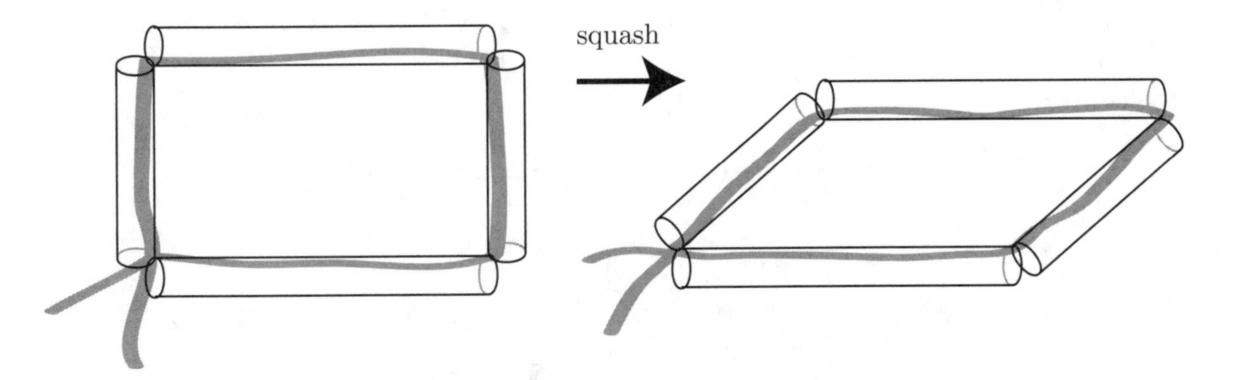

Figure 9.59: "Squashing" a Quadrilateral Made of Straws

2. Underestimate: 10.25 square centimeters (because there are 41 small squares inside the shape, and each square has area $\frac{1}{4}$ square centimeters). Overestimate: 19.25 square centimeters (77 small squares).

3. You can definitely conclude that the area of the circle is between $2r^2$ square units and $4r^2$ square units. This is because the outer square in Figure 9.57 is made up of four r by r squares, while the inner square (diamond) can be subdivided into four triangles, which can be recombined to make two r by r squares. Since the circle lies completely within the outer square and completely contains the inner square, therefore the area of the circle must be greater than $2r^2$ square units and less than $4r^2$ square units. Just by eyeballing, the area of the circle looks to be roughly half way in between these under- and over-estimates, so roughly $3r^2$ (which is pretty close to the actual πr^2!).

4. See text.

5. See Figure 9.60. ABEF and ABGH are two such examples. These parallelograms have the same area as ABCD by Cavalieri's principle, because they are just sheared versions of ABCD. In fact, if we move C and D anywhere along the line of pegs that goes through C and D, keeping the same distance between them, the resulting parallelogram will be a sheared version of ABCD and hence will have the same area as ABCD.

6. No, this "squashing" process is not the same as shearing. You can

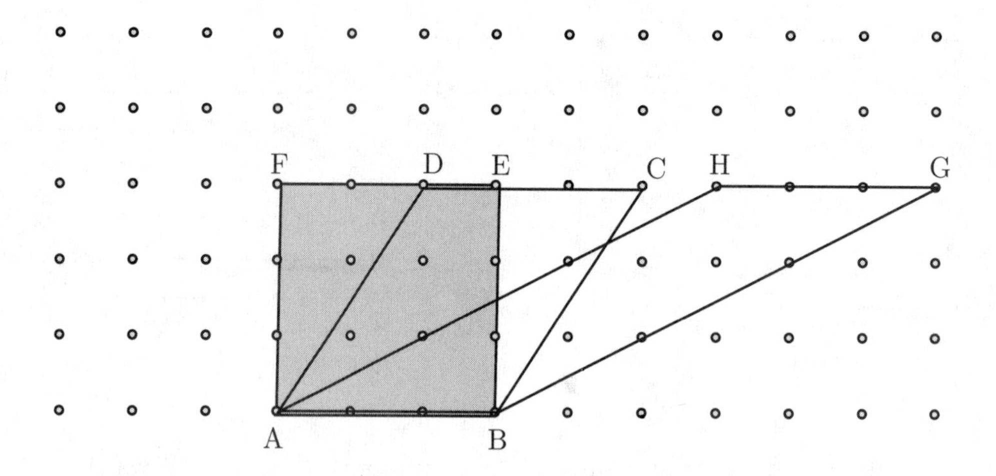

Figure 9.60: Three Parallelograms With the Same Area

tell that it's not shearing because if you made the rectangle out of very thin strips and you slid them over to make a parallelogram, they would have to become thinner in order to make the parallelogram that is formed from the straws (because the parallelogram is not as tall as the rectangle). But in the shearing process, the size of the strips does not change (either in length or in width). Therefore "squashing" is not the same as shearing.

Problems for Section 9.6 on More Ways to Determine Areas

1. Figure 9.61 shows a triangle on a pegboard. (You can think of the triangle as made out of a rubber band which is hooked around three pegs.) Describe or show pictures of at least two ways to move point C of the triangle to another peg (keeping points A and B fixed) in such a way that the area of the new triangle is the same as the area of the original triangle. Explain your reasoning.

2. The boundary between the Johnson's and the Zhang's properties is shown in Figure 9.62. The Johnsons and the Zhangs would like to change this boundary so that the new boundary is one straight line segment, and so that each family still has the same amount of land area. Describe a precise way to redraw the boundary between the

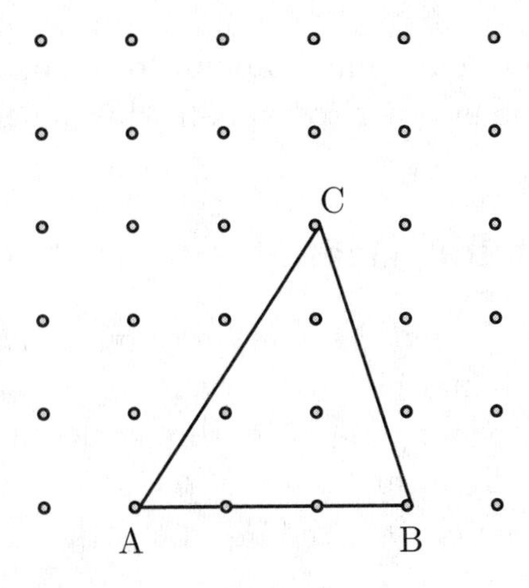

Figure 9.61: A Triangle on a Pegboard

two properties. Explain your reasoning. *Hint:* Consider shearing the triangle ABC.

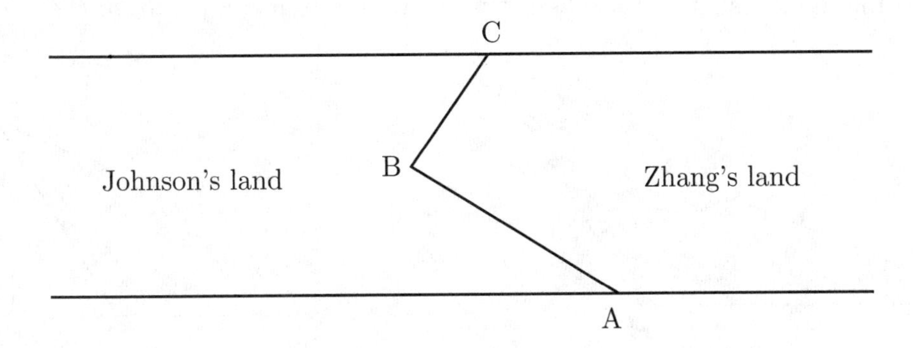

Figure 9.62: The Boundary Between Two Properties

9.7 Areas of Triangles

You probably know the familiar formula for the area of a triangle: *one half the base times the height.* Why is this formula valid? We will explain this in this section. For some triangles, we will be able to see why the area formula is valid by describing the triangle as "half of a rectangle", and by applying the *moving* and *combining* principles about area. For other triangles, we will use Cavalieri's principle in order to explain why the area formula is valid.

Base and Height for Triangles

Before we discuss the *one half the base times the height* formula for the area of a triangle, we need to know what *base* and *height* of a triangle mean. The **base** of a triangle can be any one of its three sides. In formulas, we usually write *base* when we really mean *length of the base.*

Once a base has been chosen, the **height** is the line segment that is

1. perpendicular to the base and

2. connects the base, or an extension of the base, to the corner of the triangle that is not on the base.

In formulas, we usually write *height* when we really mean *length of the height.*

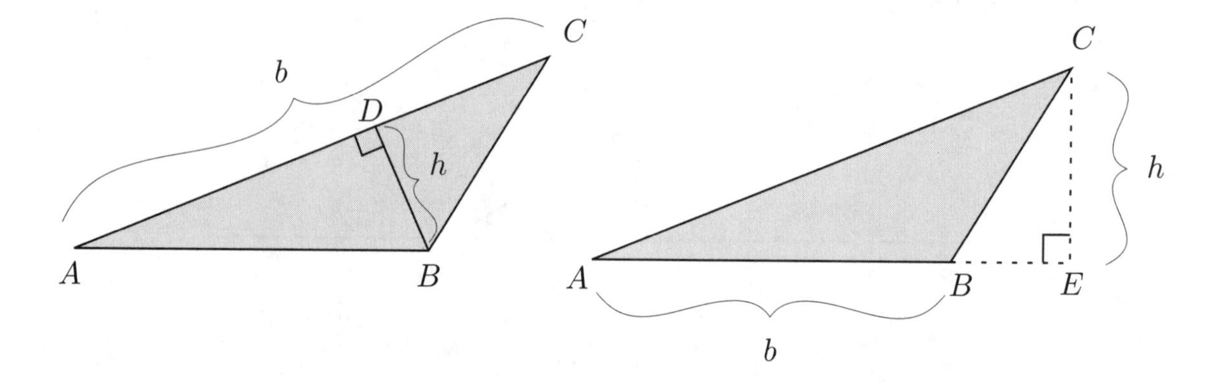

Figure 9.63: Two Ways to Pick the Base and Height of Triangle ABC

For example, Figure 9.63 shows two copies of a triangle ABC, and two of the three ways that the base b and height h could be chosen (notice that the base and height turn out to have different lengths for the different choices!). In the second case, the height is the (dashed) segment CE. In this case, even though the height h doesn't meet b itself, it does meet an *extension* of b.

Class Activity 9S: Ways to Choose the Base and Height of Triangles

The Formula for the Area of a Triangle

The familiar **formula for the area of a triangle** with base b and height h is this:

$$\text{area of triangle} = \frac{1}{2}b \times h.$$

For example, if a triangle has a base that is 5 inches long, and if the corresponding height of the triangle is 3 inches long, then the area of the triangle is

$$\frac{1}{2}5 \times 3 \text{ in}^2 = 7.5 \text{ in}^2.$$

In the formula

$$\frac{1}{2}b \times h,$$

it is assumed that both the base b and the height h are *measured in the same unit.* For example, both might be in feet, or both in centimeters, etc. But if one is in feet and the other is in inches, for example, then you need to convert both to a commmon unit. The area of the triangle is then in *square units* of whatever common unit is used for the base and height. So if the base and height were both in centimeters, then the area resulting from the formula is in square centimeters (cm^2).

Class Activity 9T: Explaining Why the Area Formula for Triangles is Valid

Explaining Why the Area Formula for Triangles is Valid

Suppose we have a triangle, and have chosen a base b and a height h for the triangle. Why is the area of the triangle equal to one half the base times the height? In some cases, such as those shown in Figure 9.64, two copies of the triangle can be subdivided and recombined, without overlapping, to form a b by h rectangle. In such cases, because of the *moving* and *combining* principles about area,

$$2 \times \text{ area of triangle} = \text{area of rectangle}.$$

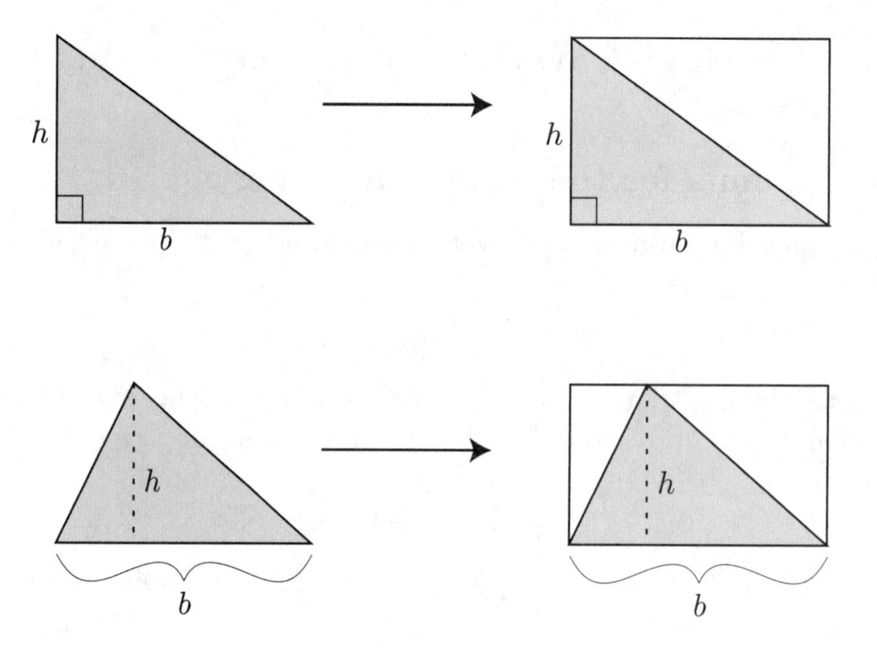

Figure 9.64: These Triangles are Half of a b by h Rectangle

But since the rectangle has area $b \times h$, therefore

$$2 \times \text{ area of triangle } = b \times h,$$

so

$$\text{area of triangle } = \frac{1}{2} b \times h.$$

But what about the triangle in Figure 9.65? For this triangle, it's not so clear how to turn it into half of a b by h rectangle. However, the formula for the area of the triangle is still valid, and we can explain why by using Cavalieri's Principle.

As shown in Figure 9.65, the triangle of Figure 9.65 can be sheared so as to form a right triangle. Recall that you can think of this shearing process as slicing the triangle into (infinitely) thin strips and giving those strips a push so as to move them over. Notice that since the shearing is done parallel to the base of the triangle, the *sheared triangle still has the same base* b *and height* h *as the original triangle.*

The new right triangle has area $\frac{1}{2}b \times h$ square units since two copies of this triangle can be combined without overlapping to form a b by h rectangle, as

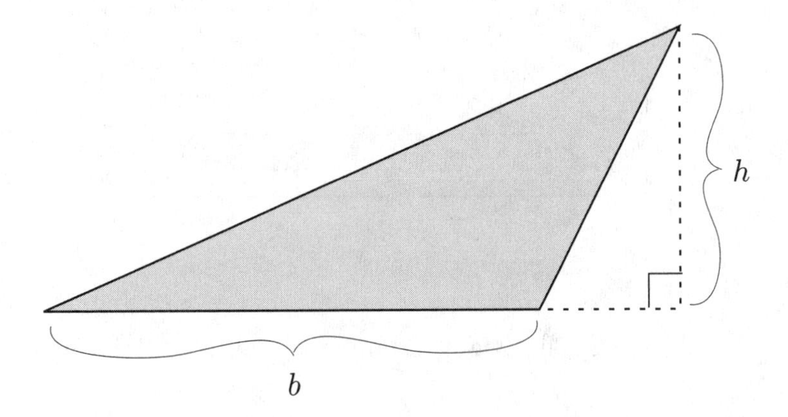

Figure 9.65: Why is the Area $\frac{1}{2}b \times h$?

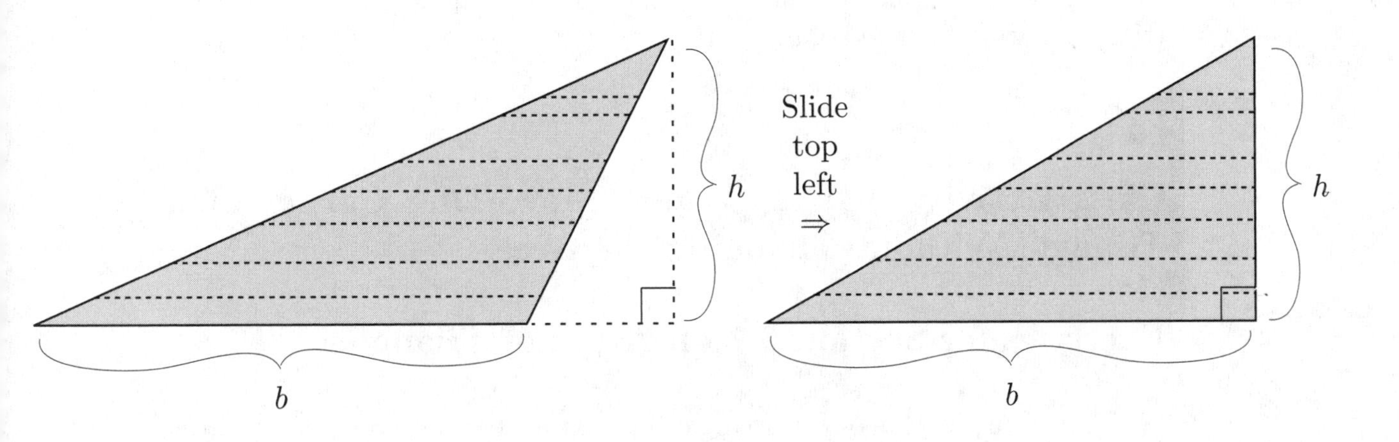

Figure 9.66: Shearing a Triangle

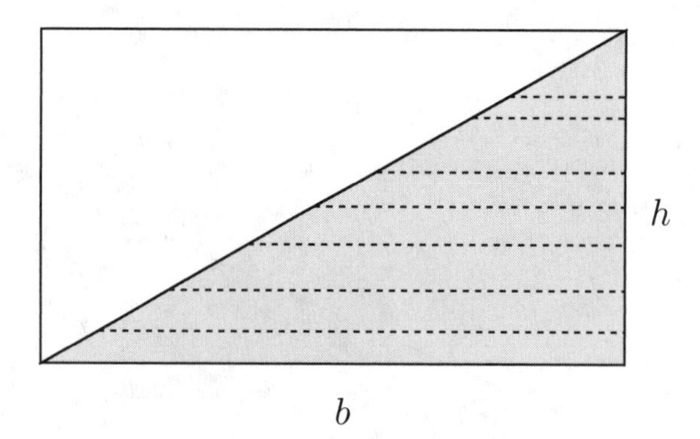

Figure 9.67: The Right Triangle is Half of a Rectangle

shown in Figure 9.67. Remember: this is the same b and h as for the original triangle. Now, according to Cavalieri's principle, the shearing process does not change the area. So the area of the original triangle is the same as the area of the new right triangle that was formed by shearing. Therefore the area of the original triangle in Figure 9.65 is also $\frac{1}{2}b \times h$ square units, which is what we wanted to show. Notice that although we worked with the particular triangle in Figure 9.65, there wasn't anything special about this triangle: the same argument applies to show that the $\frac{1}{2}b \times h$ formula for the area of a triangle is valid for *any* triangle.

Class Activity 9U: A Game: Move One Corner of a Triangle Without Changing the Area

Exercises for Section 9.7 on Areas of Triangles

1. Show the heights of the triangles in Figure 9.68 that correspond to the bases that are labeled b. Then determine the areas of the triangles.

2. Determine the areas, in square centimeters, of the shaded regions in Figures 9.69.

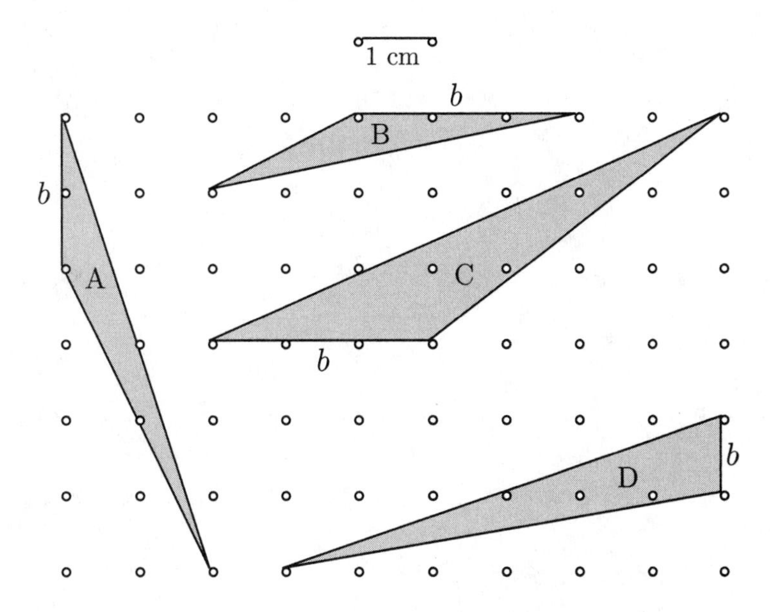

Figure 9.68: Find the Heights for the Indicated Bases

Answers to Exercises for Section 9.7 on Areas of Triangles

1. See Figure 9.70. Notice that for each of these triangles, the base must be extended in order to show the height meeting it at at right angle. Triangle A has height 2 cm and area $\frac{1}{2} \times 2 \times 2$ cm$^2 = 2$ cm^2. Triangle B has height 1 cm and area $\frac{1}{2} \times 3 \times 1$ cm$^2 = 1\frac{1}{2}$ cm^2. Triangle C has height 3 cm and area $\frac{1}{2} \times 3 \times 3$ cm$^2 = 4\frac{1}{2}$ cm^2. Triangle D has height 6 cm and area $\frac{1}{2} \times 1 \times 6$ cm$^2 = 3$ cm^2.

2. Shape A is a triangle whose base can be chosen to be the top line segment of length 2 cm. Then the height has length 7 cm, so the area of triangle A is 7 cm^2.

   As shown in Figure 9.71, triangle B and three additional triangles combine to make a 3 cm by 5 cm rectangle. The rectangle has area 15 cm^2, and the three additional triangles have areas $2\frac{1}{2}$ cm^2, and 3 cm^2, 3 cm^2, therefore

   $$(\text{area of B}) + 2\frac{1}{2} + 3 + 3 \text{ cm}^2 = 15 \text{ cm}^2.$$

   So triangle B has area $15 - 8\frac{1}{2}$ cm$^2 = 6\frac{1}{2}$cm^2.

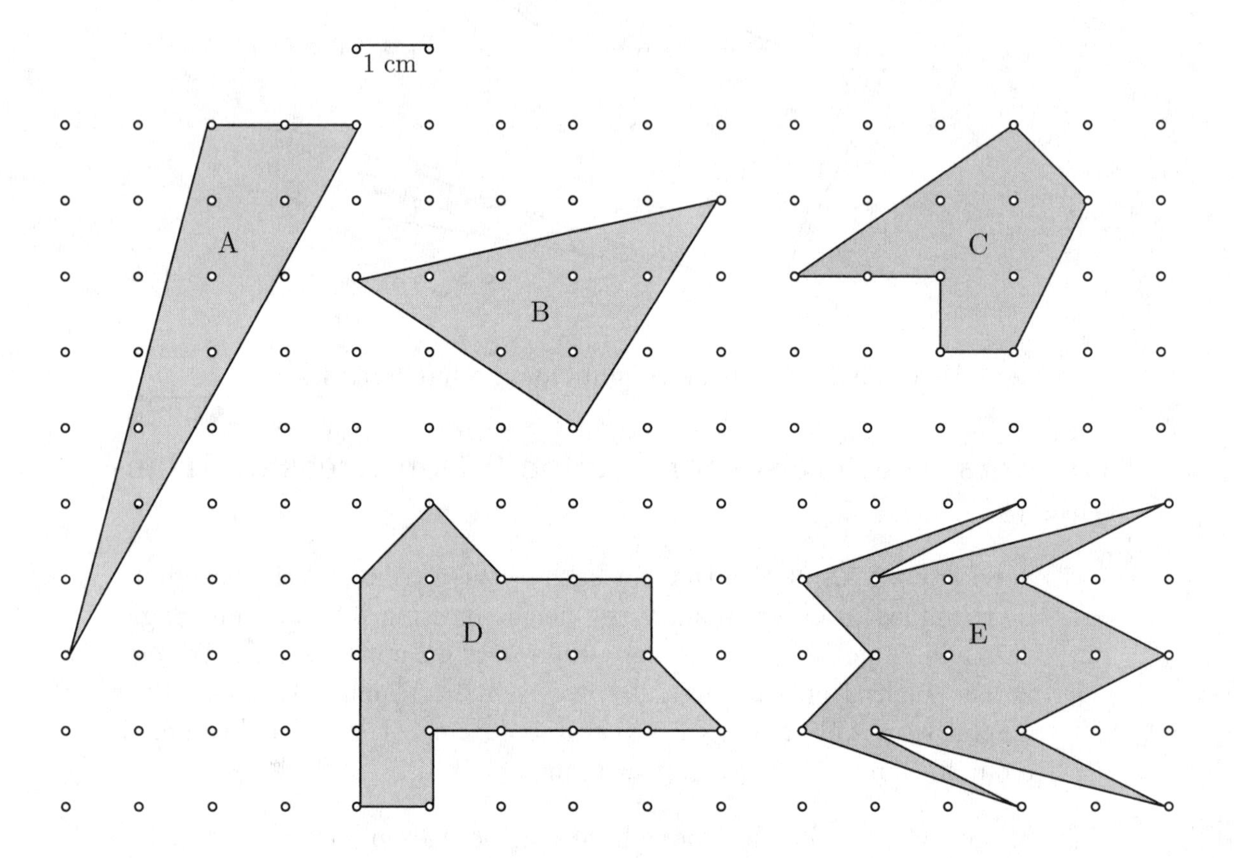

Figure 9.69: Find the Areas

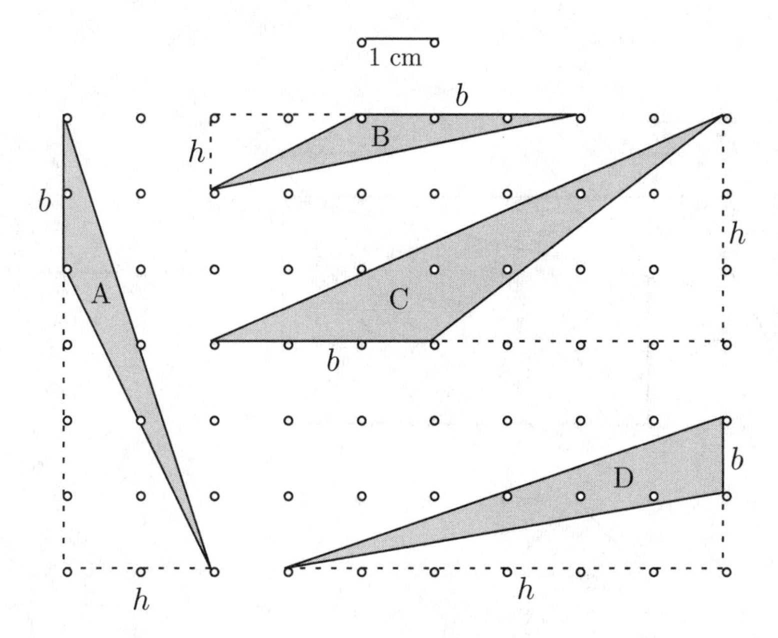

Figure 9.70: Bases and Heights

The areas of the remaining shaded shapes can be calculated by subdividing them into rectangles and triangles. Shape C has area $5\frac{1}{2}$ cm^2. Shape D has area $10\frac{1}{2}$ cm^2. Shape E has area 10 cm^2.

Problems for Section 9.7 on Areas of Triangles

1. Figure 9.72 shows a map of some land. Determine the size of this land in acres. Recall that one acre is 43,560 square feet.

2. Give a clear and thorough explanation in your own words for why the triangle in Figure 9.73 has area $\frac{1}{2}b \times h$ for the given choice of base b and height h.

3. Give a clear and thorough explanation in your own words for why the triangle in Figure 9.74 has area $\frac{1}{2}b \times h$ for the given choice of base b and height h.

4. Becky was asked to divide a rectangle into 4 equal pieces and to shade one of those pieces. Figure 9.75 shows Becky's solution. Is Becky right or not? Discuss and explain!

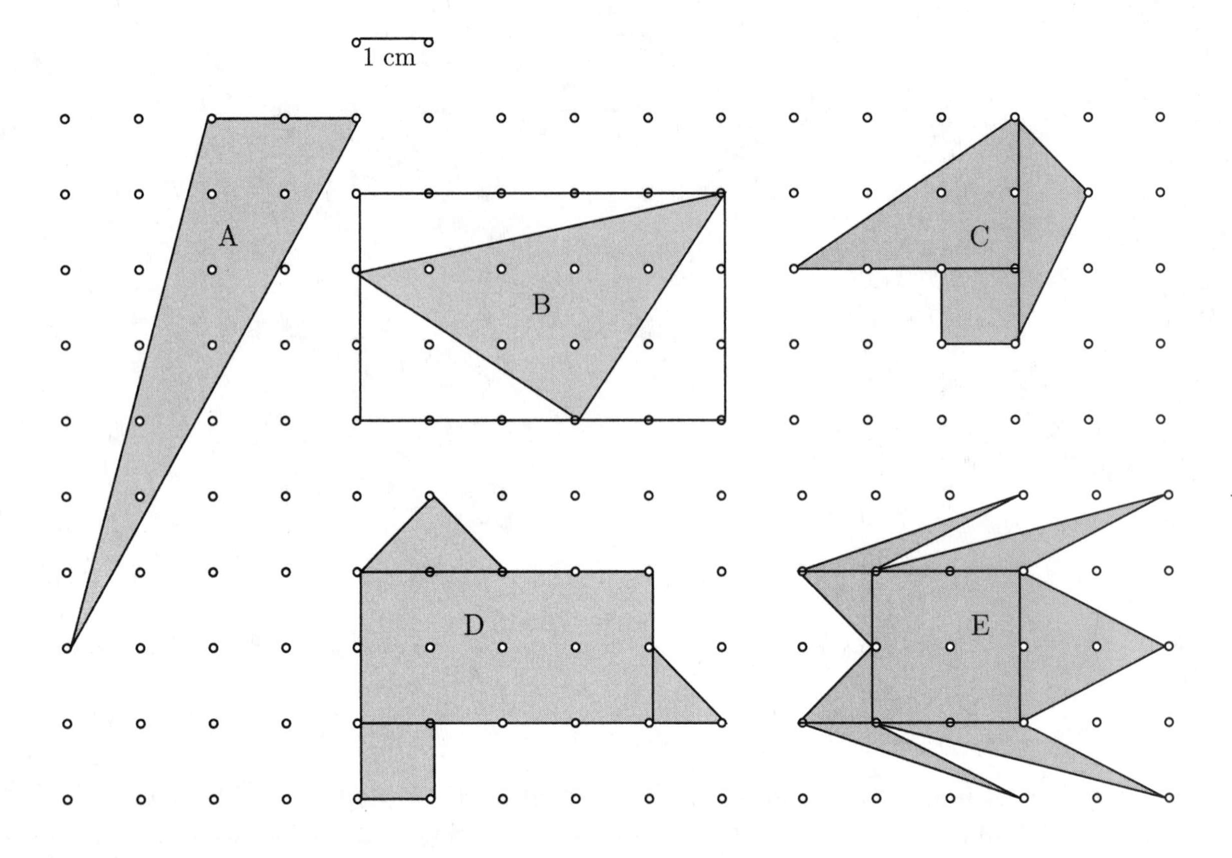

Figure 9.71: Calculate These Shapes' Areas by Subdividing

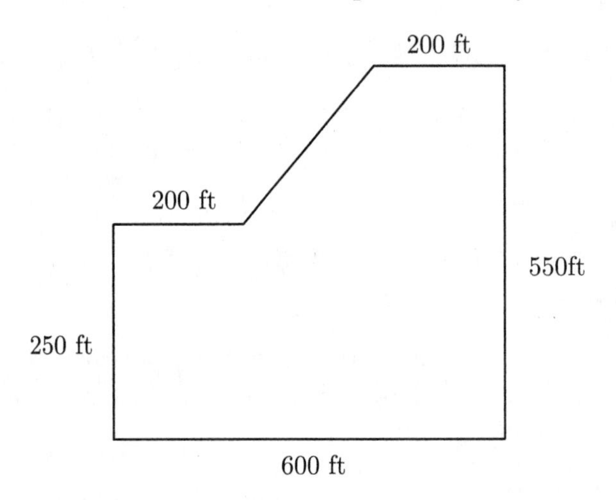

Figure 9.72: How Many Acres of Land?

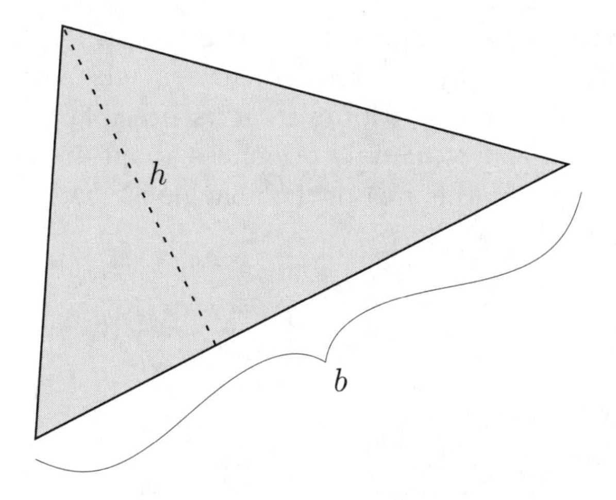

Figure 9.73: A Triangle

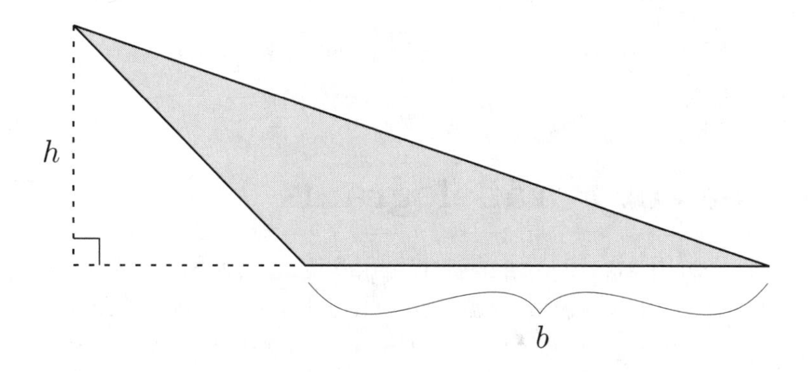

Figure 9.74: A Triangle

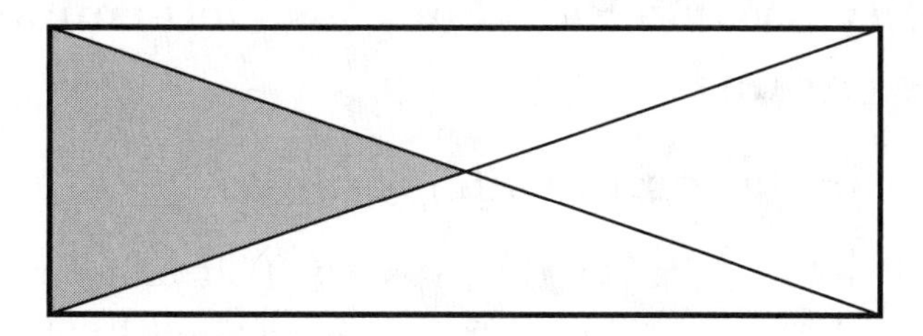

Figure 9.75: Four Equal Parts Or Not?

5. Class Activity 9T asked for a critique of an incorrect explanation based on Figure 9.76 for why the area of the triangle ABC is $\frac{1}{2}bh$ for the given choice of b and h. However, it *is* possible to use Figure 9.76 and the *moving* and *combining* principles about area to give a correct explanation for why the area of the triangle is $\frac{1}{2}bh$.

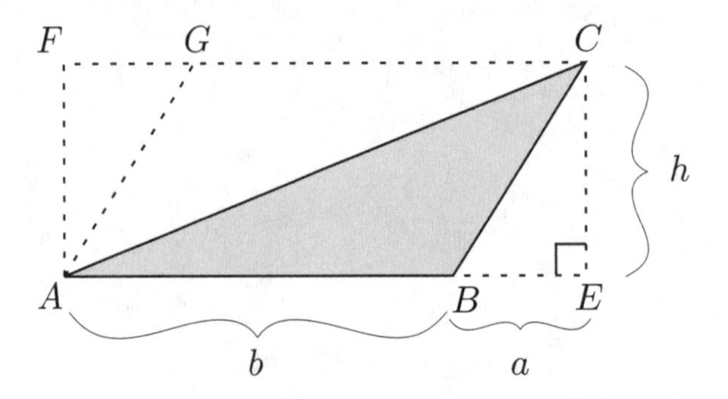

Figure 9.76: You *Can* Find the Area This Way!

9.8 Areas of Parallelograms

In this section we'll study areas of parallelograms. Since a rectangle is a special kind of parallelogram, and since the area of a rectangle is the width of the rectangle times the length of the rectangle, it would be natural to think that this same formula is true for parallelograms. But is it? The next class activity will examine this.

Class Activity 9V: Do Side Lengths Determine the Area of a Parallelogram?

An Area Formula for Parallelograms

If you did the last class activity, then you saw that unlike rectangles, it is not possible to determine the area of a parallelogram from the lengths of its sides alone. However, like triangles, there is a formula for the area of a parallelogram in terms of a base and a height.

As with triangles, the **base** of a parallelogram can be chosen to be any one of its four sides. In a formula, the word *base*, or a letter representing the base (such as b), actually means the *length* of the base.

Once a base has been chosen, the **height** of a parallelogram is a line segment that is

1. perpendicular to the base and

2. connects the base, or an extension of the base, to a corner of the parallelogram that is not on the base.

In a formula, the word *height*, or a letter representing the height (such as h), actually means the *length* of the height.

Figure 9.77 shows a parallelogram and one way to choose a base b and height h.

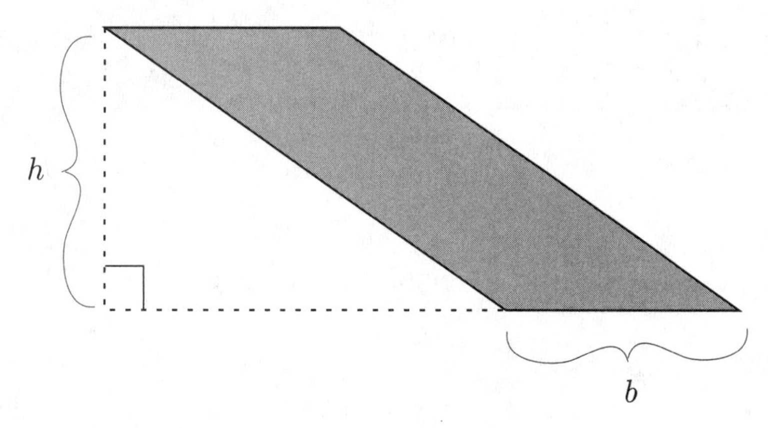

Figure 9.77: One Way to Choose the Base and Height of This Parallelogram

There is a very simple formula for the area of a parallelogram. If a parallelogram has a base that is b units long and height that is h units long then the area of the parallelogram is

$$b \times h$$

square units. In this formula we assume that b and h are measured in the same unit (for example, both in centimeters, or both in feet). If b and h are in different units, they will need to be converted to a common unit before using the formula. For example, if a parallelogram has a base that is 2 meters long and a height that is 5 cm long, then the area of the parallelogram is

$$200 \times 5 \text{ cm}^2 = 1000 \text{ cm}^2$$

because 2 meters = 200 cm.

Why is the $b \times h$ formula for areas of parallelograms valid? In some cases, such as the parallelogram on the left in Figure 9.78, we can explain why the area formula is valid by subdividing the parallelogram and recombining it to form a b by h rectangle, as shown on the right in Figure 9.78. According to the *moving* and *combining* principles about area, the area of the original parallelogram and the area of the newly formed rectangle are equal. Because the newly formed rectangle has area $b \times h$ square units, therefore the original parallelogram also has area $b \times h$ square units.

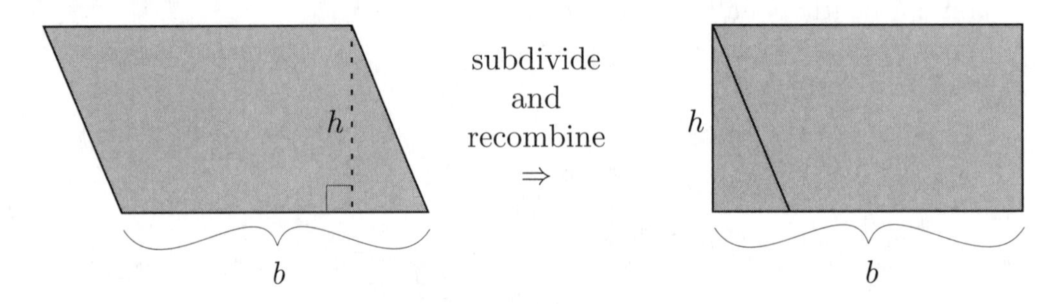

Figure 9.78: Subdividing and Recombining a Parallelogram to Make a Rectangle

In other cases, such as in the case of the parallelogram on the left in Figure 9.79, the *moving* and *combining* principles do not apply easily for the given choice of base b and height h. In this case, as well as in all other cases, the parallelogram can be sheared to form a b by h rectangle, as shown on the right of Figure 9.79. According to Cavalieri's principle, the original parallelogram and the new rectangle formed by shearing have the same area. Because the rectangle has area $b \times h$ square units, therefore the original parallelogram also has area $b \times h$ square units.

Exercises for Section 9.8 on Areas of Parallelograms

1. Every rectangle is also a parallelogram. Viewing a rectangle as a parallelogram, we can choose a base and height for it, as for any parallelogram. In the case of a rectangle, what are other names for *base* and *height*?

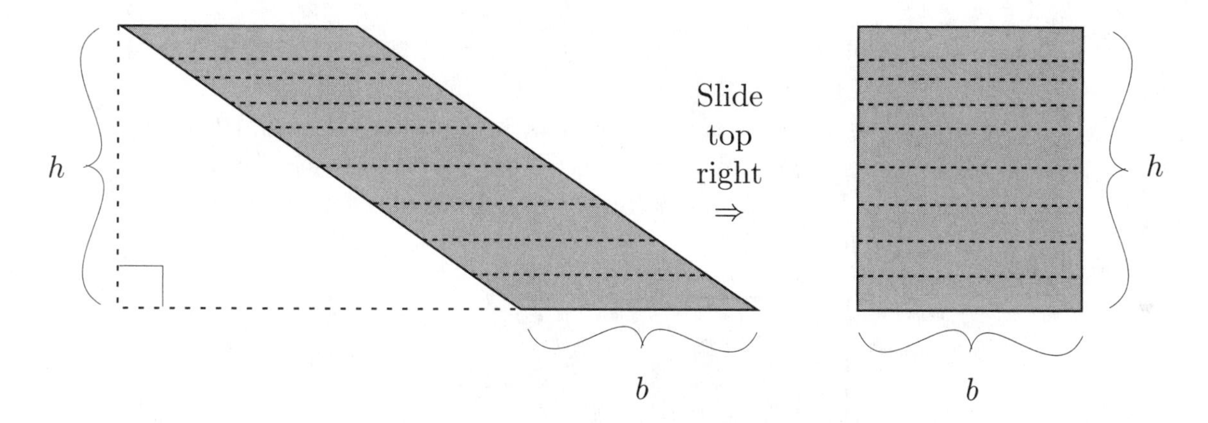

Figure 9.79: Shearing a Parallelogram Into a Rectangle

2. How are the *base*×*height* formula for areas of parallelograms and the *length*×*width* formula for areas of rectangles related?

3. Why can there not be a parallelogram area formula that is only in terms of the lengths of the sides of the parallelogram?

4. What is a formula for the area of a parallelogram, and why is this formula valid?

Answers to Exercises for Section 9.8 on Areas of Parallelograms

1. In the case of a rectangle, the *base* and *height* are the *length* and *width* of the rectangle. The base can be either the length or the width.

2. The *base*×*height* formula for areas of parallelograms generalizes the *length*×*width* formula for areas of rectangles because these two formulas are the same in the case of rectangles. This is because when a parallelogram is also a rectangle, the base can be chosen to be the length of the rectangle, and then the height is the width of the rectangle.

3. The three parallelograms in Figure 9.80 all have sides of the same length (see also Class Activity 9V; also notice that the straw rectangle in Figure 8.27 can be "squashed" to form a straw parallelogram in Figure 8.28 that has a smaller area but still has the same side lengths). If there

were a parallelogram area formula that was only in terms of the lengths of the sides of the parallelogram, then all three of these parallelograms would have to have the same area. But it is visibly clear that the three parallelograms have different areas. Therefore there can be no parallelogram area formula that is only in terms of the lengths of the sides of the parallelogram.

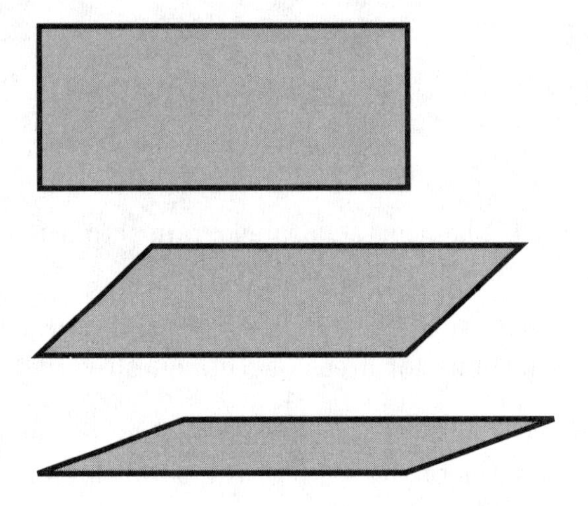

Figure 9.80: Three Parallelograms

4. Review the text.

Problems for Section 9.8 on Areas of Parallelograms

1. Figure 9.81 shows a trapezoid. This problem will help you to find a formula for the area of the trapezoid, and to explain why this formula is valid.

   (a) Using Cavalieri's Principle, explain why the trapezoid in Figure 9.81 has the same area as the new trapezoid in Figure 9.82.

   (b) Find a formula for the area of the *new trapezoid* in Figure 9.82 in terms of a, b, and h. Explain clearly why your formula gives the area of this new trapezoid.

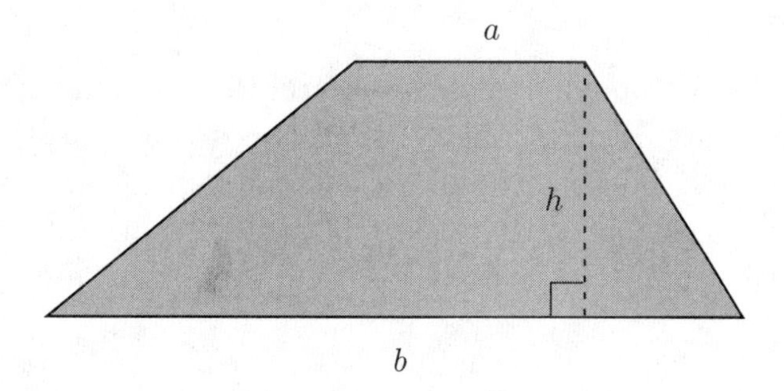

Figure 9.81: A Trapezoid

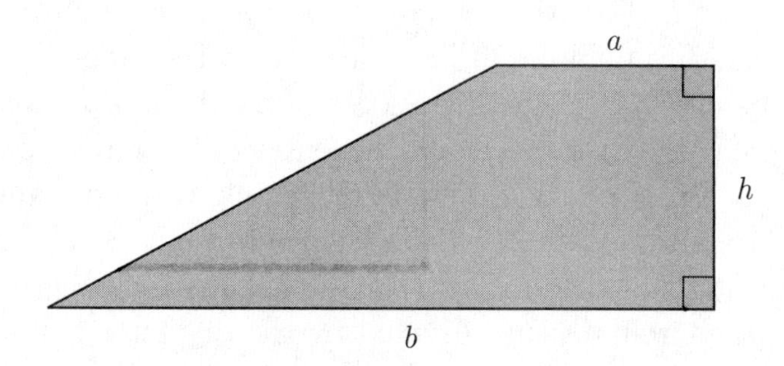

Figure 9.82: A Trapezoid With Two Right Angles

(c) Using parts (a) and (b) of this problem, give a formula in terms of a, b and h for the area of the *original trapezoid* in Figure 9.81, and explain clearly why this formula gives the area of the original trapezoid.

2. Josie has two wooden beams that are 15 feet long and two wooden beams that are 10 feet long. Josie plans to use these four beams to form the entire border around a closed garden. Josie is well known for her unusual designs. Without any other information, what is the most you can say about the area of Josie's garden? Explain.

9.9 Areas of Circles and the Number Pi

What is the area of a circle? You have probably seen the familiar formula

$$\pi r^2$$

for the area of a circle of radius r. The Greek letter π, pronounced "pie," stands for a mysterious number that is approximately equal to 3.14159. Since at least the time of the ancient Babylonians and Egyptians, nearly 4000 years ago, people have known about, and been fascinated by, the remarkable number π. In this section, we will first discuss the number π, and then we will see why the area of a circle of radius r units is πr^2 square units.

We need some terminology first. The **circumference** of a circle is the perimeter of the circle, i.e., the distance around the circle. Recall that the radius of a circle is the distance from the center of the circle to any point on the circle. Recall also that the diameter of a circle is the distance across the circle, going through the center; it is twice the radius.

It is an amazing fact that for any circle what so ever, whether huge, tiny, or in between, the circumference divided by the diameter is always equal to the same number; this number is called **pi** and is written with the Greek letter π. That the circumference of a circle divided by its diameter always results in the same number can be explained by establishing that all circles are similar, so that every circle is a scaled version of one fixed circle. In scaling, both the circumference and the diameter are multiplied by the same scale factor. Therefore, when the circumference is divided by the diameter, the same value results for all circles.

Given any circle, since

$$\text{circumference} \div \text{diameter} = \pi,$$

therefore

$$\text{circumference} = \pi \times \text{diameter}.$$

Now let r stand for the radius of a circle. Because the diameter of a circle is twice its radius, therefore we obtain the familiar expression $2\pi r$ for the circumference of a circle of radius r:

$$\text{circumference} = \pi \times 2r = 2\pi r.$$

So, for example, if you plan to make a circular garden with a radius of 10 feet, and you want to enclose the garden with a fence, then you will need

$$2\pi \times 10 \text{ ft} = 63 \text{ ft}$$

of fence.

You can do a simple experiment to determine the approximate value of π. Take a sturdy cup, can, or plate, measure the distance around the rim with a tape measure, and measure the distance across the cup, can, or plate at the widest part. When you divide these numbers you get an approximate value for π if the rim of the cup, can, or plate is a circle. But even with a big plate, an accurate tape measure, and careful measuring, the best you can probably do is to say that π is about 3.1. If you punch in π on your calculator, however, you will see many more decimal places:

$$\pi = 3.14159265\ldots.$$

Your calculator can only show a finite number of digits behind the decimal point, but it turns out that the decimal expansion of π goes on forever, and that it does not have a repeating pattern. The mathematician Johann Lambert (1728–1777) first proved this in 1761.

Why is it that such a simple and perfect shape as a circle gives rise to such a mysterious and complicated number as π? Many people are attracted to mathematics because it provides a glimpse into the mysterious, the perfect, and the infinite. Although mathematics can be used to solve a wide variety of practical problems, many people find the mystery and infinity in mathematics deeply appealing. It is not unlike the appeal of the greatest pieces of music, literature, and art. In this way, mathematics is not only a practical subject, but also informs us about what it means to be human, as music, literature, and art do.

How do we know the decimal expansion of the number π? There are many known formulas for π. One pretty one comes from the following equation:

$$\frac{\pi}{4} = 1 - \frac{1}{3} + \frac{1}{5} - \frac{1}{7} + \frac{1}{9} - \frac{1}{11} + \ldots.$$

This equation was discovered by Indian mathematicians in the 15th century. The ellipsis in the expression to the right of the equal sign indicates that this expression goes on forever, continuing the pattern of adding and subtracting fractions. Unfortunately, it is well beyond the scope of this book to explain why this formula is true or where it comes from.

Not only is the circumference of a circle related to the number π, but the area of a circle is as well. A circle of radius r units has area

$$\pi r^2 \text{ square units.}$$

For example, what is the area of a circular patio of diameter 30 feet? If the diameter is 30 feet, then the radius is 15 feet, and so the area of the patio is

$$\pi \times 15^2 \text{ ft}^2 = \pi \times 225 \text{ ft}^2 = 707 \text{ ft}^2.$$

When you use the πr^2 area formula, be sure that you only square the value of r, and not the value of πr. For example, if you multiply π times r first, and then square that result, your answer will be π times too large.

Class Activity 9W: How Big is the Number π?

Class Activity 9X: Why the Area Formula for Circles Makes Sense

Exercises for Section 9.9 on Circles and Pi

1. Using Figure 9.83 (see also Class Activity 9X), explain why it makes sense that a circle of radius r units has area πr^2 units.

2. Suppose you take a rectangular piece of paper, roll it up and tape two ends together, without overlapping them, to make a tube. If the tube is 12 inches long and has a diameter of $2\frac{1}{2}$ inches, then what were the length and width of the original rectangular piece of paper?

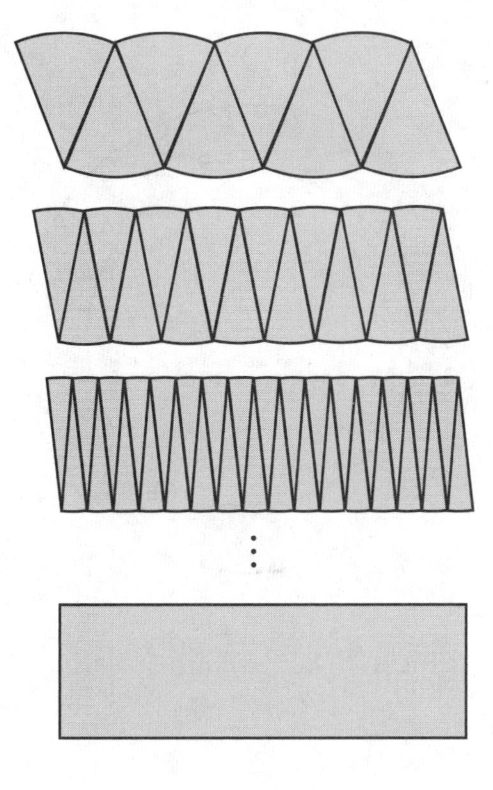

Figure 9.83: Rearranging a Circle

3. Some trees in an orchard need to have their trunks wrapped with a special tape in order to prevent an attack of pests. Each tree's trunk is about 1 foot in diameter and must be covered with tape from ground level up to a height of 4 feet. The tape is 3 inches wide. Approximately how long a piece of tape will be needed for each tree?

4. A 5 foot wide garden path is to be built around a circular garden of diameter 25 feet, as shown in Figure 9.84. What is the area of the garden path?

5. Find a formula for the surface area of a cylinder of radius r units and height h units. Include the top and bottom of the cylinder.

6. What is the area of the four-petal flower in Figure 9.85? The square is 6cm by 6cm.

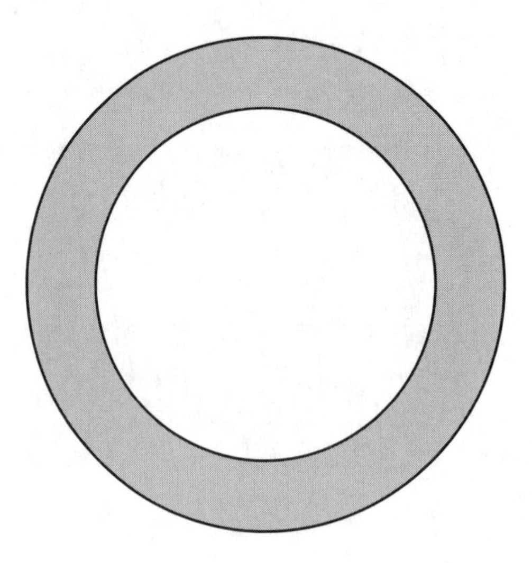

Figure 9.84: A Garden Path

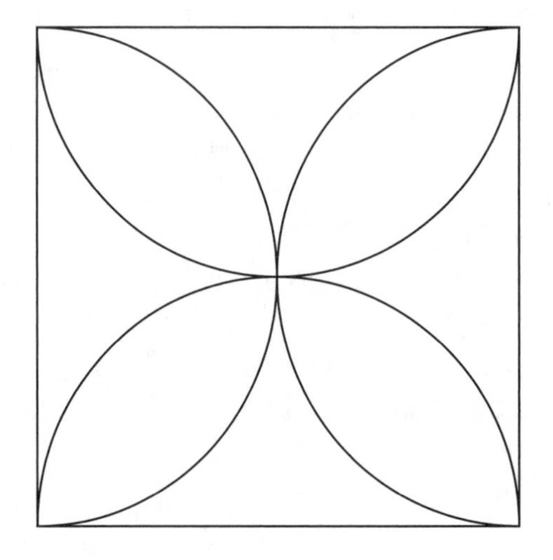

Figure 9.85: A Flower

Answers to Exercises for Section 9.9 on Circles and Pi

1. If you cut a circle into 8 pie pieces and rearrange them as at the top of Figure 9.83, then you get a shape that looks something like a rectangle. If you cut the circle into 16 or 32 pie pieces and rearrange them as in the middle of Figure 9.83, then you will get shapes that look even more like rectangles. If you could keep cutting the circle into more and more pie pieces, and keep rearranging them as before, you would get shapes that look more and more like the rectangle shown at the bottom of Figure 9.83.

   The height of the rectangle in Figure 9.83 is the radius r of the circle. To determine the width of the rectangle (in the horizontal direction), notice that in the rearranged circles at the top and in the middle of Figure 9.83, half of the pie pieces point up, and half point down. Therefore the circumference of the circle is divided equally between the top and bottom sides of the rectangle. Since the circumference of the circle is $2\pi r$, therefore the width (in the horizontal direction) of the rectangle is πr. So the rectangle is r units by πr units, and therefore has area $\pi r \times r = \pi r^2$ square units. Since the rectangle is basically a cut up and rearranged circle of radius r, the area of the rectangle ought to be equal to the area of the circle. Therefore it makes sense that the area of the circle is also πr^2.

2. The two edges of paper that are rolled up make circles of diameter 2.5 inches. Therefore the lengths of these edges are $\pi \times 2.5$ inches, which is about 9 inches. The other two edges of the paper run along the length of the cylinder, therefore they are 12 inches long. Thus the original piece of paper was about 9 inches by 12 inches.

3. One "wind" of tape all the way around a tree trunk makes an approximate circle. Because the tree trunk has diameter 1 foot, each wind around the trunk uses about π feet of tape. The tape is 3 inches wide, so it will take 4 winds for each foot of trunk height to be covered. Therefore it will take 16 winds to cover the desired amount of trunk. This will use about $16 \times \pi$ ft $= 50$ ft of tape.

4. Since the diameter of the circular garden is 25 ft, its radius is 12.5 feet. The garden together with the path form a larger circle of radius $12.5+5$

feet = 17.5 feet. By the *combining* principle about areas,

$$\text{area of garden} + \text{area of path} = \text{area of larger circle}.$$

Therefore

$$\begin{aligned}\text{area of path} &= \text{area of larger circle} - \text{area of garden}\\ &= \pi 17.5^2 - \pi 12.5^2 \text{ ft}^2\\ &= 471 \text{ ft}^2.\end{aligned}$$

5. The surface of the cylinder consists of two circles of radius r units (the top and the bottom), and a tube. Imagine slitting the tube open along its length and unrolling it, as indicated in Figure 9.86. The tube then becomes a rectangle. The height, h, of the tube becomes the length of two sides of the rectangle. The circumference of the tube, $2\pi r$, becomes the length of two other sides of the rectangle. Therefore the rectangle has area $2\pi rh$. According to the *moving* and *combining* principles about area, the surface area of the cylinder is equal to sum of the areas of the two circles (from the top and bottom), and the area of the rectangle (from the tube), which is

$$2\pi r^2 + 2\pi rh \text{ units}^2.$$

6. You can make the design by covering the square with four half-circles of tissue paper; then the flower petals are exactly the places where *two* pieces of tissue paper overlap. The remaining parts of the square are only covered with *one* layer of tissue paper. Therefore,

$$\text{area of four half} - \text{circles} = \text{area of flower} + \text{area of square}.$$

Therefore

$$\begin{aligned}\text{area of flower} &= \text{area of two circles} - \text{area of square}.\\ &= 18\pi - 36 \text{ cm}^2\\ &= 20.5 \text{ cm}^2.\end{aligned}$$

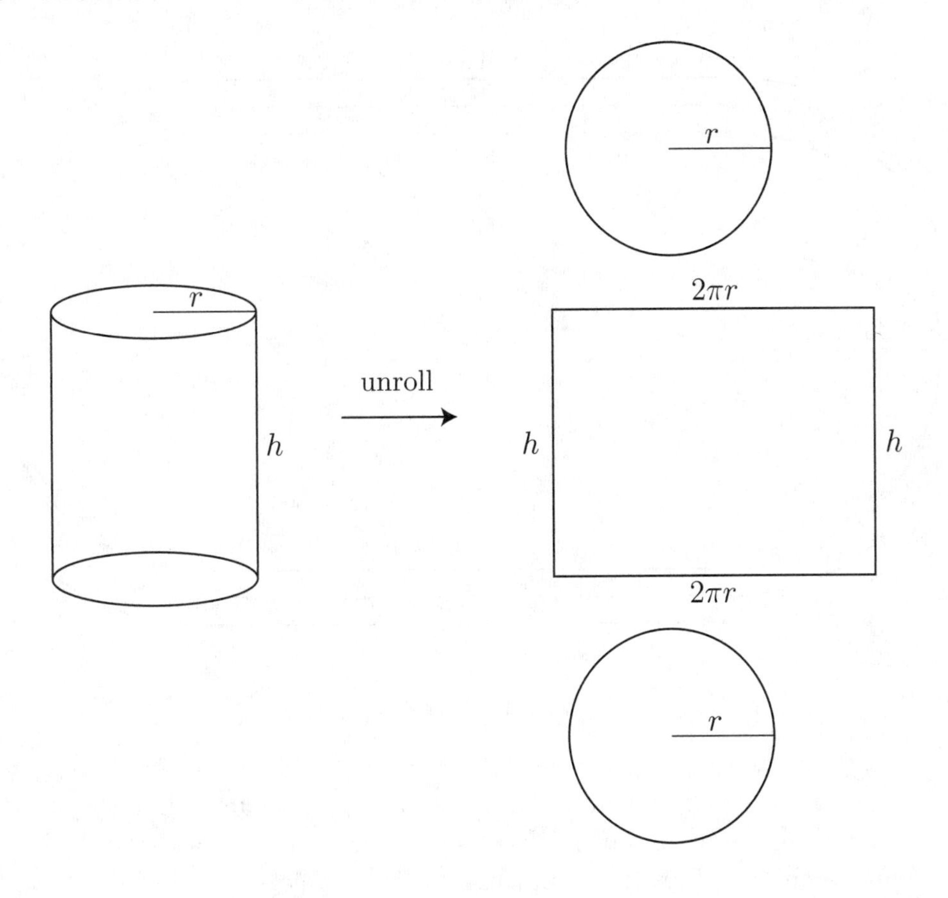

Figure 9.86: Taking a Cylinder Apart

Problems for Section 9.9 on Circles and Pi

1. A large running track is constructed to have straight sections and two semicircular sections with dimensions given in Figure 9.87. Assume that runners always run on the inside line of their track. A race will consist of one full counterclockwise revolution around the track plus an extra portion of straight segment, to end up at the finish line shown. What should the distance x between the two starting blocks be in order to make a fair race?

2. Suppose you have a large spool for winding rope onto (just like a spool of thread), such as the ones shown in Figures 9.88 and 9.89. Suppose that the spool is 1 meter long and has an inner diameter of 20 cm and an outer diameter of 60 cm. Approximately how long a piece of 5 cm

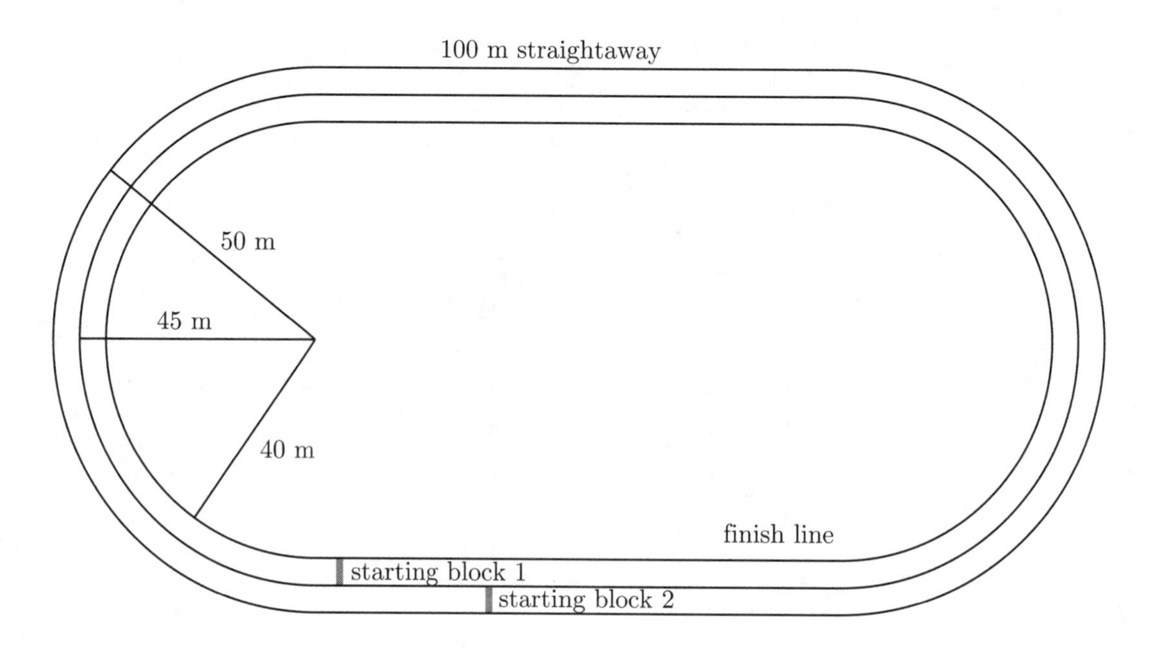

Figure 9.87: A Track

thick rope can be wound onto this spool? (Assume that the rope is wound on neatly, in layers. Each layer will consist of a row of "winds", and each "wind" will be approximately a circle.)

3. A city has a large cone-shaped Christmas tree that stands 20 feet tall and has a diameter of 15 feet at the bottom. The lights will be wound around the tree in a spiral fashion, so that each "row" of lights is about 2 feet higher than the previous row of lights. Approximately how long a strand of lights (in feet) will the city need? To answer this, it might be helpful to think of each "row" (one "wind") of lights around the tree as approximated by a circle.

4. Jack has a truck that is meant to take tires that are 26 inches in diameter. (Looking at a tire from the side of a car, a tire looks like a circle. The diameter of the tire is the diameter of this circle.) Jack puts tires on his truck that are 30 inches in diameter.

   (a) A car's spedometer works by detecting how fast the car's tires are rotating. Spedometers do not detect how big a car's tires are. When Jack's speedometer reads 60 miles per hour is that accurate,

Figure 9.88: A Large Spool of Wire

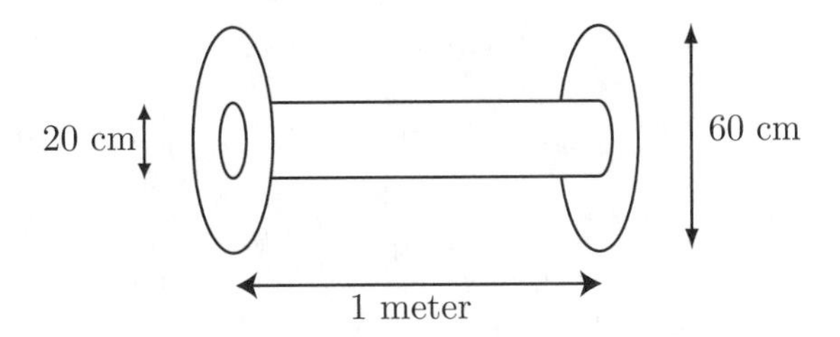

Figure 9.89: A Large Spool

or is Jack actually going slower or faster? Explain your reasoning. An exact determination of Jack's speed is not needed.

(b) Determine Jack's speed when his speedometer reads 60 mph. Explain your answer thoroughly.

5. The text gives the formula

$$\frac{\pi}{4} = 1 - \frac{1}{3} + \frac{1}{5} - \frac{1}{7} + \frac{1}{9} - \frac{1}{11} + \cdots.$$

This formula can be used to give better and better approximations to π by using more and more terms from the expression to the right of the equal sign. Here's how this works. Read the symbol $\approx$ as *is approximately equal to.*

One term: $\frac{\pi}{4} = 1$, so $\pi \approx 4$

Two terms: $\frac{\pi}{4} = 1 - \frac{1}{3}$, so $\pi \approx 2.667$.

Three terms: $\frac{\pi}{4} = 1 - \frac{1}{3} + \frac{1}{5}$, so $\pi \approx 3.467$.

(a) Use four, five, six, seven, and eight terms of the expression to the right of the equal sign above to find five more approximations to π. Also find the 20th and the 21st approximations to π.

(b) Looking at the three examples given above and your results in part (a), describe a pattern to the approximations to π obtained by this method. (Look at the sizes of your answers.) How do these approximations compare to the actual value of π?

(c) You can get better approximations to π by taking averages of successive approximations. For example, the average of the first and second approximations to π is

$$\frac{4 + 2.667}{2} = 3.334;$$

the average of the second and third approximations to π is

$$\frac{2.667 + 3.467}{2} = 3.067.$$

Find the average of the 3rd and 4th approximations to π, the 4th and 5th approximations, the 5th and 6th, the 6th and 7th and the 7th and 8th approximations to π. Also find the average of the 20th and the 21st approximations. How close is this last average to the actual value of π?

6. Tim works on the following exercise.

For each radius, r, below, find the area of a circle of that radius.

$r = 2$ in, $r = 5$ ft, $r = 8.4$ m.

Tim gives the following answers.

39.48, 246.74, 696.399

Identify the errors that Tim has made. How did Tim likely calculate his answers? Discuss how to correct the errors, including a discussion on the proper use of a calculator in solving Tim's exercise. Be sure to discuss the appropriate way to write the answers to the exercise.

7. A popular brand of soup comes in cans that are $2\frac{5}{8}$ inches in diameter and $3\frac{3}{4}$ inches tall. Each such can has a paper label that covers the entire side of the can (but not the top or the bottom).

   (a) If you remove a label from a soup can you'll see that it's made from a rectangular piece of paper. How wide and how long is this rectangle (ignoring the small overlap where two ends are glued together)? Use mathematics to solve this problem, even if you have a can to measure.

   (b) Ignoring the small folds and overlaps where the can is joined, determine how much metal sheeting is needed to make the entire can, including the top and bottom. Use the *moving* and *combining* principles about area to explain why your answer is correct. Be sure to use an appropriate unit to describe the amount of metal sheeting.

8. Let r units denote the radius of each circle in Figure 9.90. For each shaded pie portion in Figure 9.90, find a formula for its area in terms of r. Use the *moving* and *combining* principles about areas to explain why your formulas are valid.

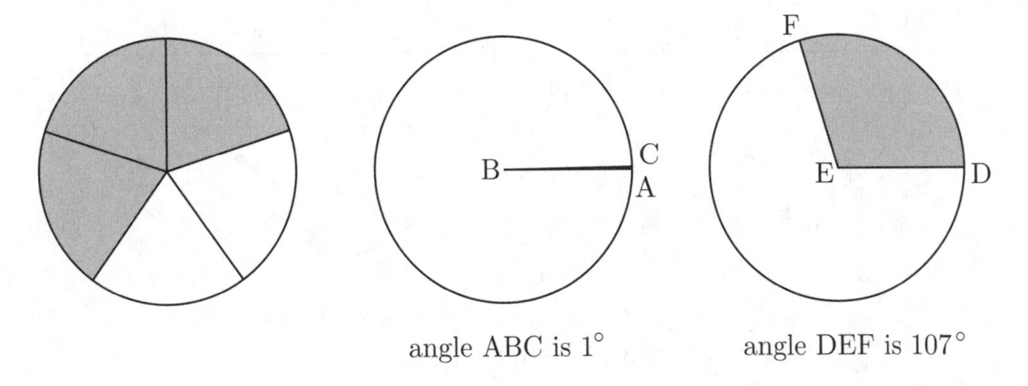

Figure 9.90: Parts of Pies

9. Suppose that when pizza dough is rolled out it costs 25 cents per square foot, and that sauce and cheese, when spread out on a pizza, have a combined cost of 60 cents per square foot. Let's say sauce and cheese are always spread out to within one inch of the edge of the pizza. Compare the sizes and costs of: a circular pizza of diameter 16 inches and a 10 inch by 20 inch rectangular pizza.

10. Lauriann and Kinsey are in charge of the annual pizza party. In the past, they've always ordered 12 inch diameter round pizzas, and each 12 inch pizza has always served 6 people. This year, the jumbo 16 inch diameter round pizzas are on special, so Lauriann and Kinsey decide to get 16 inch pizzas instead. Lauriann and Kinsey figure that since a 12 inch pizza serves 6 (which is half of 12), a 16 inch pizza should serve 8 (which is half of 16). But when Lauriann and Kinsey see a 16 inch pizza, they think it ought serve even more than 8 people. Suddenly, Kinsey realizes the flaw in their reasoning that a 16 inch pizza should serve 8. Kinsey has an idea for figuring out how many people a 16 inch pizza will serve. What mathematical reasoning might Kinsey be thinking of, and how many people should a 16 inch diameter pizza serve if a 12 inch diameter pizza serves 6? (See Figure 9.91.)

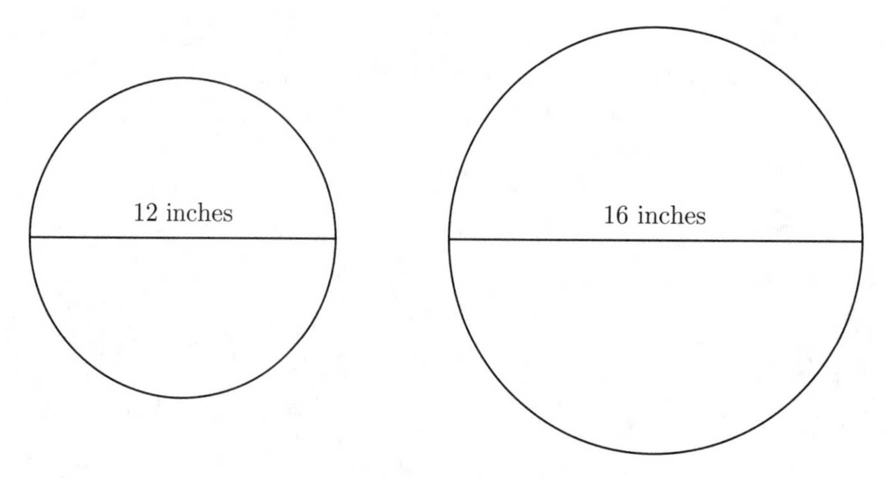

Figure 9.91: Pizzas

11. Penguins are known to huddle together in order to stay warm in very cold weather. Suppose that a certain type of penguin has a circular

cross-section approximately 14 inches in diameter (so that if you looked down on the penguin from above, the shape you would see would be a circle, 14 inches in diameter). Suppose that a group of this type of penguin is huddling in a large circular cluster, about 20 feet in diameter. (All the penguins are still standing upright on the ground, they are not piled on top of each other.)

(a) Assuming that the penguins are packed together tightly, estimate how many penguins are in this cluster (you might use areas to do this). Is this an overestimate or an underestimate?

(b) The coldest penguins in the cluster are the ones around the circumference. Approximately how many of these cold penguins are there at any given time?

(c) So that no penguin gets too cold, the penguins take turns being at the circumference. How many minutes per hour does each penguin spend at the circumference if each penguin spends the same amount of time at the circumference?

9.10 How are Perimeter and Area Related?

If you know the distance around a shape, can you determine its area? If you know the area of a shape, can you determine the distance around the shape? We will study these questions in this section.

Recall that the perimeter of a shape is the distance around a shape. For relatively small shapes you can determine the perimeter by placing a piece of string around the shape, and cutting the string so that it goes around exactly one time. The length of the string is then the perimeter of the shape. For example, the perimeter of a circle is its circumference. The shape in Figure 9.92 has perimeter

$$4 + 6 + 2 + 3 + 5 + 1 + 3 + 2 \text{ cm} = 26 \text{ cm}.$$

Note the difference between perimeter and area. The shape in Figure 9.92 has perimeter 26 cm, but it has area 21 cm^2. The perimeter of a shape is described by a unit of length, such as centimeters, whereas the area of a shape is described by a unit of area, such as square centimeters. The perimeter of

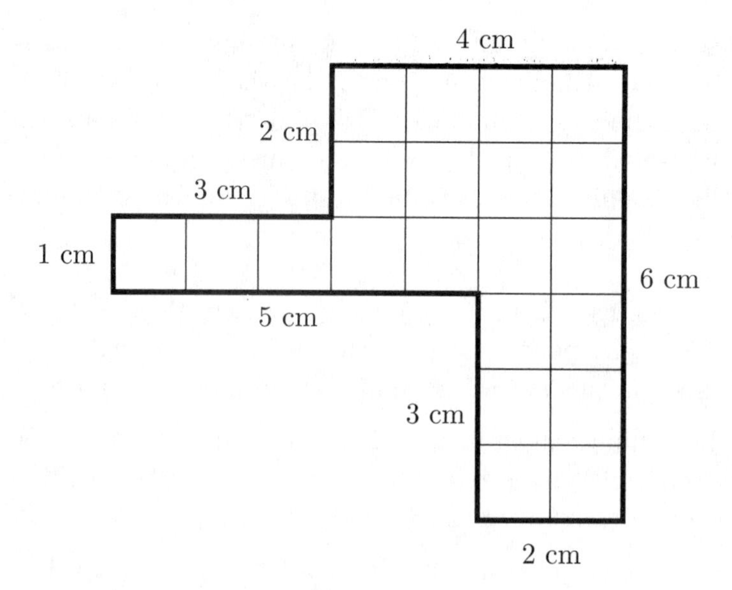

Figure 9.92: Determine the Perimeter

the shape in Figure 9.92 is the number of 1 cm segments it takes to go all the way *around the shape*, whereas the area of the shape in Figure 9.92 is the number of 1 cm by 1 cm squares it takes to *cover the shape.*

Class Activity 9Y: A Misconception about Perimeter

Class Activity 9Z: How are Perimeter and Area Related?

Class Activity 9AA: Can We Determine Area by Measuring Perimeter?

Shapes With a Fixed Perimeter

If you did Class Activity 9Z, then you probably discovered that for a given, fixed perimeter, there are many shapes that can have that perimeter, and these shapes can have different areas. By placing a loop of string on a flat surface, and by arranging the loop in different ways, you can show a variety of shapes that have the same perimeters but different areas (see Figure 9.93).

Therefore, perimeter does not determine area. But, for a given, fixed perimeter, which areas can occur?

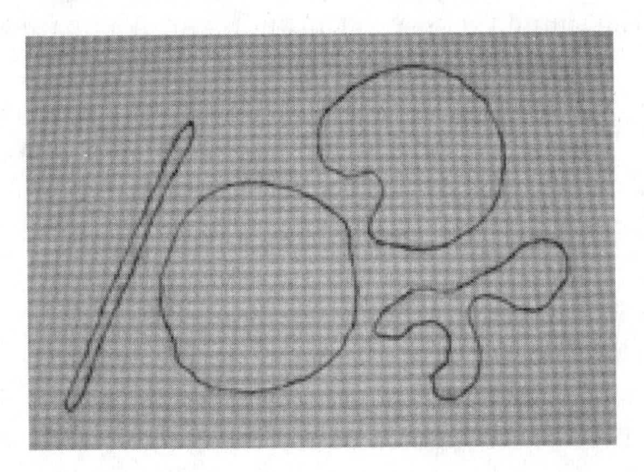

Figure 9.93: Four Strings of Equal Length Make Shapes of Different Areas

Your intuition probably tells you that of all shapes having a given, fixed perimeter, the circle is the one with the largest area. For example, of all shapes with perimeter 15 inches, a circle of circumference 15 inches is the shape that has the largest area. Since the circumference of this circle is 15 inches, its radius is

$$15 \div 2\pi \text{ in } = 2.4 \text{ in},$$

and therefore its area is

$$\pi \times 2.4^2 \text{ in}^2 = 18 \text{ in}^2.$$

By moving a 15 inch loop of string into various positions on a flat surface, you will probably find it plausible that *every* positive number less than 18 is the area, in square inches, of *some* shape of perimeter 15 in. So, for example, there is a shape that has perimeter 15 in and area 17.35982 in^2, and there is a shape that has perimeter 15 in and area 2.7156 in^2. Notice that we can say this *without actually finding shapes* that have those areas and perimeters, which could be quite a challenge.

The above is true in general: among all shapes of a given, fixed perimeter P, the circle of circumference P has the largest area, and every positive number that is less than the area of that circle is the area of some shape of perimeter P. Explanations for why these facts are true are beyond the scope of this book.

What can one say about areas of *rectangles* of a given, fixed perimeter? By stretching the loop of string between four thumbtacks pinned to cardboard, and by moving the thumbtacks, you can show various rectangles that all have the same perimeter, but have different areas. If you did Class Activity 9Z, then you probably discovered that of all rectangles of a given, fixed perimeter, the one with the largest area is a square. For example, among all rectangles of perimeter 24 inches, a square that has four sides of length 6 inches has the largest area, and this area is $6 \times 6 \text{ in}^2 = 36 \text{ in}^2$. By moving a 24 inch loop of string to form various rectangles, you can probably tell that *every* positive number less than 36 is the area, in square inches, of *some* rectangle of perimeter 24 inches. So, for example, there is a rectangle that has perimeter 24 in and area 35.723 in^2, and there is a rectangle that has perimeter 24 in and area 3.72 in^2, even though it would take some work to find the exact lengths and widths of such rectangles.

The above is true in general: among all rectangles of a given, fixed perimeter P, the square of perimeter P has the largest area, and every positive number that is less than the area of that square is the area of some rectangle of perimeter P. Although we will not do this here, these facts can be explained with algebra.

Exercises for Section 9.10 on How Perimeter and Area are Related

1. A piece of property is described as having a perimeter of 4.7 miles. Without any additional information about the property, what is the most you can say about its area? If you assume that the property is shaped like a rectangle, then what is the most you can say about its area?

Answers to Exercises for Section 9.10 on How Perimeter and Area are Related

1. Among all shapes that have perimeter 4.7 miles, the circle with circumference 4.7 miles has the largest area. This circle has radius

$$4.7 \div 2\pi \text{ miles } = .75 \text{ miles},$$

and therefore it has area

$$\pi .75^2 \text{ miles}^2 = 1.8 \text{ miles}^2.$$

So without any additional information, the best we can say about the property is that its area is at most 1.8 square miles, and that the actual area could be anywhere between 0 and 1.8 square miles.

Now suppose that the property is shaped like a rectangle. Among all rectangles of perimeter 4.7 miles, the square of perimeter 4.7 miles has the largest area. This square has 4 sides of length $4.7 \div 4$ miles $= 1.175$ miles, and therefore this square has area 1.175×1.175 miles^2 $=$ 1.4 miles^2. So if the property is know to be in the shape of a rectangle, then the best one can say about the property is that its area is at most 1.4 square miles, and that the actual area could be anywhere between 0 and 1.4 square miles.

Problems for Section 9.10 on How Perimeter and Area are Related

1. A forest has a perimeter of 210 miles, but no information is given about the shape of the forest. Justify your answers to the following (in all parts of this problem, the perimeter is still 210 miles):

   (a) Is it possible that the area of the forest is 3000 square miles? Explain.

   (b) Is it possible that the area of the forest is 3600 square miles? Explain.

   (c) If the forest is shaped like a rectangle, then is it possible that the area of the forest is 3000 square miles? Explain.

   (d) If the forest is shaped like a rectangle, then is it possible that the area of the forest is 2500 miles? Explain.

2. Write a short essay discussing whether or not Nick's idea for estimating the area of an irregular shape given in Class Activity 9AA provides a valid way to estimate the area of the irregular shape. If Nick's method is not valid, is there a way that Nick could use the string to get some sort of information about the area of the shape?

3. Consider all rectangles whose *area* is 4 square inches, including rectangles that have sides whose lengths are not whole numbers (notice that this problem says *area*, not perimeter). What are the possibilities for the perimeters of these rectangles? Is there a smallest perimeter? Is there a largest? Explore this question either by pure thought or by actual examination of rectangles. Then write a paragraph describing your exploration (including pictures, if relevant) and stating your conclusions clearly.

9.11 Principles for Determining Volumes

As with areas, there are fundamental principles that help us determine the volumes of a wide variety of solid shapes. These principles for volumes are almost identical to their counterparts for areas, and we will use them in the same ways we used the principles for determining areas.

The *Moving* and *Combining* Principles About Volumes

As with shapes in a plane, there are fundamental principles about how volumes behave when solid shapes are moved or combined.

1. If you move a solid shape rigidly without stretching or shrinking it, then its volume does not change.

2. If you combine (a finite number of) solid shapes *without overlapping* them, then the volume of the resulting solid shape is the sum of the volumes of the individual solid shapes.

We have already used these principles implicitly, in explaining why we can determine the volume of a box by multiplying its height times its width times its depth: we thought of the box as subdivided into layers, and each layer as made up of 1 unit by 1 unit by 1 unit cubes. Each small cube has volume 1 cubic unit, and the volume of the whole box (in cubic units) is the sum of the volumes of the cubes, which is just the number of cubes.

If you have a solid lump of clay, then you can mold it into various different shapes. Each of these different shapes is made of the same volume of clay. From the point of view of the *moving and combining* principles, it is as if the clay had been subdivided into many tiny pieces, and then these tiny pieces were recombined in a different way to form a new shape; therefore the new shape is made of the same volume of clay as the old shape.

Similarly, if you have water in a container and if you pour the water into another container, the volume of water stays the same, even though its shape changes.

Class Activity 9BB: Determining a Volume by Submersing in Water

Class Activity 9CC: Floating Versus Sinking: Archimedes's Principle

Cavalieri's Principle about Shearing and Volumes

As with a shape in a plane, we can shear a solid shape and obtain a new solid shape that has the same volume.

To illustrate the shearing of a solid shape, start with a polyhedron, pick one of its faces, and then imagine slicing the polyhedron into extremely thin (really, infinitely thin) slices that are parallel to the chosen face—this is rather like slicing a salami with a meat slicer. Now imagine giving those thin slices a push from the side, so that the chosen slice remains in place, but so that the other thin slices slide over, remaining parallel to the chosen face, and remaining the same distance from the chosen slice throughout the sliding process. Then you will have a new solid shape, as indicated in Figure 9.94. This process of "sliding infinitely thin slices" is called **shearing**.

You can show shearing nicely with a stack of paper. Give the stack of paper a push from the side, so that sheets of paper slide over as shown in Figure 9.94. In order to understand shearing of solid shapes, it may help you to think of the thin slices as made out of paper.

It is important to note that in the shearing process, each thin slice remains unchanged: each slice is just slid over, and is not compressed or stretched.

Cavalieri's principle for volumes says that when you shear a solid shape as described above, the volume of the original and sheared solid shapes are equal.

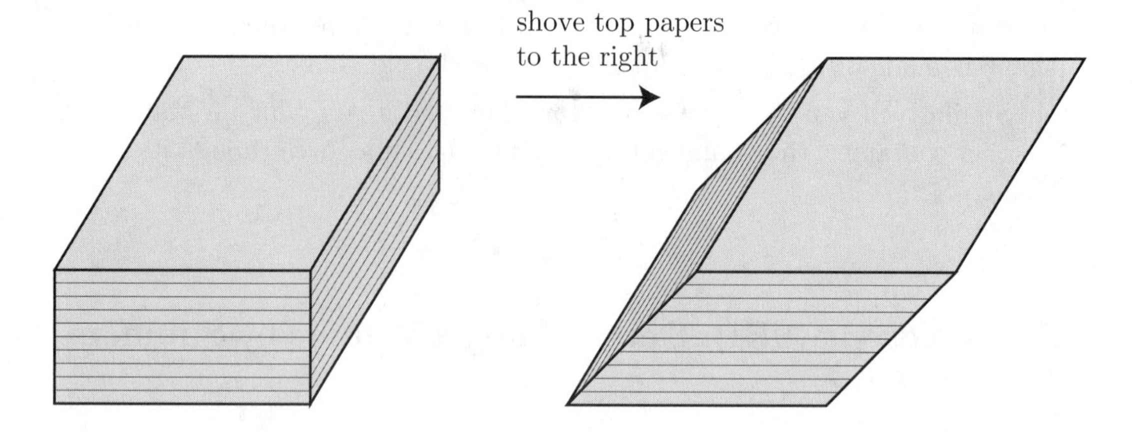

Figure 9.94: Shearing a Stack of Paper

9.12 Volumes of Prisms, Cylinders, Pyramids, and Cones

In this section we will study volume formulas for prisms, cylinders, pyramids, and cones, and we will see why these formulas make sense.

Volume Formulas for Prisms and Cylinders

Before introducing the simple volume formula for prisms and cylinders, recall that prisms and cylinders can be thought of as formed by joining two parallel, congruent bases. The **height** of a prism or cylinder is the distance between the planes containing the two bases of the prism or cylinder, measured in the direction perpendicular to the bases, as indicated in Figure 9.95. *The height is measured in the direction perpendicular to the bases, not on the slant.*

There is a very simple formula for volumes of prisms and cylinders:

$$(\text{height}) \times (\text{area of base}).$$

In the volume formula it is understood that if the height is measured in some unit, then the area of the base is measured in square units of the same unit. The volume of the prism or cylinder resulting from the formula is then in cubic units of the same basic unit.

For example, what is the volume of a 4 inch tall can that has a circular

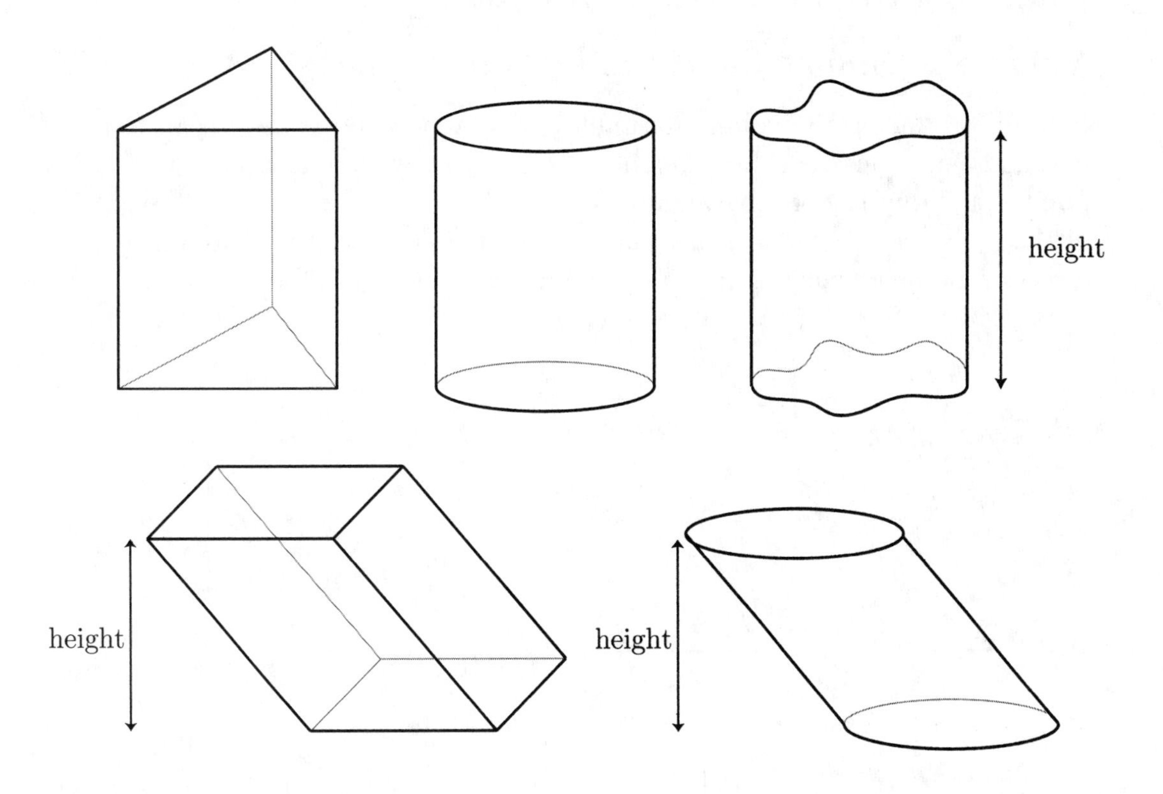

Figure 9.95: Volume = (Height)×(Area of Base)

base of radius 1.5 inches? The area of the base is $\pi(1.5)^2$ in^2 = 7.07 in^2, therefore the volume of the can is 4×7.07 in^3 = 28 in^3.

Class Activity 9DD: Why the Volume Formula for Prisms and Cylinders Makes Sense

Class Activity 9EE: Using Volume Formulas

Volume Formulas for Pyramids and Cones

Before introducing the volume formula for pyramids and cones, recall that pyramids and cones can be thought of as formed by joining a base with a point. The **height** of a pyramid or cone is the distance between the point of the pyramid or cone and the plane containing the base, measured in the direction perpendicular to the base, as indicated in Figure 9.96. *The height is measured in the direction perpendicular to the base, not on the slant.*

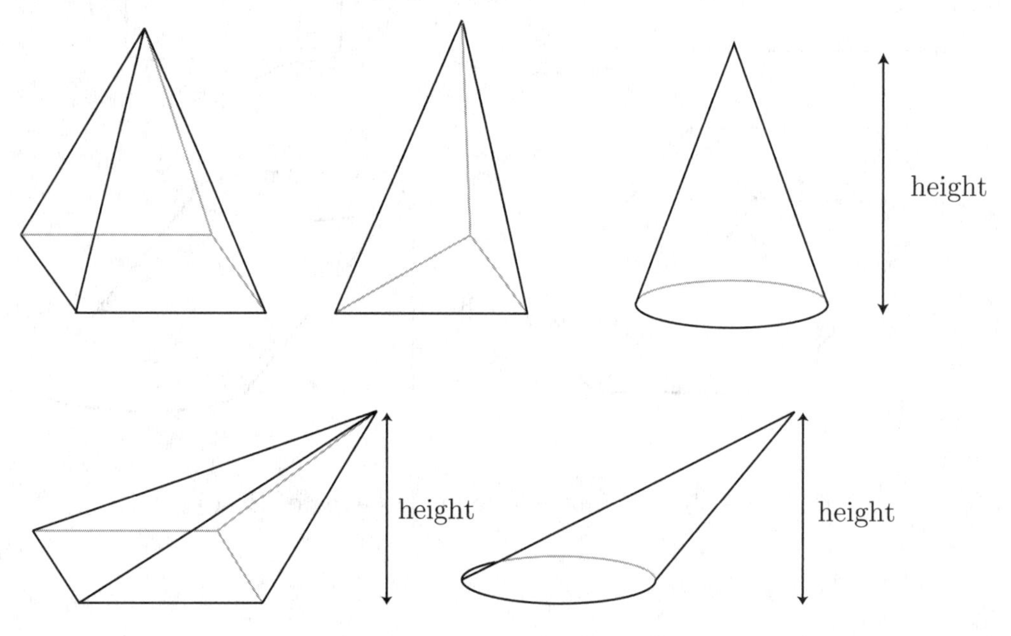

Figure 9.96: Volume = $\frac{1}{3}$(Height)×(Area of Base)

The formula for volumes of pyramids and cones is this:

$$\frac{1}{3}(\text{height}) \times (\text{area of base}).$$

In the volume formula it is understood that if the height is measured in some unit, then the area of the base is measured in square units of the same unit. The volume of the pyramid or cone resulting from the formula is then in cubic units of the same basic unit.

For example, what is the volume of sand in a cone-shaped pile that is 15 feet high and has a radius at the base of 7 feet? According to the volume formula, the volume of sand is

$$\frac{1}{3} \times 15 \times \pi(7)^2 \text{ ft}^3,$$

which is about 770 cubic feet of sand.

Where does the $\frac{1}{3}$ in the volume formula for pyramids and cones come from? The next class activity will look at this.

Class Activity 9FF: The $\frac{1}{3}$ in the Volume Formula for Pyramids and Cones

Class Activity 9GG: The Volume of a Rhombic Dodecahedron

Exercises for Section 9.12 on Volumes of Prisms, Cylinders, Pyramids, and Cones

1. The water in a full bathtub is roughly in the shape of a rectangular prism that is 54 inches long, 22 inches wide, and 9 inches high. Use the fact that 1 gallon = .134 cubic feet to determine how many gallons of water are in the bathtub.

2. A concrete patio will be made in the shape of a 15 foot by 15 foot square with half-circles attached at two opposite ends, as pictured in Figure 9.97. If the concrete will be 3 inches thick, how many cubic feet of concrete will be needed?

3. Let's suppose that a tube of toothpaste contains 15 cubic inches of toothpaste and that the circular opening where the toothpaste comes out has a diameter of $\frac{5}{16}$ of an inch. Suppose that every time you brush your teeth you squeeze out a $\frac{1}{2}$ inch long piece of toothpaste. How many times can you brush your teeth with this tube of toothpaste?

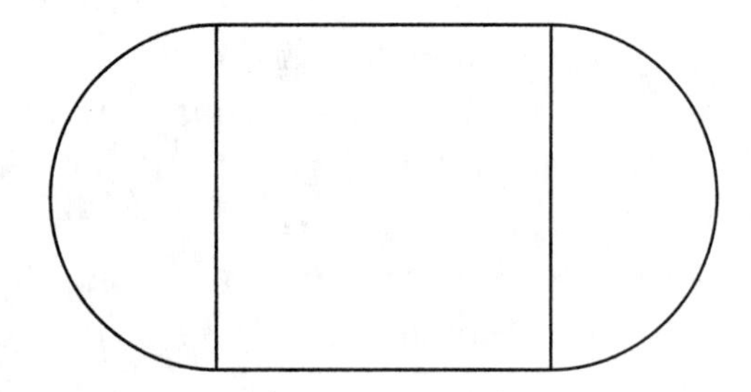

Figure 9.97: A Patio

4. The front (and back) of a greenhouse have the shape and dimensions shown in Figure 9.98. The greenhouse is 40 feet long and the angle at the top of the roof is 90°. The entire roof of the greenhouse will be covered with screening in order to block some of the light entering the greenhouse. How many square feet of screening will be needed?

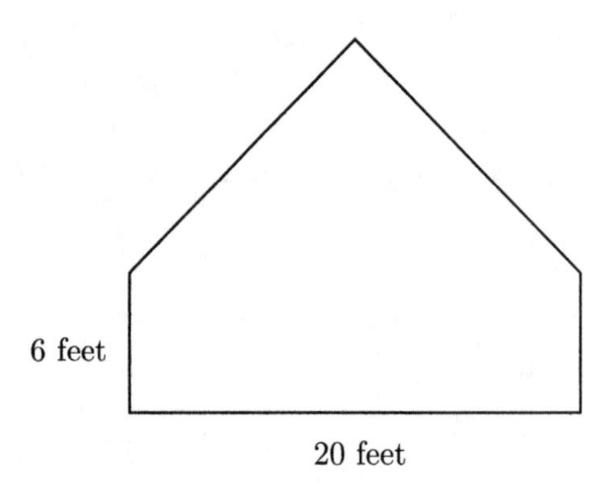

Figure 9.98: A Greenhouse

5. A typical ice cream cone is $4\frac{1}{2}$ inches tall and has a diameter of 2 inches. How many cubic inches does it hold (just up to the top)? How many fluid ounces is this? (One fluid ounce is 1.8 cubic inches.)

Answers to Exercises for Section 9.12 on Volumes of Prisms, Cylinders, Pyramids, and Cones

1. There are about 46 gallons of water in the bathtub. Because the water in the tub is $54 \div 12$ ft $= 4.5$ ft long, $22 \div 12$ ft $= 1.83$ ft wide, and $9 \div 12$ ft $= .75$ ft tall, therefore the volume of water in the tub is $4.5 \times 1.83 \times .75$ ft^3 $= 6.1875$ ft^3. Since each gallon of water is $.134$ ft^3, therefore the number of gallons of water in the tub is the number of $.134$ ft^3 in 6.1875 ft^3, which is $6.1875 \div .134$ gallons, and this is about 46 gallons.

2. 100 cubic feet of concrete will be needed for the patio. To explain why, notice that the patio is a kind of cylinder, so its volume can be calculated with the (height)$\times$(area of base) formula. The base, which is the surface of the patio, consists of a square and two half-circles. Therefore, by the *moving* and *combining* principle for areas, the area of the base is the area of the square plus the area of the circle that is created by putting the two half-circles together. So all together, the area of the patio (the base) is $15^2 + \pi(7.5)^2 = 401.7$ square feet. The height of the concrete patio is $\frac{1}{4}$ of a foot—notice that must convert 3 inches to feet in order to have consistent units. Therefore, according to the (height)$\times$(area of base) formula, the volume of concrete needed for the patio is $\frac{1}{4} \times 401.7$ cubic feet, which is about 100 cubic feet of concrete.

3. You will be able to brush your teeth 391 times. Each time you brush, the amount of toothpaste you use is the volume of a cylinder that is $\frac{1}{2}$ inch high and has a radius of $\frac{5}{32}$ of an inch. According to the volume formula for cylinders, this volume is $\frac{1}{2} \times \pi \frac{5}{32}^2$ in^3 $= .0383$ in^3. The number of times you can brush is the number of $.0383$ in^3 in 15 in^3, which is $15 \div .0383 = 391$ times.

4. 1131 square feet of screening will be needed. The roof is made out of two rectangular pieces, each of which is 40 feet long and A feet wide, where A is shown in Figure 9.99. Because the angle at the top of the roof is $90°$, we can use the Pythagorean theorem to determine A:

$$A^2 + A^2 = 20^2,$$

therefore $2A^2 = 400$, and so $A^2 = 200$, which means that $A = 14.14$ feet. Therefore each rectangular piece of roof needs $40 \times 14.14\ \text{ft}^2 = 565.69\ \text{ft}^2$ of screening. The two pieces of roof require twice as much, which is 1131 square feet of screening.

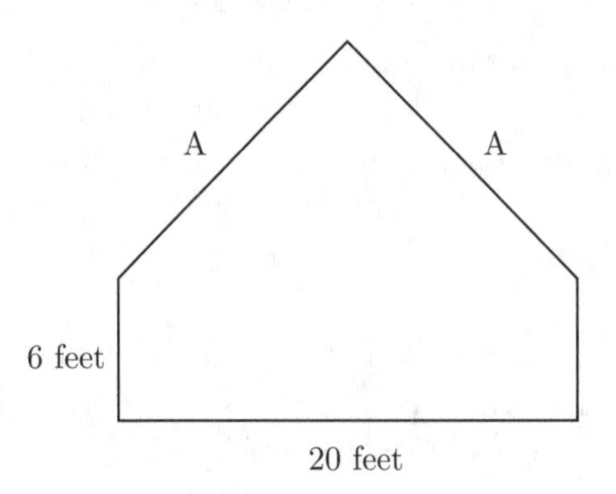

Figure 9.99: A Greenhouse

5. The cone holds 4.7 cubic inches, which is 2.6 fluid ounces. Since the diameter of an ice cream cone is 2 in, therefore its radius is 1 in, and so the area of the base of an ice cream cone (the circular hole that holds the ice cream) is $\pi \times 1^2\ \text{in}^2 = 3.14\ \text{in}^2$. According to the volume formula for cones, the volume of an ice cream cone is therefore $\frac{1}{3} \times 4.5 \times 3.14\ \text{in}^3 = 4.7\ \text{in}^3$. Since each fluid ounce is $1.8\ \text{in}^3$, therefore the number of fluid ounces the cone holds is the number of $1.8\ \text{in}^3$ in $4.7\ \text{in}^3$, which is $4.7 \div 1.8 = 2.6$ fluid ounces.

Problems for Section 9.12 on Volumes of Prisms, Cylinders, Pyramids, and Cones

1. (a) Measure how fast water comes out of some faucet of your choice. Give your answer in gallons per minute and explain how you arrived at your answer. (Notice that you do not have to fill up a gallon container to do this.)

   (b) Figure 9.100 shows a bird's eye view of a swimming pool in the shape of a cross. The pool is 4 feet deep but doesn't have any water in it. There is a small faucet on one side with which to fill the pool. How long would it take to fill up this pool, assuming that the water runs out of this faucet at the same rate as water from your faucet? Use the fact that 1 gallon = .134 cubic feet.

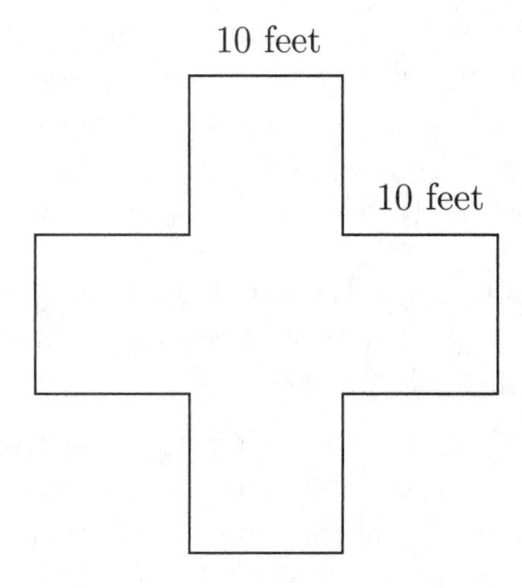

Figure 9.100: A beautiful swimming pool

2. Find either a gallon container, a half-gallon container, a quart container, a pint container, or a one-cup measure. Measure the lengths of various parts of your chosen container and use these measurements to determine the volume of your container in cubic inches. Use your answer to estimate how many gallons are in a cubic foot. (There are about $7\frac{1}{2}$ gallons in a cubic foot.)

3. One gallon is 3.79 liters, and 1 cm^3 holds 1 milliliter of liquid. Use these facts to determine the number of gallons in a cubic foot.

4. A cake recipe will make a round cake that is 6 inches in diameter and 2 inches high.

   (a) If you use the same recipe but pour the batter into a round cake pan that is 8 inches in diameter, how tall will the cake be?

   (b) Now suppose you want to use the same cake recipe to make a rectangular cake. If you use a rectangular pan that is 8 inches wide and 10 inches long, and if you want the cake to be about 2 inches tall, then how much of the recipe should you make? (For example, should you make twice as much of the recipe, half as much, three-quarters as much, or some other amount?) Give an approximate but *practical* answer.

5. A recipe for gingerbread makes a 9 in $\times$ 9 in $\times$ 2 in pan full of gingerbread. Suppose that you want to use this recipe to make a gingerbread house. You decide that you can either use a 10 in $\times$ 15 in pan or a 11 in $\times$ 17 in pan, and that you can either make a whole recipe or $\frac{1}{2}$ of the recipe of gingerbread.

   (a) Which pan should you use, and should you make the whole recipe or just half a recipe, if you want the gingerbread to be between $\frac{1}{4}$in and $\frac{1}{2}$in thick? Explain.

   (b) Draw a careful diagram showing how you would cut the gingerbread into parts to assemble into a gingerbread house. Indicate the different parts of the house (front, back, roof, etc.). When assembled, the gingerbread house should look like a real 3-dimensional house—but be as creative as you like in how you design it. You do not have to use every bit of the gingerbread to make the house, but try to use as much as possible.

   (c) If you want to put a solid, 2-inch tall fence all the way around your gingerbread house so that the fence is 6 inches away from the house, then what will be the perimeter of this fence? Explain.

   (d) To make the fence in part (c), how many batches of gingerbread recipe will you need, and what pan (or pans) will you use? Indicate how you will cut the fence out of the pan (or pans).

6. The front (and back) of a greenhouse have the shape and dimensions shown in Figure 9.101. The greenhouse is 40 feet long and the angle at the top of the roof is 90°. A fungus has begun to grow in the greenhouse, so a fungicide will need to be sprayed. The fungicide is simply sprayed into the air. To be effective, one tablespoon of fungicide is needed for every cubic yard of volume in the greenhouse. How much fungicide should be used? Give your answer in terms of units that are practical. (For example, it would not be practical to have to measure 100 tablespoons, nor would it be practical to have to measure 3.4 quarts. But it would be practical to measure 1 quart and 3 fluid ounces.)

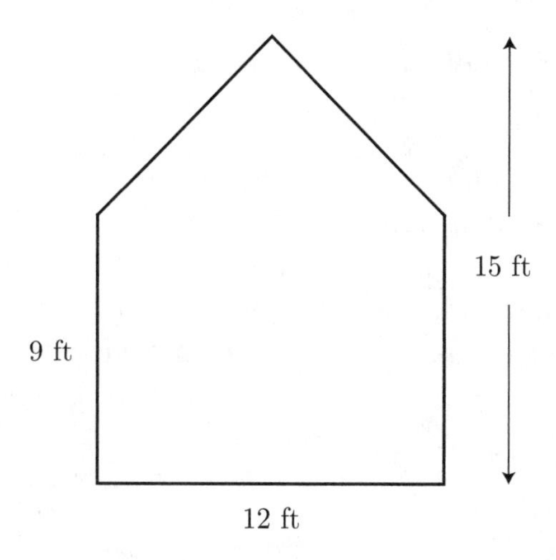

Figure 9.101: A Greenhouse

7. 2500 cubic yards of gravel will be dumped from a conveyor belt to form a cone shaped pile. How high could this pile of gravel be, and what could the circumference of the pile of gravel be at ground level? Give *two different realistically possible pairs of answers* for the height and circumference of the pile of gravel (both piles of volume 2500 cubic yards), and compare how the piles of gravel would look in the two cases.

8. A construction company wants to know how much sand is in a cone shaped pile. The company measures that the distance from the edge of the pile at ground level to the very top of the pile is 55 feet. The company also measures that the distance around the pile at ground level is 220 feet.

   (a) How much sand is in the pile? (Be sure to say what units you are measuring this in.)

   (b) The construction company has trucks that carry 10 cubic yards in each load. How many loads will it take to move the pile of sand?

9. One of the Hawaiian volcanoes is $30,000$ feet high (measured from the bottom of the ocean), and has volume $10,000$ cubic miles. Assuming that the volcano is shaped like a cone with a circular base, find the distance around the base of the volcano (at the bottom of the ocean). In other words, if a submarine were to go all the way around the base of

the volcano, at the bottom of the ocean, how far would the submarine go?

10. Suppose that a gasoline-powered engine has a gas tank in the shape of an inverted cone, with radius 10 inches and height 20 inches, as shown in Figure 9.102. The gas flows out through the tip of the cone, which points down. The shaded region represents gas in the tank. The height of this "cone of gas" is 10 inches, which is half the height of the gas tank.

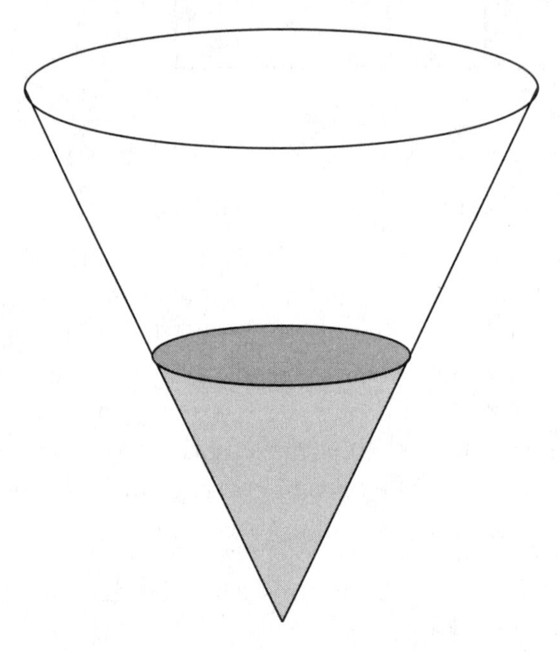

Figure 9.102: A Cone-Shaped Gas Tank

(a) What do you think: in terms of volume, is the tank half full, less than half full or more than half full? Draw a picture of the gas tank and mark approximately where you think the gas would be when the tank is $\frac{3}{4}$ full, $\frac{1}{2}$ full, and $\frac{1}{4}$ full in terms of volume.

(b) Determine the radius of the "cone of gas". *Hint:* think back to an earlier chapter.

(c) Using part (b), find the volume of the gas in the tank. Is this half, less than half, or more than half the volume of the tank?

(d) Now find the volume of gas if the height of the gas was $\frac{3}{4}$ of the height of the tank. Is this more or less than half the volume of the tank? Is this surprising?

11. Make a pattern for a cone-shaped cup that will hold about 120 ml when filled to the rim. Use a pattern like the one for a cone-shaped cup in Figure A.12 on page 387. Indicate measurements on your cup's pattern, and explain why the pattern will produce a cup that holds 120 ml. A centimeter tape measure may be helpful—you may wish to cut out the one on page 385.

12. Cut out the pattern in Figure A.12 on page 387 and tape the two straight edges together to make a small cone-shaped cup. You might want to leave a small "tab" of paper on one of the edges, especially if you use glue instead of tape. In any case, make sure that the two straight edges are joined.

    (a) Determine approximately how many fluid ounces the cone-shaped cup holds by filling it with a dry, pourable substance (such as rice, sugar, flour, or even sand), and pouring the substance into a measuring cup.

    (b) The pattern for the cone-shaped cup indicates that the two straight edges you taped together are 10 cm long, and the curved edge that makes the rim of the cup is 25 cm long. Use these measurements to determine the volume of the cone-cup in milliliters. (Remember that 1 cubic centimeter holds 1 milliliter.) Explain your method.

    (c) If you look on a can of soda, you'll see that 12 fluid ounces is 355ml. Use this relationship, and your answer to part (b), to determine how many fluid ounces the cone-cup holds. Compare your answer to your estimate in part (a).

9.13 Areas, Volumes, and Scaling

When two objects are similar, there is a scale factor such that corresponding *lengths* on the objects are related by multiplying by the scale factor (see Section 8.4). Does this mean that the *surface areas* of the objects are related by multiplying by the scale factor? Does this mean that the *volumes* of the objects are related by multiplying by the scale factor? We will examine these questions in this section.

Figure 9.103: Two Similar Cups

Class Activity 9HH: Areas and Volumes of Similar Boxes

Class Activity 9II: Areas and Volumes of Similar Cylinders

Class Activity 9JJ: Determining Areas and Volumes of Scaled Objects

How Areas and Volumes of Objects Behave Under Scaling

If you did Class Activities 9HH and 9II, then you probably discovered that if an object is scaled with a scale factor k, then even though the *lengths* of various parts of the object scale by the factor k, the *surface area* of the object scales by the factor k^2, and the *volume* of the object scales by the factor k^3. In other words, if one object (object 1) is the same as another object (object 2) except that it is k times as wide, k times as deep, and k times as tall as object 2, then *the surface area of object 1 is k^2 times as big* as the surface area of object 2, and *the volume of object 1 is k^3 times as big* as the volume of object 2.

For example, if you make a scale model of a building, and if the scale factor from the model to the building is 150 (so that all lengths on the building are 150 times as long as the corresponding length on the model), then the surface area of the building is $150^2 = 22,500$ times as large as the surface area of the model, and the volume of the building is $150^3 = 3,375,000$ times as large as the volume of the model.

Notice how nicely these scale factors for surface area and volume fit with their dimensions, and with the units used to measure surface area and volume: The surface of an object is 2-dimensional, it is measured in units of in^2, cm^2, ..., and when an object is scaled with scale factor k, the surface area scales by k^2. Similarly, the size of a full, 3-dimensional object can be measured by volume, which is measured in units of in^3, cm^3, ..., and when an object is scaled with scale factor k, the volume scales by k^3.

Exercises for Section 9.13 on Areas, Volumes, and Scaling

1. If there were a new Goodyear blimp that was $2\frac{1}{2}$ times as long, $2\frac{1}{2}$ times as wide, and $2\frac{1}{2}$ times as tall as the current one, then how much material would be needed to make the new blimp, compared to the current one? How much gas would be needed to fill the new blimp (at the same pressure), compared to the current one? (See Figure 9.104.)

Figure 9.104: A Blimp and the Blimp Scaled With Scale Factor $2\frac{1}{2}$

2. Suppose you make a scale model for a pyramid using a scale of 1:100 (so that the scale factor from the model to the actual pyramid is 100).

If your model pyramid is made out of 4 cardboard triangles, using a total of 60 square feet of cardboard, then what is the surface area of the actual pyramid?

3. If a giant was 12 feet tall, but proportioned like a typical 6 foot tall man, about how much would you expect the giant to weigh? (See Figure 9.105.)

Figure 9.105: A 6 Foot Tall Man and a 12 Foot Tall Giant

4. Suppose that a larger box is 3 times as wide, 3 times as deep, and 3 times as high as a smaller box. Use a formula for the surface area of a box to explain why the larger box's surface area is 9 times the smaller box's surface area. Use a formula for the volume of a box to explain why the larger box's volume is 9 times the smaller box's volume.

5. Suppose a large cylinder has twice the radius and twice the height of a small cylinder. Use a formula for the surface area of a cylinder to explain why the surface area of the large cylinder is 4 times the surface area of the small cylinder. Use the formula for the volume of a cylinder to explain why the volume of the large cylinder is 8 times the volume of the small cylinder.

Answers to Exercises for Section 9.13 on Areas, Volumes, and Scaling

1. According to the way surface area behaves under scaling, the new blimp's surface area would be

$$(2\frac{1}{2})^2 = \frac{25}{4} = 6\frac{1}{4}$$

times as large as the current blimp's surface area. According to the way volume behaves under scaling, the new blimp's volume would be

$$(2\frac{1}{2})^3 = \frac{125}{8} = 15\frac{5}{8}$$

times as large as the current blimp's volume.

2. According to the way surface areas behave under scaling, the actual pyramid's surface area is $100^2 = 10,000$ times as large as the surface area of the pyramid. The model is made out of 60 square feet of cardboard, so this is its surface area. Therefore the surface area of the actual pyramid is $10,000 \times 60$ square feet, which is $600,000$ square feet.

3. Since the giant is twice as tall as a typical 6 foot tall man, and since the giant is proportioned like a 6 foot tall man, the giant should be a scaled version of a 6 foot tall man, with scale factor 2. Therefore the volume of the giant should be $2^3 = 8$ times the volume of the 6 foot tall man. Assuming that weight is proportional to volume, the giant's weight should be 8 times the weight of a typical 6 foot tall man. If a typical 6 foot tall man weighs between 150 and 180 pounds, then the giant should weigh 8 times as much, or between 1200 and 1440 pounds.

4. A box that is w units wide, d units deep, and h units high has a volume of

$$wdh$$

cubic units. A box that is 3 times as wide, 3 times as deep, and 3 times as high is $3w$ units wide, $3d$ units deep, and $3h$ units high, and so has volume

$$(3w)(3d)(3h) = 27wdh$$

cubic units. Therefore a box that is 3 times as wide, 3 times as deep, and 3 times as high as another smaller box has a volume that is 27 times the volume of the smaller box.

A box that is w units wide, d units deep, and h units high has a surface area of

$$2wd + 2wh + 2dh$$

square units. This is because the box has two faces that are w units by d units, two faces that are w units by h units, and two faces that are d units by h units, as you can see by looking at the pattern for a box on page **??**. If another box is 3 times as wide, 3 times as deep, and 3 times as high, then this bigger box will have surface area

$$2(3w)(3d) + 2(3w)(3h) + 2(3d)(3h)$$

square units because its width, depth, and height are $3w$, $3d$, and $3h$ units respectively. But

$$2(3w)(3d) + 2(3w)(3h) + 2(3d)(3h) = 9(2wd + 2wh + 2dh),$$

therefore a box that is $3w$ units wide, $3d$ units deep, and $3h$ units high has 9 times the surface area of a box that is w units wide, d units deep, and h units high.

5. Let's call the radius of the small cylinder r units, and the height of the small cylinder h units. Then the volume of the small cylinder is

$$h\pi r^2$$

cubic units, according to the *(height)* × *(area of base)* volume formula. Because the larger cylinder has twice the radius and twice the height of the smaller cylinder, therefore the larger cylinder has radius $2r$ units and height $2h$ units, and so the larger cylinder has volume

$$(2h)\pi(2r)^2 = 8(h\pi r^2)$$

cubic units. Because $8h\pi r^2$ is 8 times $h\pi r^2$, therefore the volume of the large cylinder is 8 times the volume of the small cylinder.

The surface area of the small cylinder of radius r and height h is

$$2\pi r^2 h + 2\pi rh$$

square units (see the answer to Exercise 5 of Section 9.9 on page 250). Using the same formula again, but now with $2r$ substituted for r and $2h$ substituted for h, we obtain that the surface area of the big cylinder is

$$2\pi(2r)^2 + 2\pi(2r)(2h) = 8\pi r^2 + 8\pi rh = 4(2\pi r^2 + 2\pi rh)$$

square units. Because $4(2\pi r^2 + 2\pi rh)$ is 4 times $2\pi r^2 h + 2\pi rh$, therefore the surface area of the large cylinder is 4 times the surface area of the small cylinder.

Problems for Section 9.13 on Areas, Volumes, and Scaling

1. In the triangle $\triangle ADE$ pictured in Figure 9.106, the point B is halfway from A to D and the point C is halfway from A to E. Compare the areas of $\triangle ABC$ and $\triangle ADE$. Explain your reasoning clearly.

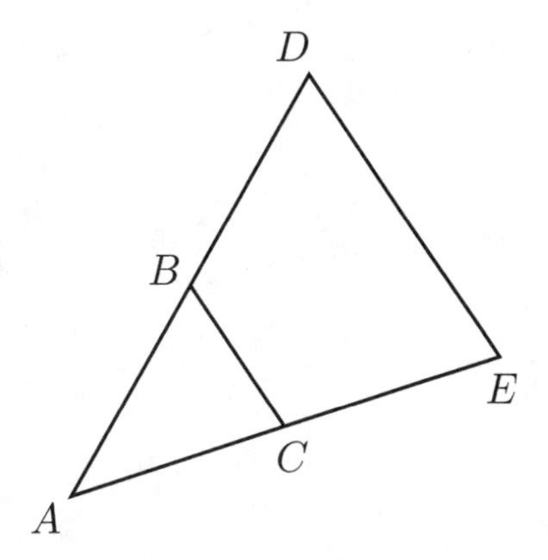

Figure 9.106: Triangles

2. A scale model is constructed for a domed baseball stadium using a scale of 1 foot to 100 feet. The model's dome is made out of 40 square feet of cardboard. The model contains 50 cubic feet of air. How many square feet of material will the actual stadium's dome be made out of? How many cubic feet of air will the actual stadium contain?

3. An artist plans to make a large sculpture of a person out of solid marble. She first makes a small scale model out of clay, using a scale of 1 inch = 2 feet. The scale model weighs 1.3 pounds. Assuming that a cubic foot of clay weighs 150 pounds and a cubic foot of marble weighs 175 pounds, how much will the large marble sculpture weigh? Explain your reasoning.

4. According to one description, King Kong was 19 feet, 8 inches tall and weighed 38 tons. Typical male gorillas are about 5 feet, 6 inches tall and weigh between 300 and 500 pounds. Assuming that King Kong was proportioned like a typical male gorilla, does his given weight of 38 tons agree with what you would expect? Explain.

5. If you know the volume of an object in cubic inches, can you find its volume in cubic feet by dividing by 12? Discuss!

Chapter 10

Functions and Algebra

10.1 Patterns, Sequences, Formulas, and Equations

Most people, including young children, enjoy discovering and experimenting with patterns. But patterns aren't just fun, by thinking about how to describe them generally we begin to develop algebraic reasoning. In this section we will see that some general formulas and equations can arise very naturally from patterns.

In order to work with changing patterns, it will be convenient to use the term *sequence*. A **sequence** is a list of items occuring in a specified order. In this section we will see that certain sequences of physical objects or pattern designs can give rise to sequences of numbers or sequences of equations.

The following example shows a simple way that a sequence can arise. Austin is making "train cars" out of snap cubes. He makes lots of train cars, each of which is made of 3 yellow snap cubes, 2 red snap cubes, and 1 blue snap cube. Austin snaps 2 train cars together to make a train. Then Austin snaps 3 train cars together to make a train, then 4, then 5, and so on. In this way there is a natural sequence of trains: the 1st in the sequence is 1 car long, the 2nd in the sequence is 2 cars long, the 3rd in the sequence is 3 cars long, and so on, as indicated in Figure 10.1.

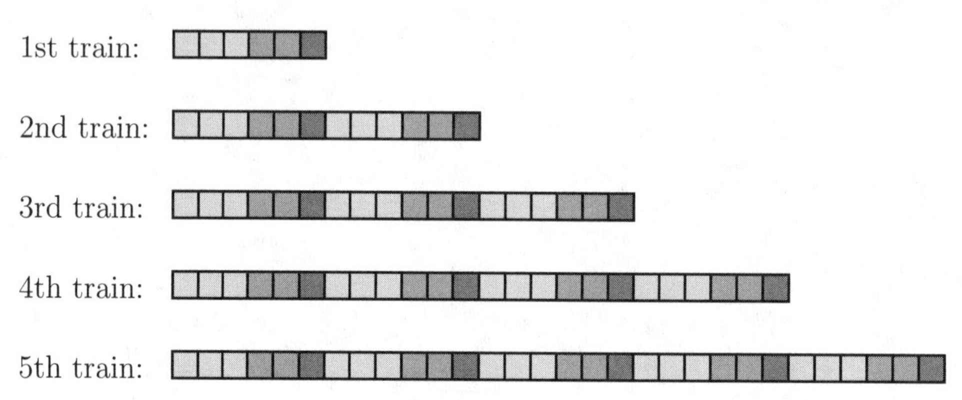

Figure 10.1: A Sequence of Snap Cube "Trains"

Austin's sequence of trains gives rise to several sequences of numbers by asking the following natural questions:

How many blue snap cubes are in the trains? How many red snap cubes are in the tains? How many yellow snap cubes are in the trains? In all, how many snap cubes are in the trains?

Because each train car has 1 blue snap cube, therefore the number of blue snap cubes goes up by 1 every time a train car is added. Thus the following sequence shows the number of blue snap cubes in the first train, in the second train, in the third train, and so on:

$$1,\ 2,\ 3,\ 4,\ 5,\ldots.$$

Because each train car has 2 red snap cubes, therefore the number of red snap cubes goes up by 2 every time a train car is added. Thus the following sequence shows the number of red snap cubes in the first train, in the second train, in the third train, and so on:

$$2,\ 4,\ 6,\ 8,\ 10,\ldots.$$

Because each train car has 3 yellow snap cubes, therefore the number of yellow snap cubes goes up by 3 every time a train car is added. Thus the following sequence shows the number of yellow snap cubes in the first train, in the second train, in the third train, and so on:

$$3,\ 6,\ 9,\ 12,\ 15,\ldots.$$

Because each train car has 6 total snap cubes, therefore the total number of snap cubes goes up by 6 every time a train car is added. Thus the following sequence shows the total number of snap cubes in the first train, in the second train, in the third train, and so on:

$$6,\ 12,\ 18,\ 24,\ 30\ldots.$$

We can record all this data about snap-cube trains succinctly in the following table:

# of train cars	# of blue cubes	# of red cubes	# of yellow cubes	total # of cubes
1	1	2	3	6
2	2	4	6	12
3	3	6	9	18
4	4	8	12	24
5	5	10	15	30
⋮	⋮	⋮	⋮	⋮

Eventually, Austin will run out of train cars and won't be able to make trains beyond a certain length. However, we can still imagine the process of snapping together more and more train cars to make longer and longer trains. We can imagine a train consisting of 100 train cars, ..., a train consisting of 1000 train cars, ..., and so on; for each counting number N, we can imagine a train consisting of N train cars.

Working generally with a train consisting of N train cars we can develop simple formulas for the numbers of snap cubes. In a train consisting of N train cars, how many blue snap cubes are there? how many red snap cubes are there? how many yellow snap cubes are there? in all, how many snap cubes are there? Because each train car has 1 blue snap cube, 2 red snap cubes, 3 yellow snap cubes, and 6 total snap cubes, there are

$$N$$

blue snap cubes,

$$2N$$

red snap cubes,

$$3N$$

yellow snap cubes, and

$$6N$$

total snap cubes in a train consisting of N train cars. Adding this information to the previous table we obtain the following table:

# of train cars	# of blue cubes	# of red cubes	# of yellow cubes	total # of cubes
1	1	2	3	6
2	2	4	6	12
3	3	6	9	18
4	4	8	12	24
5	5	10	15	30
⋮	⋮	⋮	⋮	⋮
N	N	$2N$	$3N$	$6N$

The following equations use the formulas above to express the fact that the total number of snap cubes in a train is equal to the number of blue snap cubes, plus the number of red snap cubes, plus the number of yellow snap cubes:

$$N + 2N + 3N = 6N.$$

Notice that we can consider this equation as illustrating the distributive property because

$$\begin{aligned} N + 2N + 3N &= 1N + 2N + 3N \\ &= (1 + 2 + 3)N \\ &= 6N \end{aligned}$$

Class Activity 10A: Patterns and Sequences

Class Activity 10B: Creating Sequences of Patterns

Class Activity 10C: Patterns and Equations I

Class Activity 10D: Patterns and Equations II

Class Activity 10E: Patterns and Equations III

Class Activity 10F: Creating Sequences of Patterns

Class Activity 10G: A Sequence of Three-Dimensional Block Patterns

Class Activity 10H: Patterns and Formulas

Class Activity 10I: Repeating Patterns

Exercises for Section 10.1 on Patterns, Sequences, Formulas, and Equations

1. Figure 10.2 shows a sequence of patterns made of small squares. Create a corresponding sequence of numbers by writing the total number of small squares in each pattern. Imagine the sequence of patterns continuing forever, so that for each counting number N, there is an Nth pattern. Write a formula for the number of small squares in the the Nth pattern.

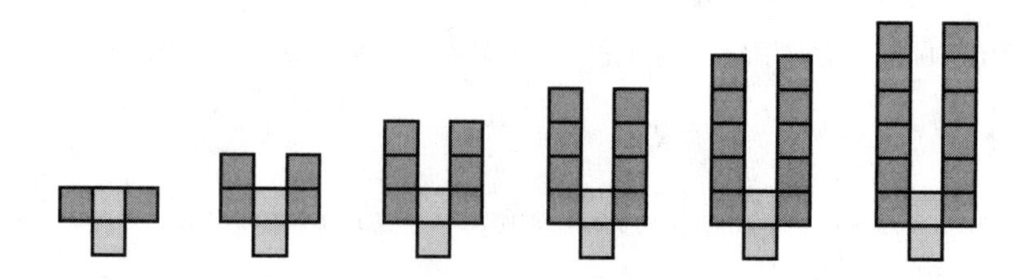

Figure 10.2: A Sequence of Patterns

Answers to Exercises for Section 10.1 on Patterns, Sequences, Formulas, and Equations

1. The number of squares in the patterns are:

$$4,\ 6,\ 8,\ 10,\ 12,\ 14, \ldots.$$

If the sequence of patterns were to continue forever, then the Nth pattern should be made of $2N + 2$ small squares. We can show this information in tabular form as follows:

# of pattern	# of squares in pattern
1	4
2	6
3	8
4	10
5	12
6	14
$\vdots$	$\vdots$
N	$2N + 2$

Problems for Section 10.1 on Patterns, Sequences, Formulas, and Equations

1. Draw a sequence of patterns made of small squares so that the Nth pattern is made of $4N + 3$ small squares.

2. Draw a sequence of rectangular patterns that illustrate the equation

$$2 \cdot (1 + 2 + 3 + \ldots N) = N \cdot (N + 1).$$

Explain why your sequence of patterns illustrates this equation.

3. (a) Create a sequence of numbers that correspond to the sequence of rectangular patterns shown in Figure 10.3 by determining the number of small squares that each rectangular pattern is made of (remember that squares are rectangles too!).

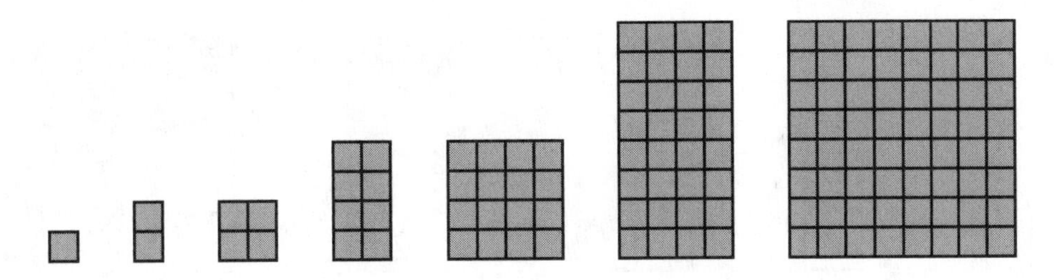

Figure 10.3: A Sequence of Rectangles Made of Squares

(b) Now imagine the sequence of rectangular patterns in Figure 10.3 continuing forever. In this sequence of patterns there is a 1st pattern, a 2nd pattern, a 3rd pattern, ..., there is a 100th pattern, ...there is a 1000th pattern, ...; for each counting number N, there is an Nth pattern in the sequence of rectangular patterns. What is a formula for the number of small squares making up the Nth rectangular pattern?

(c) Explain how the sequence in Figure 10.4 illustrates the equation

$$1 + 1 + 2 + 2^2 + 2^3 + \ldots + 2^N = 2^{(N+1)}.$$

4. Create a design that illustrates the formula

$$M + 4N + 3P$$

by imagining different portions of the design filled with different numbers of dots. Explain why your design illustrates the formula $M+4N+3P$.

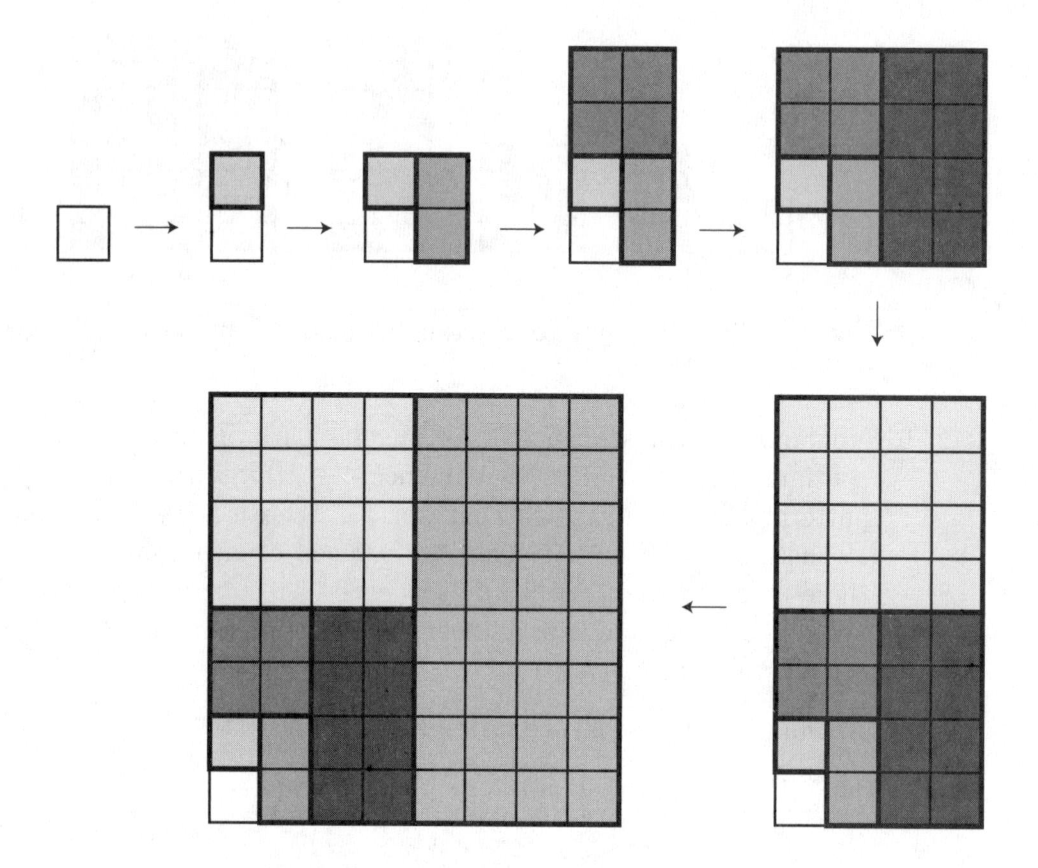

Figure 10.4: A Sequence of Rectangular Patterns

10.2 Functions

In the last section we studied sequences. Sequences are a special kind of mathematical *function.* In this section, we will define the concept of function and we will study the three main ways that functions can be represented: by formulas, by tables, and by graphs. Functions are one of the cornerstones of mathematics and science: by studying and using functions, mathematicians and scientists have been able to describe and predict a variety of phenomena, such as motions of planets, the distance a rocket will travel, or the dose of medicine that a patient needs. Even at the most elementary level, functions provide a way to organize, represent, and study information.

A **function** is a rule that assigns one output to each allowable input. Most commonly, these "inputs" and "outputs" are numbers. (The set of "allowable inputs" is called the **domain** of the function. The set of "allowable outputs" is called the **range** of the function.) Texts for children often portray functions as machines, as in Figure 10.5. With this imagery, it's easy to

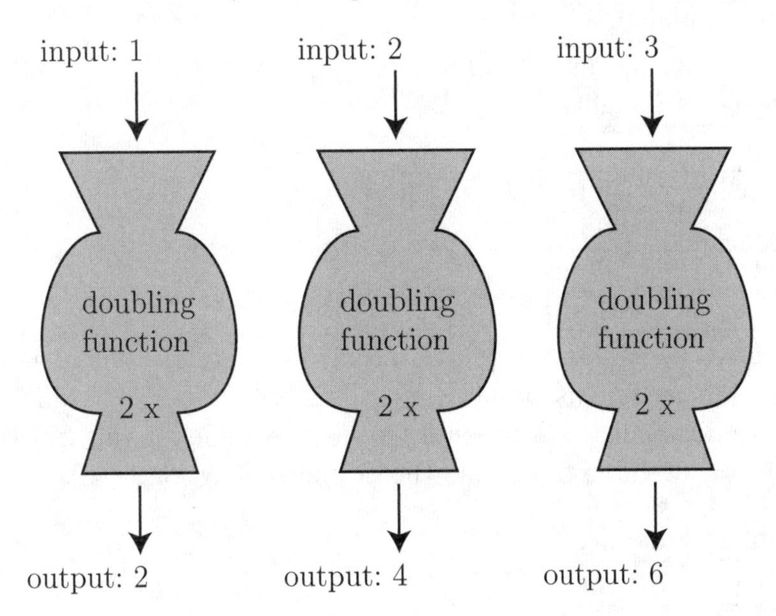

Figure 10.5: Viewing Functions as Machines

imagine putting a number into a machine; the machine somehow transforms the input number into a new output number.

For example, consider the *doubling function*: to each input number, the assigned output is twice the input. To the input 1, the doubling function assigns output 2; to the input 2, the doubling function assigns the output 4; to the input 3, the doubling function assigns the output 6; and so on.

There are several standard ways of displaying functions: with tables, with graphs and with formulas. These methods can help make functions comprehensible. We will now study these methods for representing functions.

Tables

A table provides an easily understood, organized way to display a function. To display a function in a table, make two columns: the column on the left consisting of various input values, the column on the right consisting of the corresponding output values.

For example, consider the function that assigns to each one of five girls, her height. This function is fully displayed in the following table:

Height Function	
input (girl)	output (height)
Kaitlyn	53 in
Lameisha	56 in
Sarah	49 in
Manuela	50 in
Kelli	54 in

Of course, we can also consider the height function that assigns to *each person in the world*, his or her height. In this case it would be neither informative nor realistically possible to make a table for this function.

The following table represents the doubling function:

Doubling Function	
input	output
1	2
2	4
3	6
4	8
5	10
⋮	⋮

This table does not show all possible inputs and outputs, but from the pattern we see in the table, we can determine the outputs that correspond to inputs other than the ones that are displayed.

In the case of the doubling function we can make different choices for the inputs that we want to allow. We can decide only to allow counting numbers as inputs for the doubling function. In this case, the doubling function corresponds exactly to the sequence

$$2, 4, 6, 8, 10, \ldots$$

because we assign to the Nth place in the sequence the number $2N$. In general, every sequence gives rise to a function, namely the function that assigns to a counting number, N, the value of the Nth term in the sequence.

If we decide that we want to allow all real numbers as inputs for the doubling function, then the doubling function encompasses more than just the sequence

$$2, 4, 6, 8, 10, \ldots.$$

For example, the doubling function assigns to the input 3.25, the output 6.5, but 6.5 is not in the sequence

$$2, 4, 6, 8, 10, \ldots.$$

Tables are easy to read and to understand, but they rarely display complete information about a function. Consider the function that we will call *S & P 500*, which assigns to any time since 1957, the value of the Standard and Poor's 500 Index at that time. (The Standard and Poor's 500 Index is a commonly used standard for measuring stock market performance. For further information on the S & P 500, see the website `www.spglobal.com`.) The following table displays some inputs and outputs of the S & P 500 function:

S & P 500 Function	
input (time)	output (value of S & P 500)
Jan 1, 2002, 5 pm	1148
Jan 1, 2001, 5 pm	1320
Jan 1, 2000, 5 pm	1469
Jan 1, 1999, 5 pm	1229
Jan 1, 1998, 5 pm	970
Jan 1, 1997, 5 pm	740
Jan 1, 1996, 5 pm	615
Jan 1, 1995, 5 pm	459
Jan 1, 1994, 5 pm	466
Jan 1, 1993, 5 pm	435

The table above only shows the value of the S & P 500 on January 1st of each year over a 10 year period, but it doesn't show any information about the S & P 5001's value in the intervening months. If we wanted to display more information, we could make a longer table, but that might become difficult to read and comprehend.

Another good way to display a function is to use a *graph*.

Coordinate Planes and Graphs

The graph of a function displays the function in a picture. By graphing a function we can often display more inputs and outputs than we could display in a table, and we can often discern trends and patterns in the function that are less obvious in a table.

Coordinate Planes

We graph a function that has numerical inputs and outputs in a *coordinate plane.* A **coordinate plane** is a plane, together with two perpendicular number lines in the plane that (usually) meet at the location of 0 on each number line. The two number lines can have different scales. Traditionally, one number line is displayed horizontally and the other is displayed vertically. The two number lines are called the **axes** of the coordinate plane (singular: axis). The horizontal axis is often called the **x-axis**, and the vertixal axis is often called the **y-axis**.

The main feature of a coordinate plane is that *the location of every point in the plane can be specified by referring to the two axes.* This works in the following way. To a pair of numbers, such as

$$(4.5, 3),$$

corresponds the point in the plane that is located where a vertical line through 4.5 on the horizontal axis meets a horizontal line through 3 on the vertical axis, as shown in Figure 10.6. This point is designated $(4.5, 3)$; or we say that $(4.5, 3)$ are the **coordinates** of the point. Specifically, 4.5 is the **first coordinate**, or **x-coordinate**, of the point and 3 is the **second coordinate**, or **y-coordinate**, of the point.

You can use your fingers to locate the point $(4.5, 3)$ by putting your right index finger on 4.5 on the horizontal axis and your left index finger on 3 on the vertical axis, and sliding the right index finger vertically upward and the left index finger horizontally to the right until your two fingers meet.

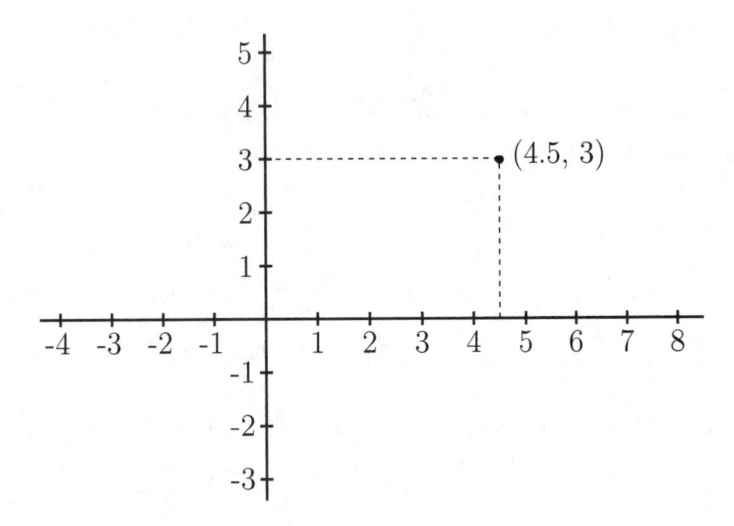

Figure 10.6: Locating $(4.5, 3)$ in a Coordinate Plane

Figure 10.7 shows the coordinates of several different points in a coordinate plane. Notice that the coordinates of a point can include negative numbers.

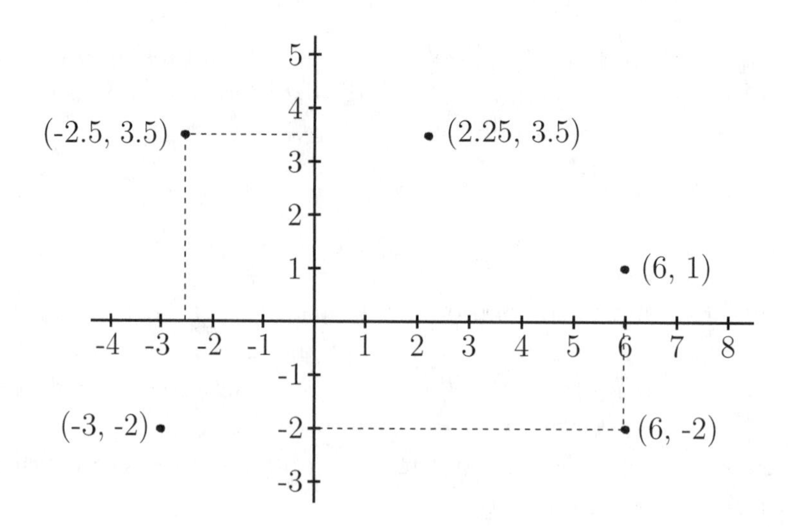

Figure 10.7: Points in a Coordinate Plane

The Graph of a Function

A function whose inputs and outputs are *numbers* has a *graph*. The **graph of a function** consists of all those points in a coordinate plane whose second coordinate is the output of first coordinate. For example, Figure 10.8 shows the graph of the doubling function when only the counting numbers are allowed to be inputs. In this case, the graph of the doubling function consists of the points

$$(1, 2),\ (2, 4),\ (3, 6),\ (4, 8),\ (5, 10), \ldots.$$

For each point on the graph, the second coordinate is twice the first coordinate.

Figure 10.9 shows the graph of the doubling function when we allow all real numbers as inputs. In this case, the graph consists of all points of the form

$$(N, 2N),$$

where N can be any real number. For example, in this case

$$(3.25, 6.5)$$

is on the graph of the doubling function, and so is

$$(-1.3, -2.6).$$

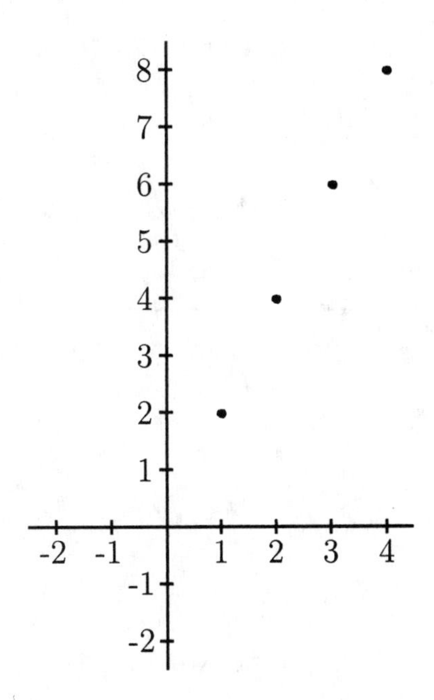

Figure 10.8: Graph of the Doubling Function With Only Counting Numbers as Inputs

The graph in Figure 10.8 consists of isolated points, but the graph in Figure 10.9 consists of so many points that the points fill in a solid line.

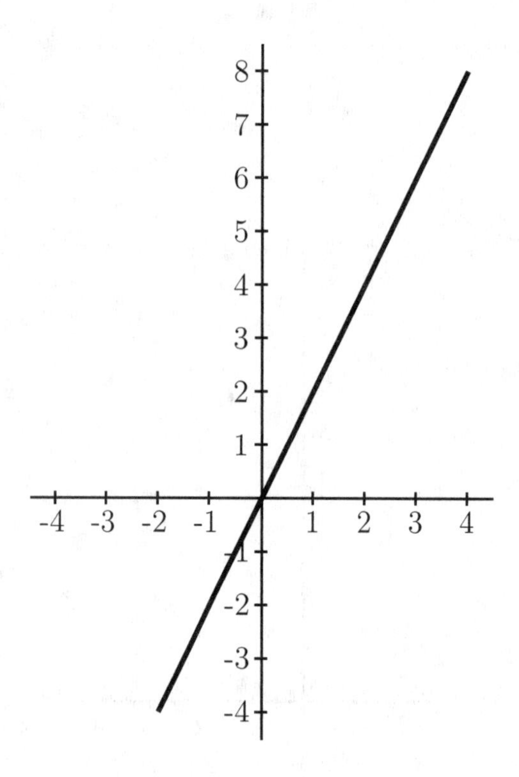

Figure 10.9: Graph of the Doubling Function When All Real Numbers are Allowed as Inputs

Drawing the Graph of a Function

In order to draw the graph of a function it often helps to display the function in a table first. For example, suppose that a seed is planted and the height of the plant growing from the seed is recorded every day after planting. This situation gives rise to a *height* function, such as the hypothetical one shown in the following table:

Plant Height Function	
input # of days since planting	output plant height in cm
1	0
2	0
3	0
4	0
5	.5
6	1.5
7	3
8	5
9	5.5
10	6
11	6
12	6

Each line in the table yields the coordinates of a point on the graph of the plant height function:

Plant Height Function		
input # of days since planting	output plant height in cm	
1	0	yields the point (1, 0)
2	0	yields the point (2, 0)
3	0	yields the point (3, 0)
4	0	yields the point (4, 0)
5	.5	yields the point (5, .5)
6	1.5	yields the point (6, 1.5)
7	3	yields the point (7, 3)
8	5	yields the point (8, 5)
9	5.5	yields the point (9, 5.5)
10	6	yields the point (10, 6)
11	6	yields the point (11, 6)
12	6	yields the point (12, 6)

The points obtained from the table for the function can then be plotted in a coordinate plane, as in Figure 10.10. Notice that in this graph, the points obtained from the table have been connected by a curve. Why does it make sense to connect the points? In between the times that the plant's height was measured, the plant continued to grow. For example, between days 7 and 8, the plant grew from 3 cm to 5 cm. So, $7\frac{1}{2}$ days after planting, the plant was probably about 4 cm tall. Therefore the point $(7.5, 4)$, or a point that is very close to this point, is also on the graph of the plant height function. For every time between 7 days after planting and 8 days after planting, the plant had some height between 3 cm and 5 cm, and this yields points on the graph that lie between the points $(7, 3)$ and $(8, 5)$, and that connect these two points with a curve.

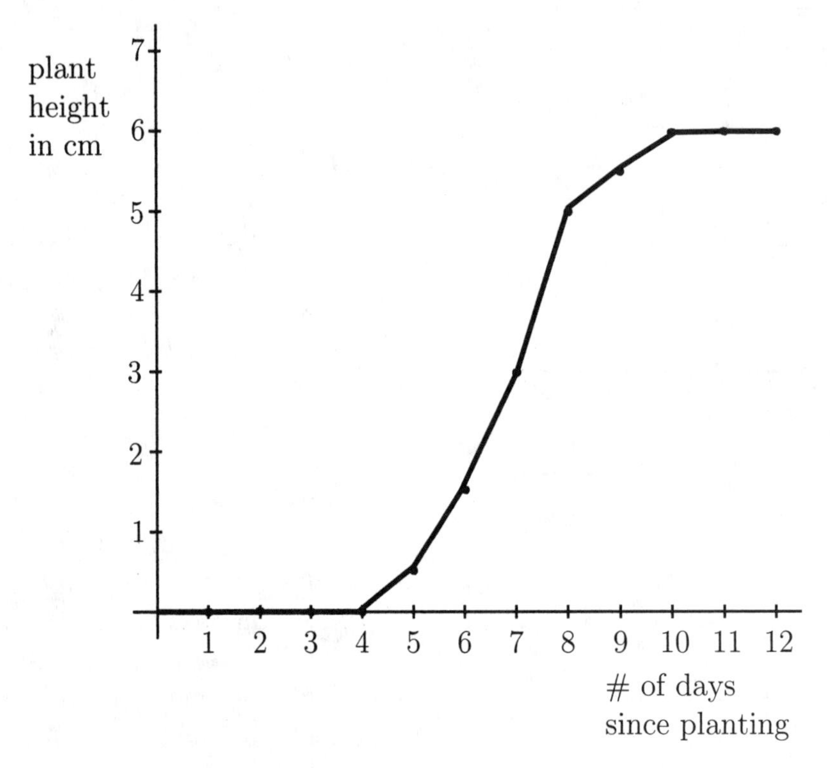

Figure 10.10: Graph of the Plant Height Function

Note two features of the graph in Figure 10.10 that help make the graph easy to see and to interpret:

- The scales on the two axes are different: the distances between numbers on the horizontal axis are smaller than the distances between numbers on the vertical axis. This is acceptable because the horizontal and vertical axes represent different kinds of quantities: the horizontal axis represents time, and the vertical axis represents height. When graphing a function, choose scales that make the function easy to see and interpret.

- The axes are labeled: the horizontal axis is labeled "# of days since planting" and the vertical axis is labeled "height in cm". Labeling the axes helps a reader understand and interpret a graph when the function arises from a real or realistic situation.

Interpreting Graphs of Functions

Once you learn how to interpret graphs of functions, you can quickly and easily see qualitative aspects of functions. Look at the graph in Figure 10.10: "read" it from left to right. Moving from left to right along the graph represents time passing: how far a point on the graph is to the right corresponds to how many days have passed since the seed was planted. The height of a point on the graph above the horizontal axis represents the height of the plant at that time (i.e., at the time represented by the horizontal location of the point). Thus, following the graph from left to right we see that the plant doesn't sprout until after 4 days. Between the 6th and 8th days the plant grows more quickly than at other times: we see this because the graph rises steeply during those times, meaning that over a short period of time, a lot of growth has occured. Between days 8 and 10 the the plant grows less rapidly: the height doesn't increase as much over those two days as it did over the previous two days. After 10 days, the plant has stopped growing: its height no longer increases.

Formulas for Functions

Some functions can be described by formulas; others cannot. To give a formula for a function means to give a formula for the output of the function in terms of the input of the function. For example, consider the doubling function. The output is always 2 times the input. Therefore, if the input is x, the output is $2x$, and so the doubling function is described by the formula

$$2x.$$

If we give the doubling function the name "D", then a standard way to describe the formula for the doubling function is to write:

$$D(x) = 2x.$$

This is shorthand for saying that when x is put into the function D, the output is $2x$.

Not all functions have formulas. For example, consider the plant height function we studied above. There is no formula to describe the height of the plant in terms of the number of days since the seed was planted.

If you have a formula for a function, then you can make a table and a graph for the function by using the formula. For example, consider the function, f, which is given by the formula

$$f(x) = x^2 - x - 2.$$

To make a table for the function, remember what it means to say that $f(x) = x^2 - x - 2$. It means that when x is put into the function, the output is $x^2 - x - 2$. So if 1 is put into the function, the output is

$$1^2 - 1 - 2 = 1 - 1 - 2 = -2.$$

If 2 is put into the function, the output is

$$2^2 - 2 - 2 = 4 - 2 - 2 = 0.$$

If 3 is put into the function, the output is

$$3^2 - 3 - 2 = 9 - 3 - 2 = 4.$$

And so on. Therefore the following is a table for the function f:

f	
input	output
-3	10
-2	4
-1	0
0	-2
.5	-2.5
1	-2
2	0
3	4
4	10

We can use this table to graph the function f, as in Figure 10.11. Assuming that all real numbers are allowed as inputs for f, the graph of f will be a curve that connects the points from the table. If you were to plot many more points, you would see that this curve is nice and smooth, as shown in Figure 10.11.

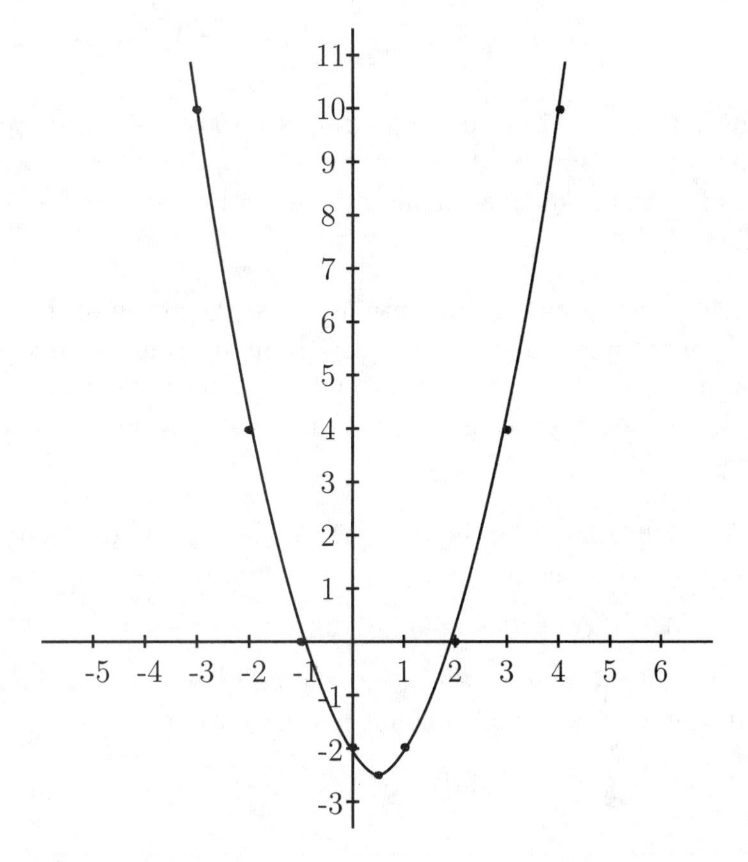

Figure 10.11: The Graph of $f(x) = x^2 - x - 2$

In more advanced mathematics, a central topic of study is the relationship between the nature of a formula for a function and the nature of the graph of the function.

Class Activity 10J: Which Points are on the Graph?

Class Activity 10K: Interpreting Graphs of Functions

Class Activity 10L: Find the Function's Formula

Exercises for Section 10.2 on Functions

1. Plot the following points in a coordinate plane:

$$(4, -2.5),\ (-2.5, 4),\ (3, 2.75),\ (-3, -2.75).$$

2. A nightlight costs \$7.00 to buy and uses \$.75 of electricity per year to operate. What function does this situation give rise to? Make a table for this function, sketch a graph of this function, and find a formula for this function.

3. For each of the following descriptions, draw the graph of the associated pollen count function, for which the input is time elapsed since the beginning of the week, and the output is the pollent count at that time. In each case, explain why you draw your graph in the shape that you do.

   (a) At the beginning of the week, the pollen count rose sharply. Later in the week, the pollen count continued to rise, but more slowly.

   (b) The pollen count fell steadily during the week.

   (c) At the beginning of the week, the pollen count fell slowly. Later in the week, the pollen count fell more rapidly.

4. Find formulas for the functions that have the following tables:

f	
input	output
0	2
1	7
2	12
3	17
4	22

g	
input	output
0	0
2	5
4	10
6	15
8	20

h	
input	output
1	2
2	4
3	8
4	16
5	32

Answers to Exercises for Section 10.2 on Functions

1. See Figure 10.12.

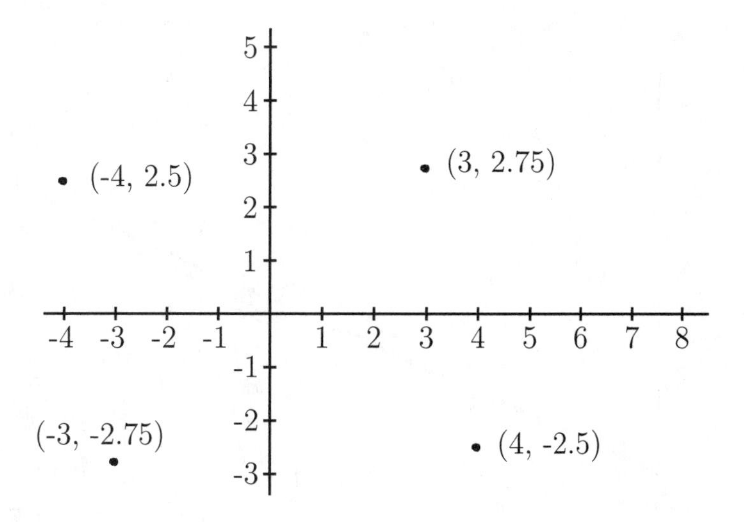

Figure 10.12: Points in a Coordinate Plane

2. The situation gives rise to a "nightlight function" for which the input is the number of years since purchase, and the output is the cost of operating the nightlight for that amount of time (where this cost includes the cost of buying the nightlight). The following is a table for the nightlight function:

Nightlight Function	
input	output
years since purchase	total cost of operating
1	\$7.75
2	\$8.50
3	\$9.25
4	\$10.00
5	\$10.75

Because the nightlight costs \$7.00 to buy and costs \$.75 for each year of operation, therefore it costs a total of

$$7 + .75y$$

dollars to operate the nightlight for y years. A sketch of the graph of the nightlight function is in Figure 10.13.

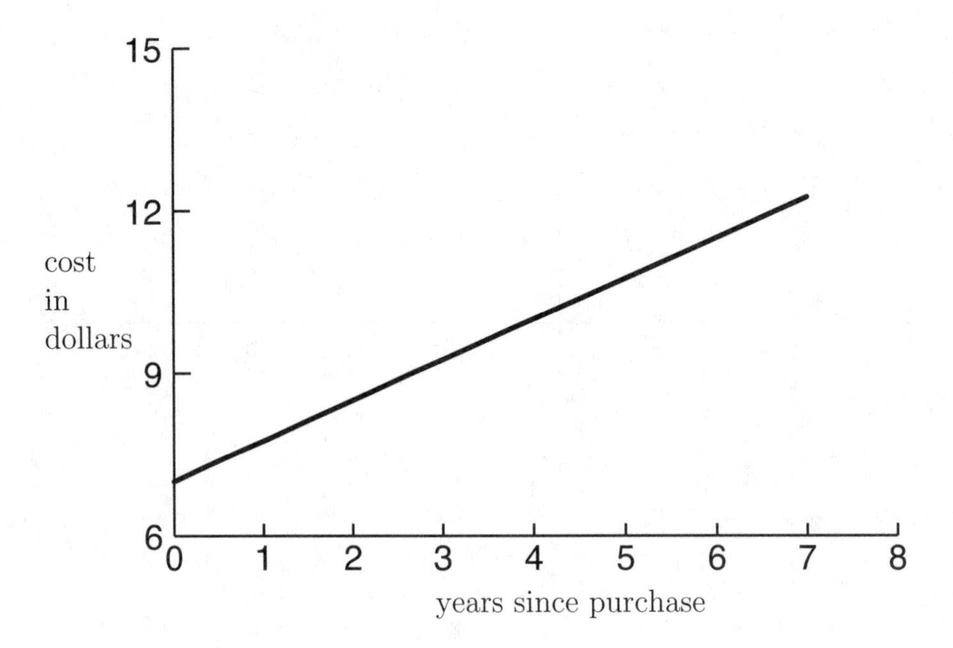

Figure 10.13: The Cost of Using a Nightlight

3. See Figure 10.14. Moving from left to right along the graph corresponds to time passing during the week.

   (a) The first graph goes up steeply at first (reading it from left to right) because for each of the first few days of the week, the pollen count is significantly higher than it was the day before. Thus the graph slopes steeply upward at first. But later in the week, toward the right of the graph, the graph must go up less steeply because for each day later in the week, the pollen count is only a little higher than it was the day before.

   (b) The second graph is a straight line that slopes downward (reading from left to right) because for each day that goes by, the pollen count drops by the same amount.

   (c) The third graph goes down slowly at first (reading from left to right) because for each of the first few days of the week, the pollen count is only a little less than it was the day before. Thus the graph slopes gently downward at first. But later in the week, toward the right of the graph, the graph must go down more steeply because for each day later in the week, the pollen count is significantly lower than it was the day before. Thus towards the right, the graph slopes steeply downward.

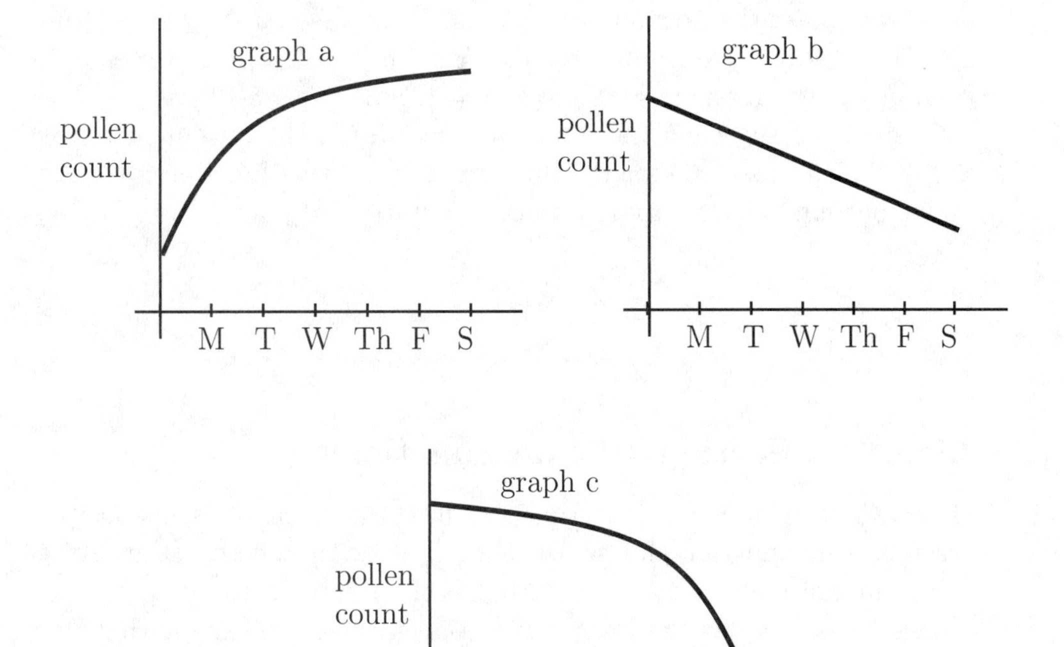

Figure 10.14: Pollen Counts

4.

$$f(x) = 5x + 2$$

You can guess this formula by noticing that when the input increases by 1, the output increases by 5. The formula $5x$ has the property that when you increase x by 1, the output, $5x$, increases by 5. But the function f "starts" at 2, i.e., $f(0) = 2$, so add 2 to $5x$ to get the proper formula for f. Notice that you can always check if your formula is correct by seeing if it gives you the correct output for each input.

$$g(x) = \frac{5}{2}x$$

You can guess this formula as follows. It is reasonable to guess that $g(1)$ is halfway between $g(0)$ and $g(2)$, that $g(3)$ is halway between $g(2)$ and $g(4)$, and so on. If so, then $g(1) = \frac{5}{2}$, $g(3) = \frac{15}{2}$, $g(5) = \frac{25}{2}$, and so on. In this case, when the input increases by 1, the output increases by $\frac{5}{2}$. This is also true for the function $\frac{5}{2}x$, and this function gives the same outputs for the inputs given in the table for g.

$$h(x) = 2^x$$

When the input increases by 1, the output doubles.

Problems for Section 10.2 on Functions

1. For each of the following descriptions, draw the graph of the associated temperature function, for which the input is time elapsed since the beginning of the day, and the output is the temperature at that time. In each case, explain why you draw your graph in the shape that you do.

   (a) At the beginning of the day, the temperature dropped sharply. Later on, the temperature continued to drop, but more gradually.

   (b) The temperature rose steadily at the beginning of the day, remainded stable in the middle of the day, and then fell steadily later in the day.

   (c) The temperature rose quickly at the beginning of the day until it reached its peak. Then the temperature fell gradually throughout the rest of the day.

2. Draw *two different* graphs of two different functions that both have the following properties:

   - When the input is 1, the output is 3.
   - When the input is 2, the output is 6.
   - When the input is 3, the output is 9.

3. Find *two different* formulas for two different functions that both have the following two properties:

   - When the input is 0, the output is 1.
   - When the input is 1, the output is 2.

4. Find formulas for the following functions.

f	
input	output
0	3
1	7
2	11
3	15
4	19

g	
input	output
0	1
2	4
4	7
6	10
8	13

h	
input	output
0	-1
1	0
2	3
3	8
4	15

Chapter 11

Statistics

The field of statistics provides tools for studying questions that can be addressed with data. Increasingly, a variety of data is available to tell us about population, health, financial and business concerns, the environment, and any number of other aspects of society and the world around us. Statistical concepts can help us interpret this data.

Concerning the learning of statistics, the National Council of Teachers of Mathematics recommends the following (see [3]):

In prekindergarten through grade 2 all students should—

- Formulate questions that can be addressed with data and collect, organize, and display relevant data to answer them:
  - pose questions and gather data about themselves and their surroundings;
  - sort and classify objects according to their attributes and organize data about the objects;
  - represent data using concrete objects, pictures, and graphs.
- Select and use appropriate statistical methods to analyze data:
  - describe parts of the data and the set of data as a whole to determine what the data show.
  - describe the shape and important features of a set of data and compare related data sets, with an emphasis on how the data are distributed;
  - use measures of center, focusing on the median, and understand what each does and does not indicate about the data set;
  - compare different representations of the same data and evaluate how well each representation shows important aspects of the data.
- Develop and evaluate inferences and predictions that are based on data
  - discuss events related to students' experiences as likely or unlikely.

In grades 3–5 all students should—

- Formulate questions that can be addressed with data and collect, organize, and display relevant data to answer them:

  - design investigations to address a question and consider how data-collection methods affect the nature of the data set;
  - collect data using observations, surveys, and experiments;
  - represent data using tables and graphs such as line plots, bar graphs, and line graphs;
  - recognize the differences in representing categorical and numerical data.

- Select and use appropriate statistical methods to analyze data:

  - describe the shape and important features of a set of data and compare related data sets, with an emphasis on how the data are distributed;
  - use measures of center, focusing on the median, and understand what each does and does not indicate about the data set;
  - compare different representations of the same data and evaluate how well each representation shows important aspects of the data.

- Develop and evaluate inferences and predictions that are based on data:

  - propose and justify conclusions and predictions that are based on data and design studies to further investigate the conclusions or predictions.

11.1 Designing Investigations and Gathering Data

We humans are naturally curious about each other and about the world. We want to know what others around us think about the political questions of the day, about the latest movies, TV shows, and music, and about many other topics of interest. Local, state, and national governments want to know facts about population, employment, income, and education in order to design appropriate development plans, or to take needed actions. Scientists want to know how to cure and prevent diseases, as well as to know about the world around us. Therefore news organizations, government bodies, and scientists regularly design investigations to address questions, and they gather data to help answer those questions.

Some investigations use surveys to gather data. In a survey, people respond to questions or provide information either by phone, in person, or in writing. Some investigations use observations to gather data, such as by counting the number of turtle eggs in a certain area, or by measuring the height of a plant every day as it grows. Some investigations use controlled experiments to generate data. For example in determining whether a new drug is safe and effective, a standard experiment is for half of a group of volunteers to be given the drug, and the other half to be given a placebo. The results are observed to see how well the drug worked.

Class Activity 11A: Difficulties in Conducting Good Surveys

It may seem like a simple matter to determine people's opinions or facts about people by surveying. But if you did Class Activity 11A, they you probably began to realize that it is not.

In gathering data by surveying, one issue is how to choose the people to be surveyed. In most cases, it is not possible to ask all relevant people the survey question. In this case, a sample of the population of interest must be surveyed. But unless the sample is chosen carefully, the sample's answers may not accurately reflect the answers of the population of interest. For example, if you wanted to know how people were going to vote in an upcoming election, and if you only asked your friends how they plan to vote, you would probably not get an accurate picture of how others in your county or state would vote. We tend to be friends with like-minded people. The best way to pick a representative sample for a survey is to pick a *random* sample. A random sample can be chosen using a list of random numbers.

Another issue in surveying is the nature of the questions. Different people may interpret the same question in different ways. For example, even a relatively simple question such as "how many children are in your family?" can be interpreted in different ways. Does it mean how many people under the age of 18 are in your family? Does it mean how many siblings you have, plus yourself? Should step-children be included? A good survey question must be clear.

Investigations in which data is gathered by observation must often contend with errors in measurement. Whenever we measure the length of an object or its weight, the result can never be completely accurate. In fact, even with very accurate measuring devices, the same object can be measured several different times with different results. This is just the nature of measurement.

In controlled experiments, such as in an experiment to determine if a new drug is safe and effective, a group of volunteers will be divided into two groups (more groups are used if the drug is being compared to other drugs). One group receives the new drug, the other group receives a placebo—an inert substance that is made to look like a real drug. Patients are observed to see if their symptoms improve and if there are any side-effects.

But such an experiment is only useful if it accurately predicts the outcome of the use of the drug in the population at large. Therefore it is crucial that the two groups, the one receiving the drug and the one receiving the placebo, are as alike as possible in all ways except in whether or not they receive the drug. The two groups should also be as much like the general population who would take the drug as possible. Therefore, the two groups are usually chosen *randomly* (such as by flipping a coin: heads goes to group 1, tails to group 2); in addition, such experiments are usually *double-blind*, that is, neither the patient nor the treating physician knows who is in which group.

In controlled drug experiments, the randomness of the assignment of patients to the two groups and the double-blindness ensure that the study will accurately predict the outcome of the use of new drug in the whole population. If the patients were not assigned randomly to the two groups then the person assigning the patients to a group might subconsciously pick patients with certain attributes for one or the other group, thereby making the group that is treated by the new drug slightly different from the group receiving the placebo. This could result in inaccurate predictions about the safety and effectiveness of the drug in the population at large. Similarly, if the patient knows whether or not they have received the new drug, their resulting mental state may affect their recovery. If the physician knows whether or not the new drug has been administered, their body language may be different toward the patient and the patient may again be affected differently.

Class Activity 11B: Is it a Representative Sample?

Class Activity 11C: Which Experiment is Better?

Exercises for Section 11.1 on Designing Investigations and Gathering Data

1. Kaitlyn, a third grader, asked 5 of her friends in class who their favorite singer is. All five said Britney Spears. Can Kaitlyn conclude that most of the children at her school would say Britney Spears is their favorite singer? Why or why not?

2. What are some issues that may arise in using a survey to determine people's opinion on some issue or to determine facts about people?

Answers to Exercises for Section 11.1 on Designing Investigations and Gathering Data

1. No, Kaitlyn can't conclude that most of the children at her school would say Britney Spears is their favorite singer. Kaitlyn's friends may all have a similar taste in music, and their musical taste may not be representative of the musical tastes of the whole school.

2. See text.

Problems for Section 11.1 on Designing Investigations and Gathering Data

1. A group studying violence wants to determine the attitudes towards violence of all the 5th graders in a county. The group plans to do a survey, but the group does not have the time or resources to survey all 5th graders in the county. The group can afford to survey 120 fifth graders. For each of the following methods of choosing 120 fifth graders to survey, discuss advantages and disadvantages of that method. Based on your discussion, which of the three methods will be best for determining the attitudes towards violence of all the 5th graders in the county? (There are 6 elementary schools in the county and each school has 4 fifth grades.)

   (a) Have each elementary school principal in the county select 20 fifth graders.

(b) Have each fifth grade teacher in the county select 5 fifth graders.

(c) Obtain a list of all 5th graders in the county. Obtain a list of 120 random numbers from 1 to the number of fifth graders in the county. Use the list of random numbers to select 120 random 5th graders in the county.

11.2 Displaying Data

Every day we have the opportunity to see the results of surveys in newspapers or on television news shows. Sometimes results are reported with numbers (such as a number of people or as a percentage), but other times, results are shown in a table, or in a chart or graph. Charts and graphs help us get a sense of what data mean because a chart or graph can show us the "big picture" about data in a way that we can understand quickly and easily. In this section we will study the most common ways to display data in charts and graphs. We will use data displays to help us interpret data and to help us communicate our interpretation to others. We will also look at some common errors in working with data displays.

Common Data Displays

Real Graphs and Pictographs

The most basic kinds of graphs are real graphs and pictographs. Both of these are suitable for use with young children and can be used as a springboard for understanding other kinds of graphs. A **real graph** or **object graph** displays actual objects in a graph form; a **pictograph** is like a real graph except that it uses pictures of objects instead of actual objects. In some pictographs, a single picture may represent more than one object. We can use real graphs and pictographs to show how a collection of related objects are sorted into different groups. The real graph or pictograph allows us to see at a glance which groups have more, and which groups have fewer objects in them.

Suppose we have a tub filled with small beads that are alike except that they come in a variety of colors: yellow, pink, purple, and green. We can mix up the beads and scoop some out. How many beads of each color do we have? Which color beads do we have the most of, which do we have the least of? To answer these questions at a glance, we can line the beads up, as indicated in Figure 11.1. This creates a real graph.

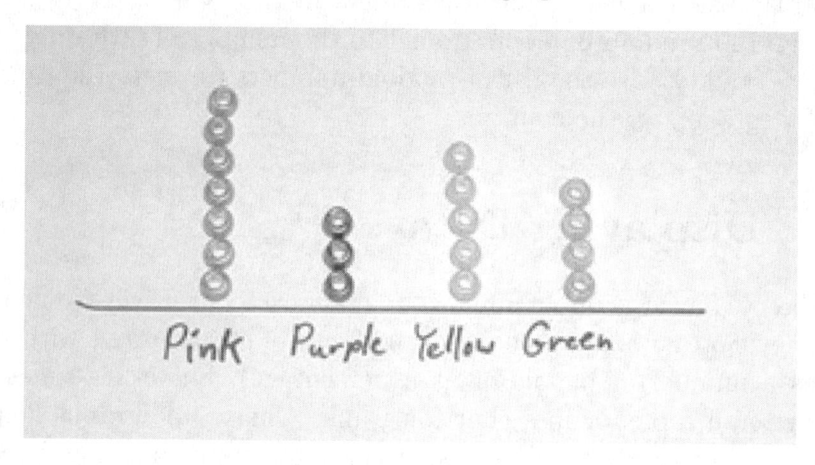

Figure 11.1: A Real Graph: Beads of Different Colors

A real graph can be turned into a pictograph by replacing the objects with pictures of the objects, as in Figure 11.2.

When we want to represent a large number of objects in a pictograph it usually makes sense to allow each picture to stand for more than one object. Each picture could stand for 2, 5, 10, 100, 1000, or some other convenient number of objects. If each picture stands for more than 1 object, then a key near the graph should show clearly how many objects the picture stands for.

Suppose a 5th grade class of students in Georgia has penpals in Wyoming, Idaho, Oregon, and Washington state. The class can look up the populations of these states on the internet, by going to the U.S. Census Bureau homepage at `www.census.gov/`:

Georgia	8,383,915
Washington	5,987,973
Oregon	3,472,867
Idaho	1,321,006
Wyoming	494,423

Although the table above shows the (estimated) populations quite precisely, to get a good feel for the relative populations of the states, the class can first round the populations to the nearest half-million, and then draw a pictograph, as in Figure 11.3. In this pictograph, the picture of a person represents 1 million people, so half a million people can be represented by

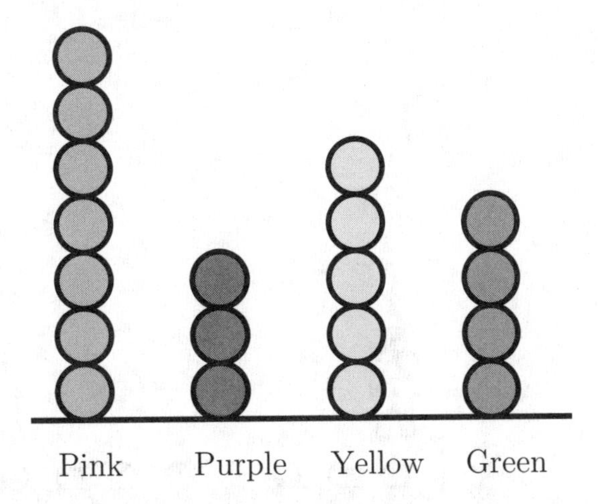

Figure 11.2: A Pictograph: Beads of Different Colors

half a person. In the pictograph we can see at a glance how much bigger in population some of these states are than others.

Pictographs can be horizontal, as in Figure 11.3, or vertical, as in Figure 11.4.

Bar Graphs

A **bar graph**, or **bar chart** is essentially just a streamlined version of a pictograph. A bar graph shows at a glance the relative sizes of different categories. The pictographs of Figures 11.3 and 11.4 have been turned into horizontal and vertical bar graphs in Figure 11.5.

A **double bar graph** is a bar graph in which each category has been subdivided into two subcategories. For example, if the bar graph is about people, then the two subcategories might be "male" and "female". Figure 11.6 shows mean annual earnings in 1989 of full-time year-round workers, 25–29 years old. The different colored bars for men and women show the different earnings of these two groups in each of the different educational levels. The data used in constructing the double bar graph was gathered by the U.S. Census Bureau and is available on the internet at
`govinfo.library.orst.edu/earn-stateis.html`

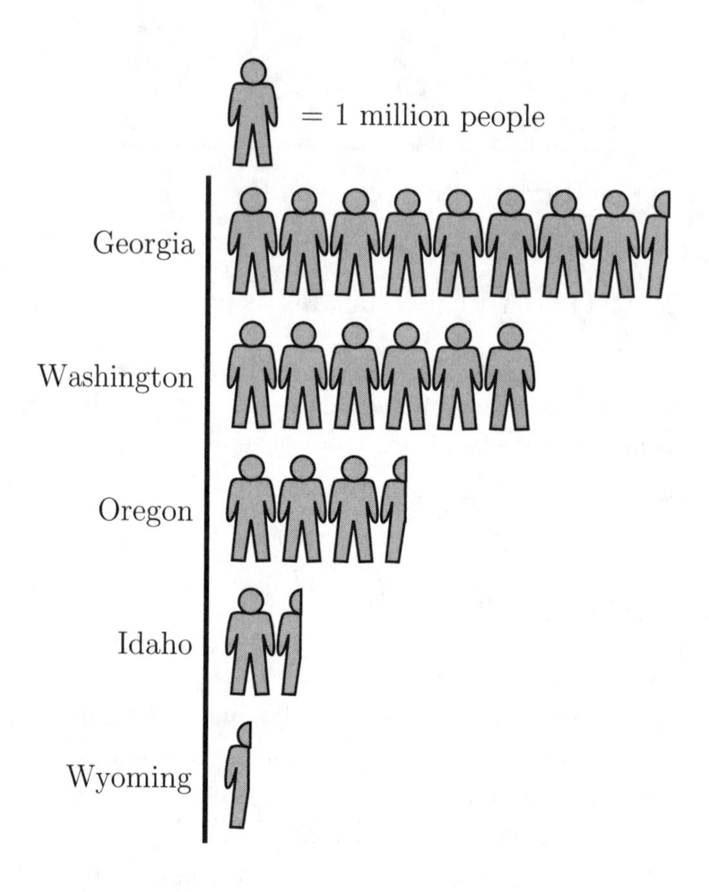

Figure 11.3: Pictograph: Current Populations of Some States

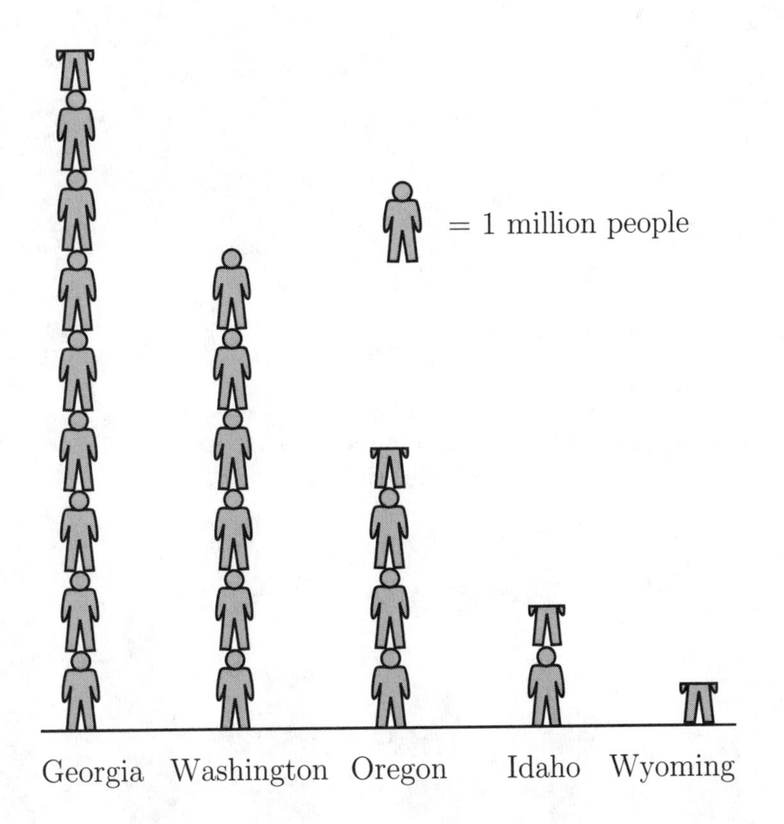

Figure 11.4: Vertical Pictograph: Current Populations of Some States

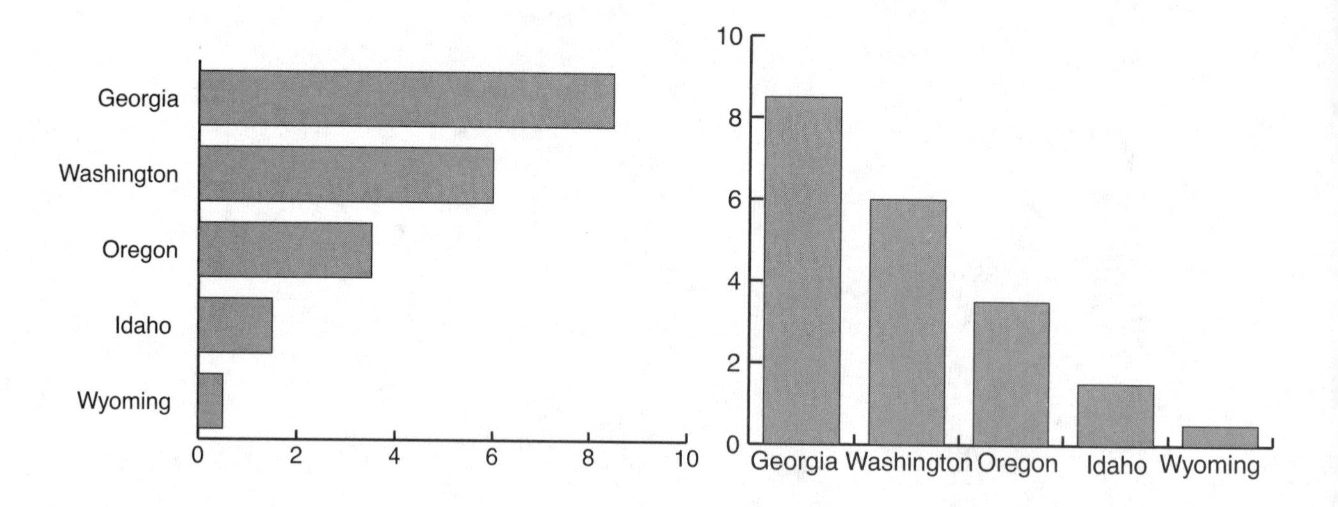

Figure 11.5: Bar Graphs: Current Populations of Some States

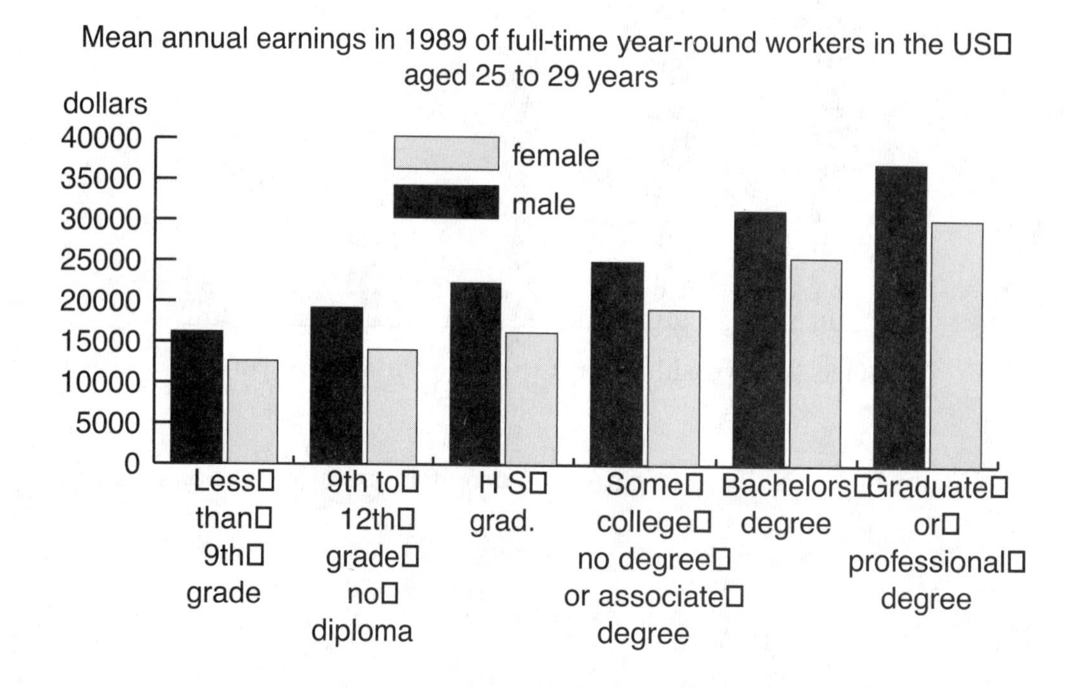

Figure 11.6: Double Bar Graph: Mean Annual Earnings by Education and Sex

Line Plots

A **line plot** is a kind of pictograph in which the categories are numbers, and the pictures are Xs. The Xs function as tally marks.

Let's say we have a bag of blackeyed peas and a measuring teaspoon. How many peas are in a teaspoon? We can scoop out a teaspoon of peas and count the number of peas. If we do it again and again, we may get different numbers of peas in a teaspoon. The following numbers could be the number of peas in a teaspoon that we get from many different trials:

$$18, 19, 19, 17, 19, 18, 20, 18, 19, 19, 20, 20, 19,$$

$$20, 18, 19, 18, 17, 18, 20, 18, 18, 16, 18, 19.$$

We can organize this data in a line plot, as shown in Figure 11.7 Each X in the line plot represents a number in the list above. We can see right away from the line plot that a teaspoon of blackeyed peas usually consists of 18 or 19 peas.

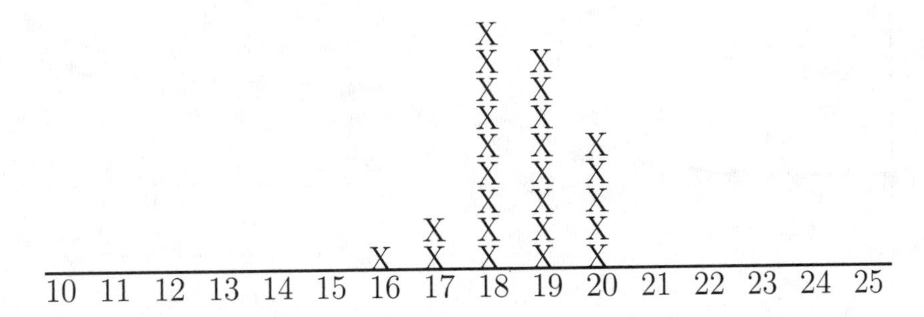

Figure 11.7: Line Plot: The Number of Blackeyed Peas in a Teaspoon

Line plots are easy to draw, and you can even draw a line plot as you collect data—you don't necessarily have to write your data down first before you draw a line plot. For this reason, line plots can be a quick and handy way to organize numerical data.

Line Graphs

A **line graph** is a graph in which adjacent data points are connected by a line. Often, a line graph is just a graph of a function, such as the graph of a population of a region over a period of time, or the graph of the temperature of something over a period of time.

Line graphs are appropriate for displaying "continuously varying" data, such as data that varies over a period of time. Figure 11.8 shows that the average math scores of age 9 children have been slowly increasing, as measured by NAEP, the National Assessment of Educational Progress. Notice that the points that are plotted in Figure 11.8 indicate that the NAEP math test was given in 1978, 1982, 1986, 1990, 1992, 1994, 1996, and 1999. It makes sense to connect these points because if the test were given at times in between, and if the children's average math scores were plotted as points, these points would probably be close to the lines drawn in the graph of Figure 11.8.

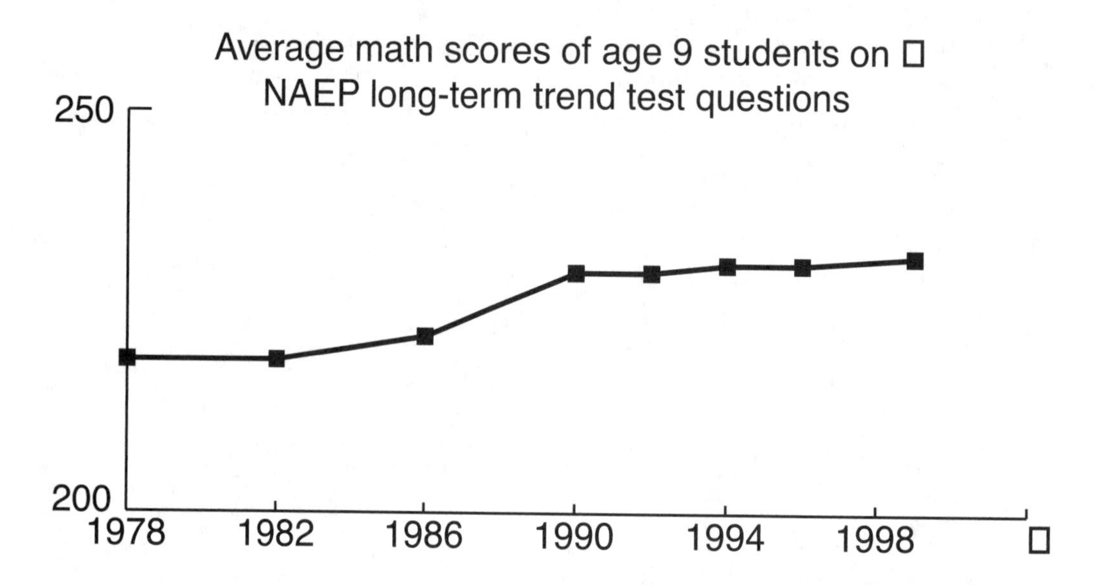

Figure 11.8: Line Graph: Math NAEP Scores of Age 9 Children

Line graphs are not appropriate for displaying data of a "categorical" nature. For example, it would not be appropriate to use a line graph to display the data on state populations that is displayed in the pictographs and bar graphs of Figures 11.4 and 11.5. Figure 11.9 shows an inappropriate line graph of this sort. It doesn't make sense to use a line graph in this case because there is no logical reason to connect the state populations.

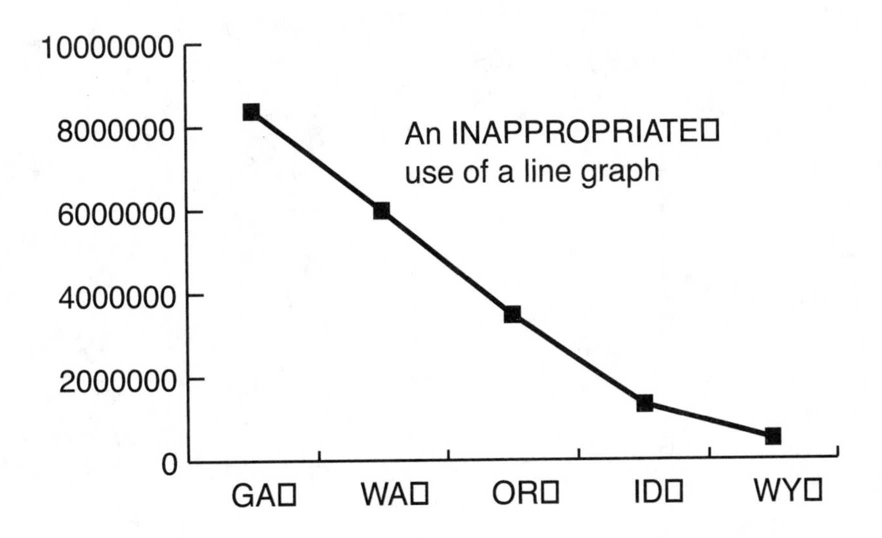

Figure 11.9: An Inappropriate Use of a Line Graph: Connecting State Populations

Pie Charts

A **pie chart** or **pie graph** uses a subdivided circle to show how data partitions into categories. Figure 11.10 shows in a pie chart how U.S. businesses in 1999 were divided according to how many employees they had. We can see at a glance that the vast majority of businesses had fewer than 20 employees. The data for this pie graph were taken from
`www.census.gov/pub/epcd/cbp/view/us99.txt`
from the U.S. Census Bureau (see `www.census.gov/`).

By their nature, pie charts involve percentages or fractions. In fact, simple activities with pie charts can be used to introduce fractions and percents to children. For example, let's say we have a small bag filled with 6 black snap cubes, 8 purple snap cubes, 6 green snap cubes, and 4 white snap cubes. We can arrange these snap cubes evenly around a paper plate, keeping like colors together, as in Figure 11.11 We can then make "pie wedges" showing $\frac{1}{2}$, $\frac{1}{3}$, $\frac{1}{4}$, and $\frac{1}{6}$ of the plate. By filling the plate with pie wedges, we can see that $\frac{1}{3}$ of the snap cubes are purple, $\frac{1}{4}$ are black, $\frac{1}{4}$ are green, and $\frac{1}{6}$ are white.

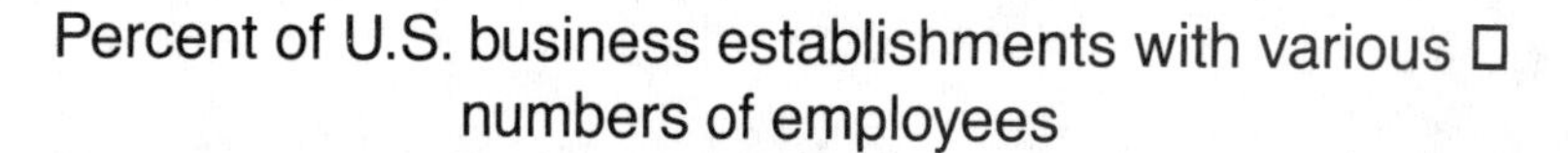

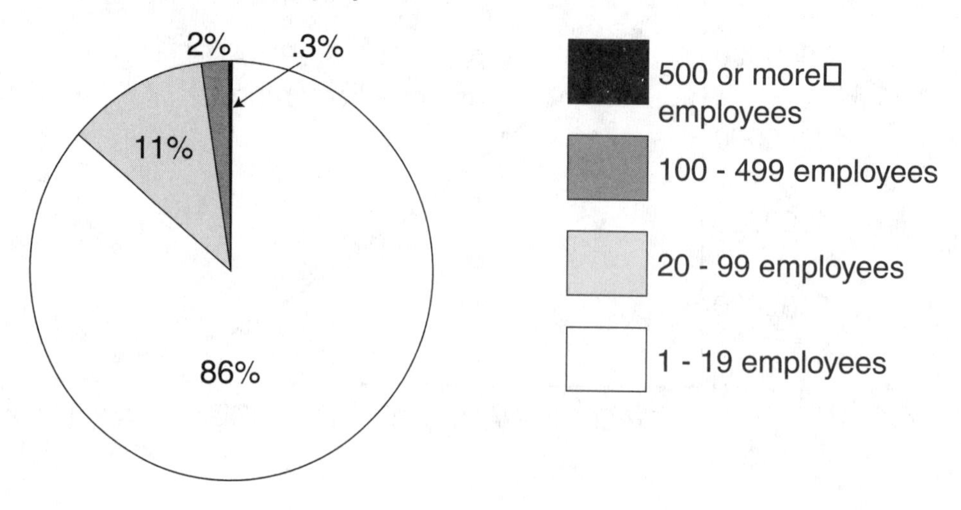

Figure 11.10: Pie Chart: Percent of U.S. Businesses With Various Numbers of Employees

Creating and Interpreting Data Displays

Data displays are created for the purpose of communicating and interpreting data. Therefore, in creating a data display, try to choose a display that will communicate the point you want to make most clearly. Give your display a title, so that the reader will know what it is about. Label your display clearly.

Read data displays carefully before interpreting them. For example, the vertical axis of the line graph in Figure 11.8 starts at 200, not at 0, which can make changes appear greater than they are. For example, you might initially think that the math scores in 1999 were almost 50% higher than they were in 1978. In fact, the math scores in 1999 were only about 6% higher than they were in 1978.

Be careful about drawing conclusions about cause and effect from a data display. For example, in Figure 11.6, we see that women earn less money than men in all categories of education. We might be tempted to conclude that this is due to discrimination against women on the basis of sex. While this could be the case, there could also be other factors affecting the difference in

Figure 11.11: A Pie Chart in a Paper Plate

earnings. For example, many of the women might have taken time off from work in order to raise children. This could cause many women to have less work experience, and therefore earn less. In any case, further study would be needed to determine the cause of the difference in earnings.

Although we can read specific numerical information from data displays, their main value is in showing qualitative information about the data. So when you read a data display, you should look for qualitative information that it conveys, and think about the implications of this information. For example, the pie chart in Figure 11.10 shows that only a tiny fraction of U.S. businesses have 100 or more employees. Therefore a law that would only apply to businesses with 100 or more employees would not affect the vast majority of businesses. On the other hand, from the display in Figure 11.10, we do not know what percent of all *employees* work in businesses with 100 or more employees. Further data would need to be gathered to determine this.

Class Activity 11D: What do You Learn From the Display?

Class Activity 11E: Display These Data

Class Activity 11F: What is Wrong With These Displays?

Class Activity 11G: Three Levels of Questions about Graphs

Class Activity 11H: Toy Car on a Ramp

Class Activity 11I: Floating on Air

Exercises for Section 11.2 on Displaying Data

1. A class plays a fishing game in which there is a large tub filled with plastic fish that are identical, except that some are red and some are white. A child is blindfolded and pulls 10 fish out of the tub. The child writes down how many of each color fish they got, and then puts the fish back in the tub. Each child takes a turn. The results are shown in the table below.

Name	Fish
Michelle	9 red, 1 white
Tyler	7 red, 3 white
Antrice	7 red, 3 white
Yoon-He	6 red, 4 white
Anne	6 red, 4 white
Peter	6 red, 4 white
Brandon	9 red, 1 white
Brittany	6 red, 4 white
Orlando	4 red, 6 white
Chelsey	7 red, 3 white
Sarah	8 red, 2 white
Adam	7 red, 3 white
Lauren	6 red, 4 white
Letitia	9 red, 1 white
Jarvis	7 red, 3 white

Show several different ways to display this data.

2. When is it appropriate to use a line graph?

3. If you have data consisting of percentages, is it always possible to display this data in a single pie graph?

Answers to Exercises for Section 11.2 on Displaying Data

1. One simple way to display the data is with a pictograph, as in Figure 11.12.

   This pictograph can be turned into a bar graph, as in the text (in this case, each bar could be part red, part white).

   We can also create a different kind of display: one which shows how many children had each of the possible arrangements of 10 fish, as does the line plot in Figure 11.13. In this line plot, we can see at a glance that most of the children picked 6 or 7 red fish (and 4 or 3 white fish).

2. See text.

Name	Fish
Michelle	● ● ● ● ● ● ● ● ● ○
Tyler	● ● ● ● ● ● ● ○ ○ ○
Antrice	● ● ● ● ● ● ● ○ ○ ○
Yoon-He	● ● ● ● ● ● ○ ○ ○ ○
Anne	● ● ● ● ● ● ○ ○ ○ ○
Peter	● ● ● ● ● ● ○ ○ ○ ○
Brandon	● ● ● ● ● ● ● ● ● ○
Brittany	● ● ● ● ● ● ○ ○ ○ ○
Orlando	● ● ● ● ○ ○ ○ ○ ○ ○
Chelsey	● ● ● ● ● ● ● ○ ○ ○
Sarah	● ● ● ● ● ● ● ● ○ ○
Adam	● ● ● ● ● ● ● ○ ○ ○
Lauren	● ● ● ● ● ● ○ ○ ○ ○
Letitia	● ● ● ● ● ● ● ● ● ○
Jarvis	● ● ● ● ● ● ● ○ ○ ○

Figure 11.12: A Pictograph

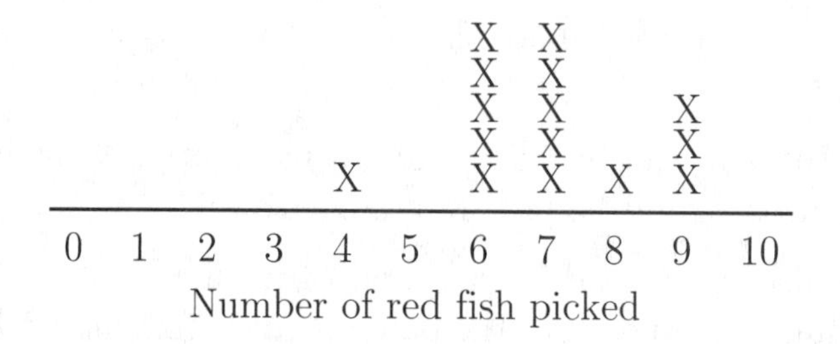

Figure 11.13: Number of Children Picking the Given Numbers of Fish

3. No, it is not always possible to display data consisting of percentages in a single pie graph. If the percentages aren't *separate parts of the same whole*, then you can't display them in a pie graph. For example, the following table on children's eating is based on the table found on the internet at
`www.childstats.gov/ac2001/ECON4Dtbl.asp`

Food	Percent of 4–6 year olds meeting the dietary recommendation for the food
Grains	27%
Vegetables	16%
Fruits	29%
Saturated fat	28%

Even though the percentages add to 100%, they probably do not represent *separate* groups of children. For example, many of the 16% of the children who meet the recommendations on vegetables probably also meet the recommendations on fruits. Therefore it would not be appropriate to display this data in a pie graph.

Problems for Section 11.2 on Displaying Data

1. The line graph in Figure 11.14 was taken from the internet at
`www.childstats.gov/ac1999/poptxt.asp#pop2`
which is maintained by the Federal Interagency Forum on Child and Family Statistics (see `www.childstats.gov`). Write a paragraph discussing what you can learn from this graph, and discussing possible implications of the information displayed in this graph. Your discussion should go beyond a simple numerical reading of the graph.

2. The National Assessment of Education Progress (NAEP), often called "the Nation's Report Card", is a national assessment of what children know and can do in various subjects. Information about NAEP is available on the internet at
`nces.ed.gov/nationsreportcard/`
Using the internet, find the NAEP 1999 Long-Term Trend Summary

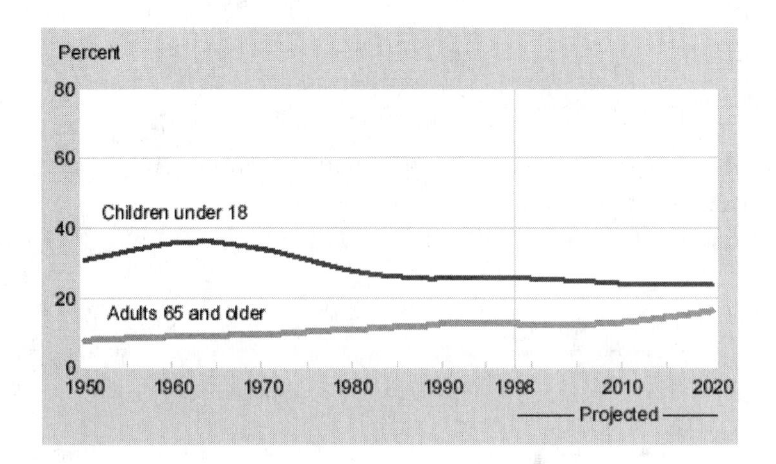

Figure 11.14: Percent Population of Children Under 18 and Adults over 65

Data Tables, which should be at
`nces.ed.gov/naep3/tables/Ltt1999/`
By clicking in the boxes for "scale scores and performance data", you can view long term trend data in mathematics, reading, and science. Browse through this data. Notice that it includes a variety of information, such as data on television watching, time spent on homework, reading practices in and out of school, and use of scientific instruments, among other items. Select some data that interests you, write a paragraph about what the data tells you, and make a display of the data to help convey your point.

3. The National Assessment of Education Progress (NAEP), often called "the Nation's Report Card", is a national assessment of what children know and can do in various subjects. Information about NAEP is available on the internet at
`nces.ed.gov/nationsreportcard/`
Using the internet, find NAEP data about states, which should be at
`nces.ed.gov/nationsreportcard/states/`
Gather some data about one or several states that you find interesting, write a paragraph about what the data shows, and make a display of the data to help convey your point.

4. (a) Describe *in detail* an activity suitable for use with elementary

school children in which the children gather data and create an appropriate display for the data.

(b) Write several questions that you could ask the children about the proposed graph in part (a), once the graph is completed. Your questions should aim to help the children learn to "read" the graph for information.

5. At a math center in a class, there is a bag filled with 40 red blocks and 10 blue blocks. Each child in the class of 25 will do the following activity at the math center: pick a block out of the bag without looking, record the block's color, and put the block back into the bag. Each child will do this 10 times in a row. Then the child will write on an index card the number of red blocks picked, and the number of blue blocks picked.

   (a) Describe a good way to display the *whole class's data* using one of the types of graphical displays discussed in this section. Your proposed display should be a realistic and practical way show *every* child's piece of data (in a class of 25).

   (b) Write several questions that you could ask the children in the class about the graph in part (a), once the graph is completed. Your questions should aim to help the children learn to "read" the graph for information.

11.3 The Center of Data: Average and Median

When we have a collection of numerical data, such as the collection of scores on a test, or the heights of a group of 6 year old children, our first questions about the data are usually "what is typical or representative of this data? what is the center of this data?". When a data set is large, it is especially helpful to know the answers to these questions. For example, if the data consists of the test scores of all students taking the SAT in a given year, no one person would want to look at the entire data set (even if they could); it would be overwhelming, and impossible to comprehend. Instead, we want ways of understanding the nature of the data.

To say what is representative, in the center, or typical of a list of numbers, we commonly use either the *average*, the *median*, or the *mode*.

To calculate the **average** (also called the **arithmetic mean** or just plain **mean**) of a list of numbers, add all the numbers and divide this sum by the number of numbers in the list. For example, the average of

$$7, \quad 10, \quad 11, \quad 8, \quad 10$$

is

$$(7 + 10 + 11 + 8 + 10) \div 5 = 46 \div 5 = 9.2.$$

We divide by 5 because there are 5 numbers in the list $7, 10, 11, 8, 10$.

To calculate the **median** of a list of numbers, organize the list from smallest to largest (or vice versa). The number in the middle of the list is the median. If there is no number in the middle, then the median is half way between the two middle numbers. For example, to find the median of

$$7, \quad 10, \quad 11, \quad 8, \quad 10,$$

rearrange the numbers from smallest to largest:

$$7, \quad 8, \quad \boxed{10}, \quad 10, \quad 11.$$

The number 10 is exactly in the middle, so it is the median. There are 6 numbers in the list

$$3, \quad 3, \quad \boxed{4, \quad 5}, \quad 5, \quad 6,$$

so there is no single number in the middle. The median is halfway between the two middle numbers 4 and 5, and therefore the median is 4.5.

To calculate the **mode** of a list of numbers, reorganize the list so as to group equal numbers in the list together. The mode is the number that occurs most frequently in the list. For example, to find the mode of

$$11, \quad 9, \quad 10, \quad 8, \quad 11, \quad 10, \quad 11, \quad 9, \quad 12$$

first reorganize the list:

$$8, \quad 9, \quad 9, \quad 10, \quad 10, \quad 11, \quad 11, \quad 11, \quad 12.$$

The number 11 occurs 3 times, which is more than any other number on the list. Therefore 11 is the mode of this list of numbers. Statisticians usually do not use the mode because relatively small changes in data can produce large changes in the mode.

Class Activity 11J: The Average as "Making Even" or "Leveling Out"

Class Activity 11K: The Average as "Balance Point"

Class Activity 11L: Same Median, Different Average

Class Activity 11M: Same Average, Different Median

Class Activity 11N: Can More than Half be Above Average?

Exercises for Section 11.3 on the Average and the Median

1. Explain how to use physical objects to describe the average of a data set as "leveling out" the data set.

2. Explain how to show the average of a data set as the "balance point" of the data set.

3. Explain why the average of a list of numbers must always be in between the smallest and largest numbers in the list.

4. Explain why the average of two numbers is exactly half way between the two numbers.

5. Juanita read an average of 3 books a day for 4 days. How many books will Juanita need to read on the 5th day so that she will have read an average of 5 books a day over 5 days? Solve this problem in several ways and explain your solutions.

6. George's average score on his math tests in the first quarter is 60. George's average score on his math tests in the second quarter is 80. George's semester score in math is the average of all the tests George took in the first and second quarters. Can George necessarily calculate his semester score by averaging 60 and 80? If so, explain why; if not, explain why not, explain what other information you would need to calculate George's semester average, and show how to calculate this average.

7. Suppose that all 4th graders in a state take a writing competency test that is scored on a 5 point scale. Is it possible for 80% of the 4th graders to score below average? If so, show how that could occur; if not, explain why not.

Answers to Exercises for Section 11.3 on the Average and the Median

1. See Class Activity 11J.

2. See Class Activity 11K.

3. If we think of the average of a list of numbers as its "balance point", then clearly the average must be in between the largest and smallest numbers on the list because the balance point must be in between those numbers.

   Similarly, if we think of the average as "leveling out" the numbers in a list, then with this point of view it's also clear that the average must be in between the smallest and largest numbers in the list.

4. If we think of the average of two numbers as the balance point between those numbers, then it's clear that the average is exactly half way between those numbers because the balance point is exactly half way between the numbers.

5. Method 1: Imagine putting the books that Juanita has read into 4 stacks, one for each day. Because she has read an average of 3 books per day, the books can be redistributed so that there are 3 books in each of the 4 stacks. This is because when you add the number of books Juanita has read and divide by 4, the result is the average, which is 3; in other words, when you put all of Juanita's books into 4 equal groups, there are 3 books in each group. Now picture a 5th stack with an unknown number of books in it—the number of books Juanita must read to make the average 5 books per day. If the books in this 5th stack are distributed among all 5 stacks so that all 5 stacks have the same number of books, then this common number of books is the average number of books Juanita will have read over 5 days. Therefore this 5th stack must have 5 books plus another $4 \times 2 = 8$ books, so as to

distribute 2 books to each of the other 4 stacks. So Juanita must read $5 + 8 = 13$ books on the 5th day.

Method 2: To read an average of 5 books a day over 5 days, Juanita must read a total of 25 books. This is because

$$\text{(total books)} \div 5 = \text{average},$$

therefore

$$\text{(total books)} \div 5 = 5,$$

and so

$$\text{total books} = 5 \times 5 = 25.$$

Similarly, because Juanita has already read an average of 3 books per day for 4 days, therefore she has read a total of $4 \times 3 = 12$ books. Therefore Juanita must read $25 - 12 = 13$ books on the 5th day.

6. No, George cannot necessarily calculate his semester score by averaging 60 and 80. This is because we don't know if George took the same number of tests in the first and second quarters. To calculate George's average score over the whole semester, we need to know how many tests he took in the first and second quarters. Suppose George took 3 tests in the first quarter and 5 tests in the second quarter. Then the sum of his first three test scores is $3 \times 60 = 180$, and the sum of his next 5 test scores is $5 \times 80 = 400$, so the sum of all 8 of his test scores is $180 + 400 = 580$. Therefore George's average score on all 8 tests is $580 \div 8 = 72.5$. Notice that this is not the same as the average of 60 and 80, which is only 70.

7. Yes, it is possible for 80% of the 4th graders to score below average. For example, suppose there are 200,000 4th graders and that:

   10,000 score 1 point,
   150,000 score 2 points,
   10,000 score 3 points,
   25,000 score 4 points, and
   5,000 score 5 points,

   as shown in the bar graph of Figure 11.15. Then you can see from the bar graph that the "balance point" is between 2 and 3, and so 160,000 children, namely 80% of the 200,000 children, score below this average.

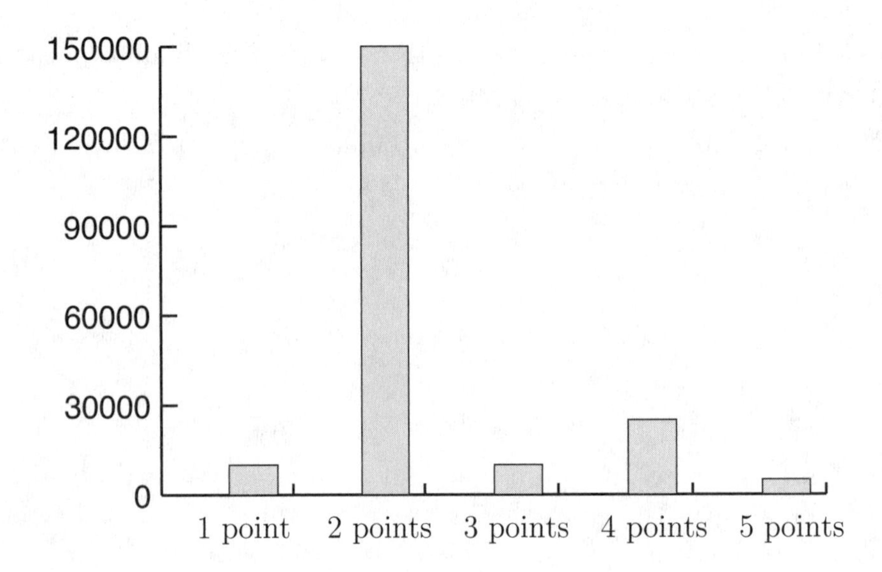

Figure 11.15: Scores on a Writing Test

Problems for Section 11.3 on the Average and the Median

1. John's average annual income over a 4-year period was \$25,000. What would John's average annual income have to be for the next 3 years so that his average annual income over the 7-year period would be \$50,000? Explain your solution clearly.

2. Julia's average on her first 3 math tests was 80. Her average on her next 2 math tests was 95. What is Julia's average on all 5 math tests? Solve this problem in two different ways and explain your solutions clearly.

3. A teacher gives a 10-point test to a class of 10 children.

   (a) Is it possible for 9 of the 10 children to score above average on the test? If so, give an example to show how, if not, explain clearly why not.

   (b) Is it possible for all 10 of the children to score above average on the test? If so, give an example to show how, if not, explain clearly why not.

(c) If the average score on the test is 8, what are the lowest and highest possible median scores on the test? Explain your answers briefly.

4. (a) In Ritzy county, a large percentage of the population has a high annual income and a small percentage of the population has a low annual income. Which would you expect to be greater, the median annual income or the average annual income in Ritzy county? Explain.

   (b) In Normal county, a large percentage of the population has a low annual income and a small percentage of the population has a high annual income. Which would you expect to be greater, the median annual income or the average annual income in Normal county? Explain.

5. In Ritzy county, the average annual household income is $100,000. In neighboring Normal county, the average annual household income is $30,000. Does it follow that in the two-county area (consisting of Ritzy county and Normal county), the average annual household income is the average of $100,000 and $30,000? If so, explain why; if not, explain why not, explain what other information you would need in order to calculate the average annual household income in the two-county area, and show how to calculate this average if you had this information.

6. In county A, the average score on the grade 5 Iowa Test of Basic Skills (ITBS) was 50. In neighboring county B, the average score on the grade 5 ITBS was 71. Can you conclude that the average score on the grade 5 ITBS in the two-county region (consisting of counties A and B) can be calculated by averaging 50 and 71? If so, explain why; if not, explain why not, explain what other information you would need to know in order to calculate the average grade 5 ITBS score in the two-county region, and show how to calculate this average if you had this information.

7. The **average speed** of a moving object during a period of time is the distance the object traveled, divided by the length of the time period. For example, if you left on a trip at 1pm, arrived at 3:15 pm and drove 85 miles, then the average speed for your trip would be

$$85 \div 2.25 \text{ mph} = 37.8 \text{ mph}.$$

(a) Jane drives 72 miles from Athens, Georgia to Atlanta, Georgia with an average speed of 48 miles per hour. When she gets to Atlanta, Jane turns around immediately, and heads back to Athens along the same route. Using this information, find several different possible average speeds for Jane's *entire trip* from Athens back to Athens.

(b) Ignoring practical issues, such as speed limits and how fast cars can go, would it be theoretically possible for Jane to average 100 miles per hour for the whole trip from Athens to Atlanta and back that is described in part (a)? If so, explain how, if not, explain why not.

(c) Realistically, what are the largest and smallest possible average speeds for Jane's entire trip from Athens back to Athens that is described in part (a)? Explain your answers clearly.

11.4 The Spread of Data: Percentiles

The average and the median of numerical data tell us about the "center" of the data. But very different looking data can have the same median or average. The line plots in Figure 11.16 represent hypothetical test scores on a 10-point quiz. In all three line plots, the median and the average are both 7. But the way the test scores are distributed is very different in the three cases. Therefore it is useful to have ways of summarizing how numerical data is "spread out". In this brief section, we will study one tool for reporting how data is spread out, namely percentiles. Statisticians often use another concept for studying the spread of data, that of **standard deviation**. We will not study standard deviations here, but almost any book on statistics, such as [1], discusses the topic.

A **percentile** is a generalization of the median. The 90th percentile of a set of numerical data is the number such that 90% of the data is less than or equal to that number and 10% is greater than or equal to that number. (If there is not one single number that satisfies this criterion, then use the midpoint of all numbers that do, as in the median.) Other percentiles are defined similarly. The median of a list of numbers is the 50th percentile, because 50% of the list is at or below the median, and 50% is at or above the median.

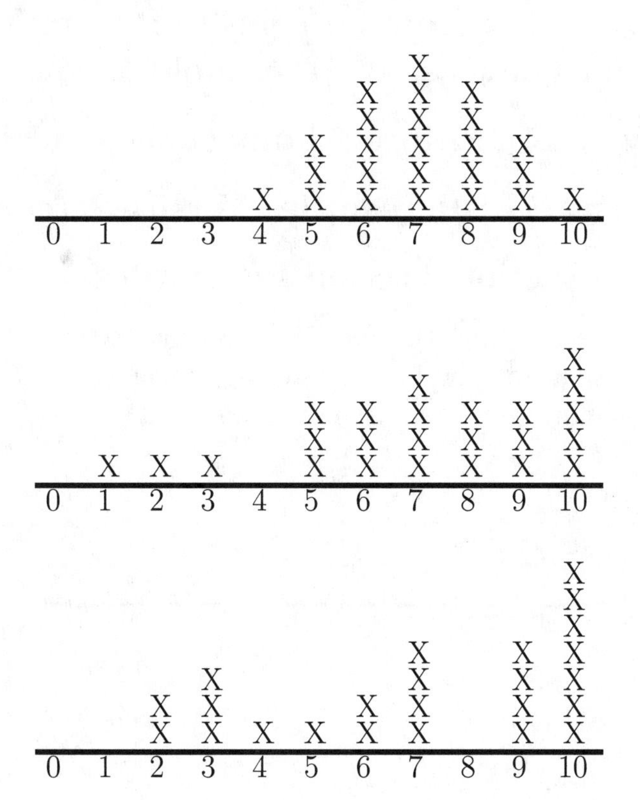

Figure 11.16: Three Different Hypothetical Test Results Each With Median 7 and Average 7

Results of standardized tests are often reported as percentiles. So if a student scores in the 75th percentile on a test, then that student has scored the same as or better than 75% of the students taking the test.

Class Activity 11O: Determining Percentiles

Class Activity 11P: How Percentiles Inform You About Data Spread: The Case of Household Income

Class Activity 11Q: More About Household Income

Class Activity 11R: Percentiles Versus Percent Correct

Exercises for Section 11.4 on Percentiles

1. Determine the 25th, 50th, and 75th percentiles for the hypothetical test scores shown in the line plots of Figure 11.17. Discuss how these percentiles indicate the way the data is spread.

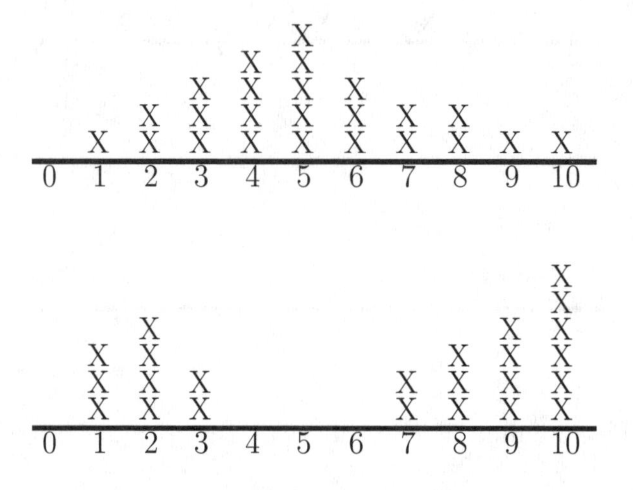

Figure 11.17: Hypothetical Test Results

2. Figure 11.18 shows bar graphs of hypothetical scores on a 400 point test given at two different schools. Determine approximately the 20th, 40th, 60th, and 80th percentiles for the test at each of the schools. Discuss how the percentiles indicate the way the data is spread.

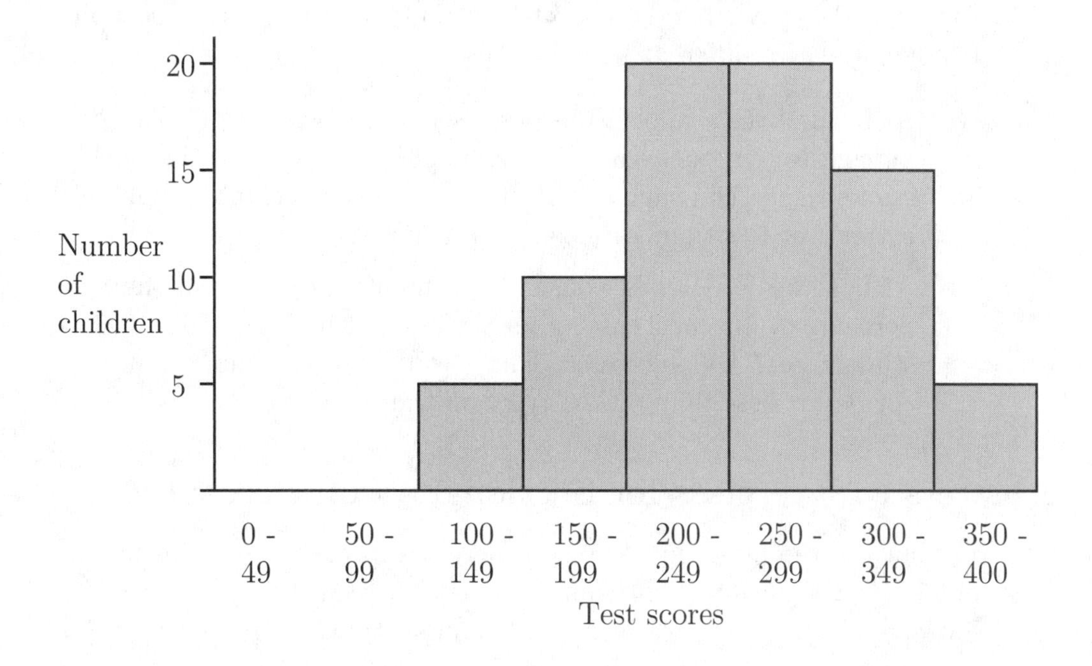

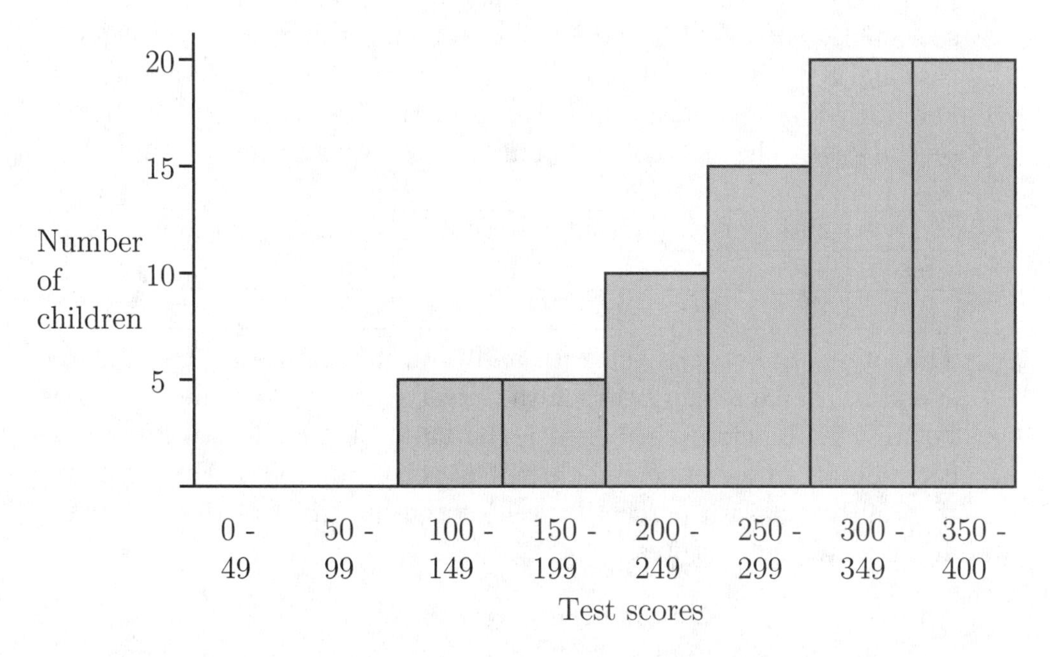

Figure 11.18: Hypothetical Test Results at Two Schools

3. Figure 11.19 shows bar graphs of hypothetical scores on a 400 point test given at two different schools.

   (a) At both of the schools, consider a student who scored at the 20th percentile. In each case, approximately what score did such a student get? Did such a student get 20% correct, more than 20% correct, or less than 20% correct on the test?

   (b) At both of the schools, consider a student who scored at the 80th percentile. In each case, approximately what score did such a student get? Did such a student get 80% correct, more than 80% correct, or less than 80% correct on the test?

Answers to Exercises for Section 11.4 on Percentiles

1. For both of the hypothetical tests, there are 24 test scores. Since 25% of 25 is 6, the 25th percentile is the score between the score of the 6th lowest and 7th lowest test scores. Therefore the 25th percentile is 3.5 in the first case, and 2 in the second case. The 50th percentile is the score between the 12th and 13th lowest scores, which is 5 in the first case and 8 in the second case. The 75th percentile is between the 18th and 19th lowest scores, which is 6.5 in the first case, and 9.5 in the second case. The table below summarizes these results.

	25th percentile	50th percentile	75th percentile
Test 1	3.5	5	6.5
Test 2	2	8	9.5

   The large gap between the 25th and 50th percentiles for test 2 correspond to the large gap between the lower scores and the higher scores. For test 1, the closeness between the 25th, 50th and 75th percentiles corresponds to the bunching of the test scores around 5. For test 2, the high 50th and 75th percentiles reflect the bunching of the test scores around the higher scores.

2. An easy way to determine the percentiles is to divide the bar graph into squares, as in Figure 11.19. Each square represents 5 students. In each bar graph, there are 15 such squares, so 20% of the students is represented by 3 squares. Therefore for each school, the 20th percentile occurs where 3 squares are to the left and the remaining 12 squares are

to the right, the 40th percentile occurs where 6 squares worth are to the left and the remaining 9 squares worth are to the right, and so on for the 60th and 80th percentiles as well.

So in the first school, the 20th percentile is 200; the 40th percentile should be about $\frac{3}{4}$ of the way between 200 and 250, which is about 240; the 60th percentile should be about half way between 250 and 300, at about 275; and the 80th percentile should be about one third of the way between 300 and 350, at about 315. Similarly for the second school. The percentiles are shown in the following table.

	Approximate values of percentiles			
	20th percentile	40th percentile	60th percentile	80th percentile
School 1	200	240	275	315
School 2	225	285	325	360

The higher values for the 20th, 40th, 60th, and 80th percentiles in the second school correspond to the the way the test scores are shifted toward the higher scores.

3. In the first school, a student scoring at the 20th percentile scored about 200 points, which is 50% of the maximum possible score of 400. At the second school, a student scoring at the 20th percentile probably scored about 225 points, or about 56% of the maximum possible score. Thus we see that scoring in the 20th percentile is not the same as scoring 20% on the test.

   In the first school, a student scoring at the 80th percentile probably scored about 315 points, or about 79% on the test. In the second school, a student scoring at the 80th percentile probably scored about 360 points, or about 90%.

Problems for Section 11.4 on Percentiles

1. What is the difference between scoring in the 90th percentile on a test and scoring 90% correct on a test. Discuss this carefully, giving examples to illustrate.

2. What is the purpose of reporting a child's percentile on a state or national standardized test? How is this purpose different from reporting the child's percent correct on a test?

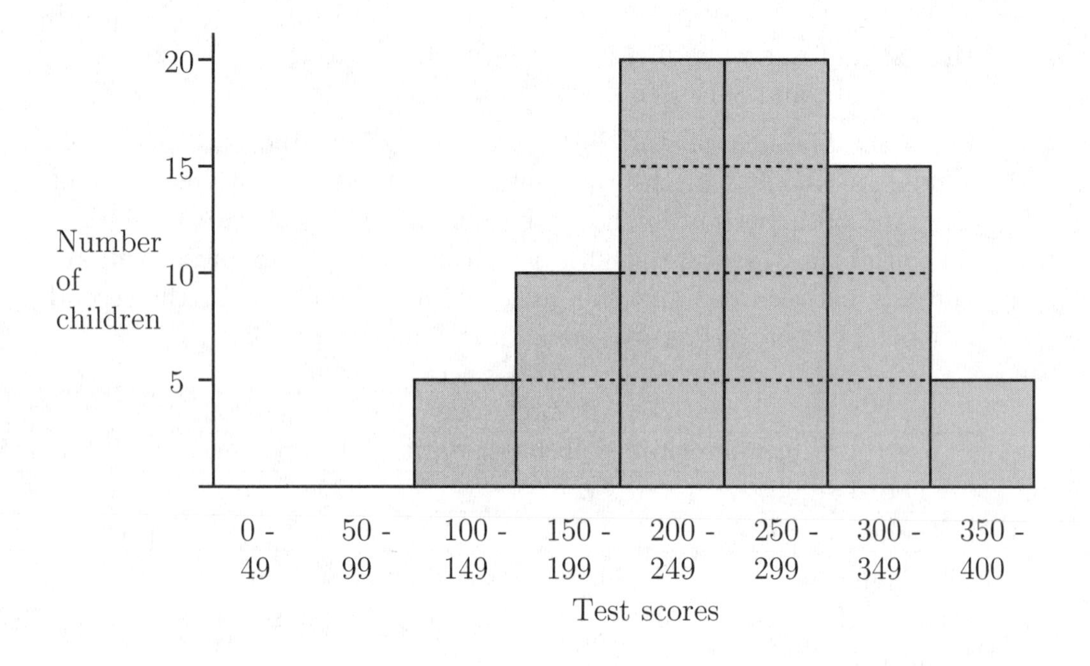

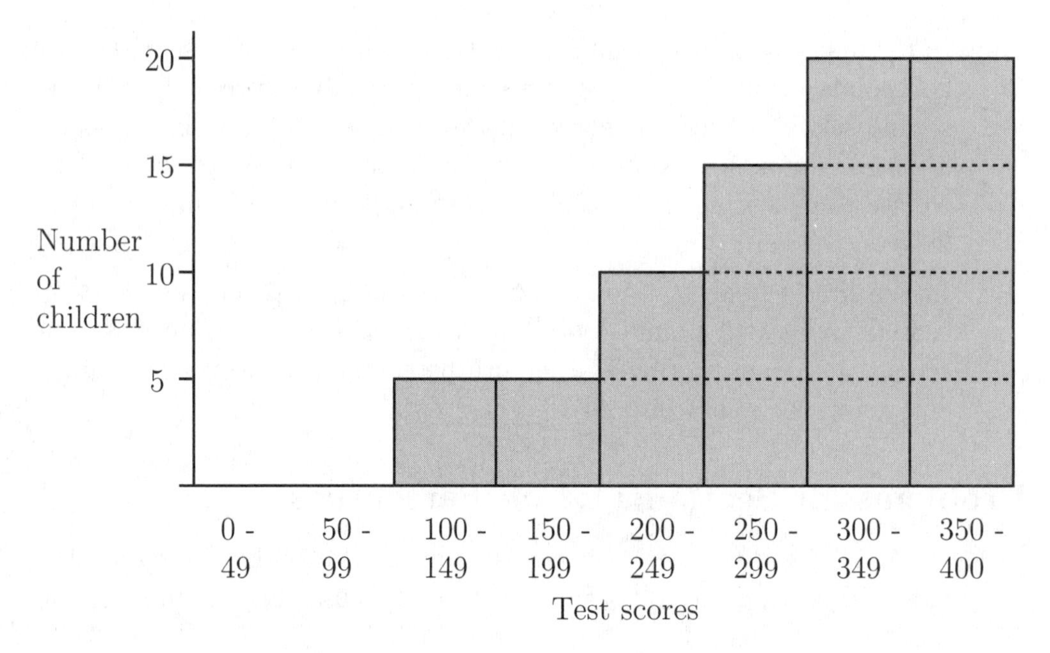

Figure 11.19: Hypothetical Test Results at Two Schools

Chapter 12

Probability

The study of probability arises naturally when we play games in which we flip coins, roll dice, spin spinners, or pick cards. These kind of games involve an element of chance: we don't know what the spinner will land on next, or how the dice will roll. This element of chance adds excitement to a game, and so games of chance can provide a fun way to introduce children to the study of probability.

If probability were only used in analyzing games of chance, it would not be a very important subject. But probability has far wider applications. In business and finance, probability can be used in determining how best to allocate assets. Farmers can use data on the probability of various amounts of rain to determine how best to plant their fields. In medicine, probability can be used to determine how likely it is that a person has a certain disease, given outcomes of test results.

The National Council of Teachers of Mathematics [3] recommends the following concerning the study of probability:

> In grades 3–5 all students should—
> Understand and apply basic concepts of probability
>
> - describe events as likely or unlikely and discuss the degree of likelihood using such words as certain, equally likely, and impossible;
> - predict the probability of outcomes of simple experiments and test the predictions;
> - understand that the measure of the likelihood of an event can be represented by a number from 0 to 1.

12.1 Some Basic Principles of Probability

In this section we will study some of the basic principles of probability and we will see some simple contexts in which to introduce basic ideas of probability. We will also use basic principles of probability to calculate probabilities in simple situations.

Given an experiment or a situation in which various outcomes are possible, the **probability** of a given outcome is a measure of how likely that outcome is. The probability of a given outcome is the fraction or percentage of times that outcome should occur "in the ideal". In reality, when you perform an experiment a number of times, the *actual* fraction of times that the given outcome occurs will usually *not* be equal to the probability of that outcome, but it is usually close to the probability if you performed the experiment many times. When you perform an experiment a number of times, the fraction of times that a given outcome occurs is called the **experimental probability** of that outcome.

Probabilities are always between 0 and 1, or equivalently, when they are given as percentages, between 0% and 100%. A probability of 0% means that outcome will not occur; a probability of 100% means that outcome is certain to occur; a probability of 50% means the outcome is as likely to occur as not to occur. For example, when we flip a coin, the probability of getting *either* a heads *or* a tails is 100% because one or the other is certain to occur. The probability of getting a heads is 50%, or $\frac{1}{2}$, because heads and tails are equally likely to occur.

If we have a coin in hand, how do we *know for sure* that a coin is *fair*, i.e., that heads and tails are equally likely to occur when we flip the coin? In fact, we usually *don't* know for sure that the coin is fair, but unless the coin has been weighted lopsidedly, there is no reason to believe that one of heads or tails should be more likely to occur than the other. If it were important to know for sure that the coin is fair, for example if the coin were to be used in a state lottery, then the coin would be flipped many times, perhaps thousands of times, to determine if the coin is likely to be fair.

There are several key principles about probability that we can use to calculate probabilities:

Principles About Probability

1. If two outcomes of an experiment or situation are equally likely, then their probabilities are equal.

2. If there are several outcomes of an experiment or situation that cannot occur simultaneously, then the probability that either one of those outcomes will occur is the sum of the probabilities of each of the individual outcomes.

3. If an experiment is performed many times, then the fraction of times that a given outcome occurs is likely to be close to the probability of that outcome occuring.

Using Principles of Probability to Calculate Simple Probabilities

Here is an example of how to use the first two principles to calculate a probability. Suppose you have an ordinary die that is in the shape of a cube and has dots on each of its six faces. Each face has a different number of dots on it, ranging from 1 dot to 6 dots. You can roll the die and count the number of dots on the face that lands up. Let's assume that the die is not weighted, so that each face is equally likely to land up. Then by principle 1, the probability of rolling a 1 is equal to the probability of rolling a 2, which is equal to the probability of rolling a 3, and so on, up to 6, in other words,

$$\text{prob. of } 1 = \text{prob. of } 2 = \text{prob. of } 3 = \text{prob. of } 4 = \text{prob. of } 5 = \text{prob. of } 6.$$

Now when you roll a die, the probability of rolling either a 1 or a 2 or a 3 or a 4 or a 5 or a 6 is 1, because *some* face has to land up. And you can't *simultaneously* roll a 2 and a 3 on one die, or any other pair of distinct numbers. Therefore by principle 2,

$$\text{prob. of } 1 + \text{prob. of } 2 + \text{prob. of } 3 + \text{prob. of } 4 + \text{prob. of } 5 + \text{prob. of } 6 = 1.$$

Because all these probabilities are equal and add up to 1, therefore each of them must be $\frac{1}{6}$, i.e.,

$$\begin{aligned}
\text{prob. of } 1 &= \frac{1}{6} \\
\text{prob. of } 2 &= \frac{1}{6} \\
\text{prob. of } 3 &= \frac{1}{6} \\
\text{prob. of } 4 &= \frac{1}{6} \\
\text{prob. of } 5 &= \frac{1}{6} \\
\text{prob. of } 6 &= \frac{1}{6}
\end{aligned}$$

In general, because of the first two principles of probability, if an experiment or situation has N distinct possible outcomes, none of which can occur simultaneously, and if all of these outcomes are equally likely, then for each possible outcome, the probability of that outcome is $\frac{1}{N}$.

Displaying all Outcomes of an Experiment with an Organized List, an Array, or a Tree Diagram to Calculate a Probability

Often, we can use techniques we learned for showing multiplicative structure to display all possible outcomes of an experiment. If all possible outcomes are equally likely, then we can calculate the probability that one of those outcomes will occur by using principles (1) and (2).

For example, suppose you have three coins: a penny, a nickel, and a dime. If you flip all three coins, what is the probability that all three come up heads? First, let's determine how many different possible outcomes there are when we flip the three coins. For example, three heads is one possible outcome. Heads on the penny and nickel, and tails on the dime is another possible outcome. We could show all possible outcomes in an organized list, or we can use a tree diagram, as in Figure 12.1.

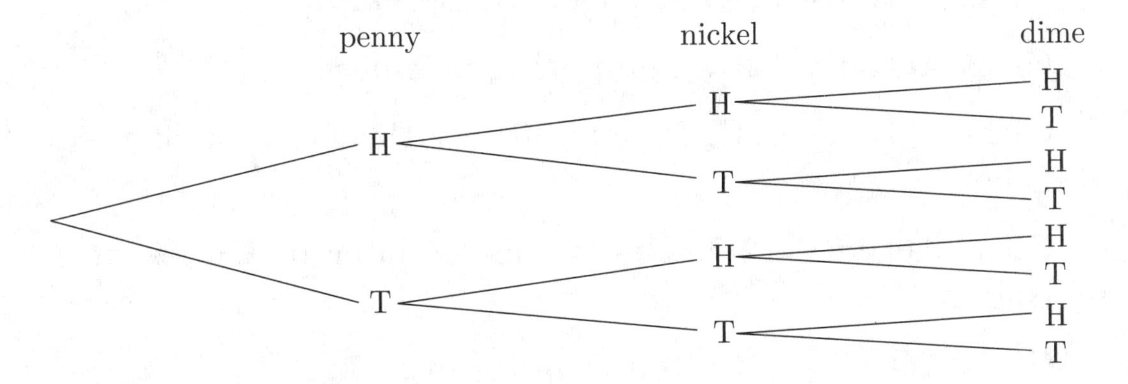

Figure 12.1: Flipping Three Coins

For each of the two outcomes for the penny, the nickel has two possible outcomes, so there are

$$2 \times 2$$

possible outcomes for the penny and nickel. For each of the 2×2 possible outcomes for the penny and nickel, there are two possible outcomes for the dime. So all together, there are

$$2 \times 2 \times 2 = 8$$

total possible outcomes when we flip the three coins.

Next, since pennies, nickels, and dimes are equally likely to come up heads as tails, all of these eight possible outcomes shown in the tree diagram are equally likely. Therefore, according to principles (1) and (2), the probability of getting three heads when we flip three coins is $\frac{1}{8}$, or 12.5%.

We can use the tree diagram of Figure 12.1 to calculate other probabilities as well. For example, what is the probability of getting two heads and one tails when we toss the three coins? Two heads and one tails occurs in three of the eight possible outcomes, namely HHT, HTH, and THH. According to principle (2), the probability that either one of these three outcomes will occur is

$$\frac{1}{8} + \frac{1}{8} + \frac{1}{8} = \frac{3}{8},$$

or 37.5%.

Class Activity 12A: A Dice Rolling Game

Class Activity 12B: Comparing Spinners

Class Activity 12C: Probability With Different Shaped Dice

Class Activity 12D: Guess the Number of Cubes in a Bag

Class Activity 12E: Ten Pennies in a Bag

Class Activity 12F: Some Probability Misconceptions

Class Activity 12G: Using a Tree Diagram to Calculate a Probability

Exercises for Section 12.1 on Some Basic Principles of Probability

1. Consider the experiment of rolling a die two times.

   (a) List all possible outcomes in a systematic way. How many possible outcomes are there?

(b) How many ways are there of getting a 1 on both rolls of the die? Since all possible outcomes are equally likely, therefore what is the probability of getting a 1 on both rolls of the die?

(c) Count the number of ways of getting a 1 on either the first roll or the second roll, or on both rolls. Therefore what is the probability of getting at least one 1 on two rolls?

(d) Count the number of ways of getting either a a 1 or a 2 on each roll (this includes the cases 1,1 and 1, 2, but not 4, 1 for example). Therefore what is the probability of getting either a 1 or a 2 on each roll?

2. What is the probability of getting four heads in a row on four tosses of a coin?

3. A children's game has a spinner that is equally likely to land on any one of four colors: red, blue, yellow, green.

   (a) What is the probability of getting a red both times on two spins?

   (b) What is the probability of getting red at least once on two spins?

   (c) What is the probability of getting red three times in a row in three spins?

4. Suppose you have four marbles in a bag: three red, and one black. If you reach into the bag without looking and randomly pick out two marbles at once, what is the probability that both of the marbles you pick will be red?

Answers to Exercises for Section 12.1 on Some Basic Principles of Probability

1. (a) To list all possible outcomes, you could draw a tree diagram, or you could make an array like this one:

$$\begin{array}{cccccc} (1,1) & (1,2) & (1,3) & (1,4) & (1,5) & (1,6) \\ (2,1) & (2,2) & (2,3) & (2,4) & (2,5) & (2,6) \\ (3,1) & (3,2) & (3,3) & (3,4) & (3,5) & (3,6) \\ (4,1) & (4,2) & (4,3) & (4,4) & (4,5) & (4,6) \\ (5,1) & (5,2) & (5,3) & (5,4) & (5,5) & (5,6) \\ (6,1) & (6,2) & (6,3) & (6,4) & (6,5) & (6,6) \end{array}$$

Here $(4,3)$ stands for a 4 on the first roll and a 3 on the second roll. There are $6 \times 6 = 36$ possible outcomes.

(b) A 1 on both rolls occurs in only one way (namely $(1,1)$ above). All 36 possible outcomes are equally likely, so by principles (1) and (2), the probability of getting a one on both rolls is $\frac{1}{36}$, or 2.8%.

(c) There are 11 ways (the first row together with the first column) of getting a 1 on either the first roll or the second roll, or both. Therefore, by principle (2), the probability of getting at least one 1 on two rolls of a die is $\frac{11}{36}$, or 30.6% (so you'd expect to get at least one 1 a little less than one third of the time).

(d) There are four ways to get either a 1 or a 2 on both rolls, namely $(1,1), (1,2), (2,1)$, and $(2,2)$. Therefore, by principle (2), the probability of getting either a 1 or a 2 each time on two rolls of a die is $\frac{4}{36} = \frac{1}{9}$, or about 11%.

2. Think of the tree diagram for four tosses of a coin: it will have four stages, two branches coming from each previous branch. So there will be $2 \times 2 \times 2 \times 2\times = 16$ possible outcomes, all of which are equally likely. Only one of those outcomes produces four heads in a row. So by principles (1) and (2), the probability of getting four heads in a row on four tosses of a coin is $\frac{1}{16}$, or 6.25%.

3. Use either a tree diagram or an organized list for the first two questions; think about what the tree diagram looks like for the third question. There is only one way to get red both times on two spins. There are seven ways to get red at least once on two spins.

   (a) $\frac{1}{16}$

   (b) $\frac{7}{16}$

   (c) $\frac{1}{64}$

4. If you think of the three red marbles as labeled *red 1*, *red 2*, and *red 3*, then the following list shows all possible ways of picking a pair of marbles from the bag:

$$\begin{array}{ccc} (R1, R2) & (R1, R3) & (R1, B) \\ & (R2, R3) & (R2, B) \\ & & (R3, B) \end{array}$$

(Notice that $(R2, R1)$ would be the same as $(R1, R2)$. Similarly for other cases.)

These 6 possible ways of picking two marbles are all equally likely. Out of the 6 ways of picking two marbles, there are 3 ways of picking two red marbles. Therefore the probability of picking 2 red marbles out of the bag containing 3 red and 1 black marble is $\frac{3}{6} = \frac{1}{2}$.

Problems for Section 12.1 on Some Basic Principles of Probability

1. Let's say you're making up a game for a fundraiser. You take five ping pong balls, number them from 1 to 5, and put all five in a brown bag. Contestants will pick the five balls out of the bag one at a time, without looking, and line the balls up in the order they were picked. If the contestant picks 1, 2, 3, 4, 5, in that order, then the contestant wins a prize of $2. Otherwise, the contestant wins nothing. Each contestant pays $1 to play.

   (a) What is the probability that a contestant will win the prize?

   (b) If 100 people play your game, then approximately how many people would you expect to win the prize? Therefore about how much money would you expect your game to make for the fundraiser?

2. (a) A waitress is serving five people at a table. She has the five dishes they ordered (all five are different), but she can't remember who gets what. How many different possible ways are there for her to give the five dishes to the five people? Explain!

   (b) Based on part (a), if the waitress just hands the dishes out randomly, what is the probability that she will hand the dishes out correctly?

3. There are 4 black marbles and 5 red marbles in a bag. If you reach in and randomly pick out two marbles, what is the probability that both are red? Explain your reasoning clearly.

4. Social security numbers have 9 digits and are presented in the form

$$xxx - xx - xxxx,$$

where the "x"s can be any digit from 0 to 9.

(a) How many different social security numbers can be made this way? Explain briefly.

(b) If the population of the U.S. is 260 million people, and if everybody living in the U.S. has a social security number, then what fraction of the possible social security numbers are in current use? (This fraction is the probability that a random number in the format of a social security number is the social security number of a living person.)

12.2 Using the Meaning of Fraction Multiplication to Calculate Probabilities

In this section we will use basic principles of probability and the meaning of fraction multiplication to calculate probabilities. This relies on the following way of thinking about the outcome of an experiment. Consider what would happen "ideally" if the experiment were repeated a very large number of times. Calculate directly the "ideal" fraction of times the outcome you are interested in occurs. This fraction is the probability of that outcome.

For example, if we roll a die two times, what is the probability that we will get two 1s in a row? Think about doing the experiment of rolling a die two times in a row many times over. Because the numbers 1, 2, 3, 4, 5, and 6 are all equally likely to come up on one roll of a die, therefore, in the ideal, the first roll will be a 1 in $\frac{1}{6}$ of the rolls. Now consider those $\frac{1}{6}$ of the experiments when the first roll is a 1. Of those double rolls in which the first roll is a 1, in the ideal the second roll will be a one $\frac{1}{6}$ of the time too, because the outcome of the first roll has no bearing on the outcome of the second roll. Therefore, in the ideal, in $\frac{1}{6}$ of $\frac{1}{6}$ of all the experiments, both rolls will be a 1. According to the meaning of multiplication,

$\frac{1}{6}$ of $\frac{1}{6}$ of all the experiments

is

$$\frac{1}{6} \cdot \frac{1}{6} = \frac{1}{36}$$

of all the experiments. Therefore, in the ideal, in $\frac{1}{36}$ of all the experiments, you will roll two ones, and so the probability of rolling two 1s in a row on two rolls of a die is $\frac{1}{36}$. (In this case, this probability can also be calculated using an array or a tree diagram.)

The Case of Spot, the Drug Sniffing Dog

Here is another example. In this example, we will consider the outcome of a large number of experiments in the ideal. We then divide the number of times the outcome we are interested in occurs by the total number of experiments in order to calculate the probability that the outcome we are interested in occurs. In this way, the fraction multiplication used to calculate a probability is more indirect than in the previous example.

Spot is a dog who has been trained to sniff luggage to detect drugs. But Spot isn't perfect. His trainers conducted careful experiments and found that in luggage that *doesn't* contain any drugs, Spot will nevertheless bark to indicate the presence of drugs for 1% of this luggage. In luggage that *does* contain drugs, Spot will bark to indicate he thinks drugs are present for about 97% of this luggage (so he fails to recognize the presence of drugs in about 3% of luggage containing drugs).

Consider the consequences of putting Spot in a busy airport where police estimate that 1 in $10,000$ pieces of luggage (or .01%) passing through that airport contain drugs. Here's a question we should answer before we put Spot in a busy airport:

> Questions: If Spot barks to indicate that a piece of luggage contains drugs, what is the probability that the luggage actually does contain drugs? (Remember, Spot is not perfect, sometimes he barks when there are no drugs.)

Before you read on, make a guess: what do you think this probability is? You might be surprised at the actual answer, which we will now calculate. It is not 97%.

Imagine that we made Spot sniff a very large number of pieces of luggage, let's say $1,000,000$ pieces of luggage. For these $1,000,000$ pieces of luggage, we will first determine in the ideal how many of these Spot will bark for, indicating that he thinks they contain drugs. Then we will determine how many of these pieces of luggage actually do contain drugs.

According to the police estimates, out of 1,000,000 pieces of luggage at this airport, about 100 will contain drugs. The remaining 999,900 will not contain drugs. Based on the information we have about Spot, of the 100 pieces of luggage that contain drugs, Spot will bark, recognizing the presence of drugs in 97 pieces of luggage, in the ideal. In the ideal, for the 999,900 pieces of luggage that don't contain drugs, Spot will still bark for 1% of these pieces of luggage, which is 9999 pieces of luggage. So, all together, Spot will bark indicating that he thinks drugs are present for

$$9999 + 97 = 10,096$$

pieces of luggage, in the ideal. But of those 10,096 pieces of luggage, only 97 actually contain drugs (because the other 9999 of the 10,096 were ones where Spot barked, but they didn't actually contain any drugs). So of the 10,096 times that Spot barks, indicating he thinks there are drugs, only 97 of those times are drugs *actually present.* Now

$$\frac{97}{10,096} = .0096\ldots = .96\%$$

so that of the times when Spot barks, indicating he thinks drugs are present, only .96% of the time are drugs actually present. So if Spot barks to indicate he *thinks* drugs are present, the probability that drugs actually *are* present is .96%, which is *less than one percent*! Even though Spot's statistics sounded pretty good at the start, we might think twice about putting him on the job, since many innocent people could be delayed at the airport.

Class Activity 12H: Probabilities With a Spinner

Problems for Section 12.2 on Using the Meaning of Fraction Multiplication to Calculate Probabilities

1. Due to its high population, China has a stringent policy on having children. In rural China, couples are allowed to have either one or two children according to the following rule: if their first child is a boy, they are not allowed to have any more children. If their first child is a girl, then they are allowed to have a second child. Let's assume that

all couples follow this policy and have as many children as the policy allows.

(a) Make a guess: do you think this policy will result in more boys, more girls, or about the same number of boys as girls being born?

(b) Now consider a random group of 100,000 rural Chinese couples who will have children. Assume that any time a couple has a child the probability of having a boy is $\frac{1}{2}$.

i. In the ideal, how many couples will have a boy as their first child, and how many will have a girl as their first child?

ii. In the ideal, of those couples who have a girl first, how many will have a boy as their second child and how many will have a girl as their second child?

iii. Therefore, overall, what fraction of the children born in rural China under the given policy should be boys and what fraction should be girls?

2. The Pretty Flower Company starts plants from seed and sells the seedlings to nurseries. They know from experience that about 60% of the calla lily seeds they plant will sprout and become a seedling. Each calla lily seed costs twenty cents, and a pot containing at least one sprouted calla lily seedling can be sold for $2.00. Pots that don't contain a sprouted seedling must be thrown out. The company figures that costs for a pot, potting soil, water, fertilizer, fungicide and labor are $0.30 per pot (whether or not a seed in the pot sprouts). The Pretty Flower Company is debating between planting one or two seeds per pot. Help them figure out which choice will be more profitable by working through the following problems.

(a) Suppose the Pretty Flower Company plants one calla lily seed in each of 100 pots. Using the information above, approximately how much profit should the Pretty Flower Company expect to make on these 100 pots? Profit is income minus expenses.

(b) Now suppose that the Pretty Flower Company plants two calla lily seeds (one on the left, one on the right) in each of 100 pots. Assume that whether or not the left seed sprouts has no influence on whether or not the right seed sprouts. So the right seed will still sprout in about 60% of the pots in which the left seed does not sprout. Explain why the Pretty Flower Company should expect about 84 of the 100 pots to sprout at least one seed.

(c) Using part (b), determine how much profit the Pretty Flower Company should expect to make on 100 pots if two calla lily seeds are planted per pot. Compare your answer to part (a). Which is expected to be more profitable: one seed or two seeds per pot?

(d) What if calla lily seeds cost fifty cents each instead of twenty cents each (but everything else stays the same). Now which is expected to be more profitable: one seed or two seeds per pot?

3. Suppose you have 100 light bulbs and one of them is defective. If you pick two light bulbs out at random (either both at the same time, or first one, then another, without replacing the first light bulb), what is the probability that neither one of your chosen light bulbs is defective?

4. Suppose that in a survey of a large, random group of people, 17% were found to be smokers and 83% were not smokers. Suppose further that .08% of the smokers and .01% of the non-smokers from the group died of lung cancer (be careful: notice the decimal points in these percentages—they are not 8% and 1%).

   (a) What percent of the group died of lung cancer?

   (b) Of the people in the group who died of lung cancer, what percent were smokers? (It may help you to make up a number of people for the large group and to calculate the number of people who died of lung cancer and the number of people who were smokers and also died of lung cancer.)

5. Suppose that 1% of the population has a certain disease. Let's also suppose that there is a test for the disease, but it is not completely accurate: it has a 2% rate of false positives and a 1% rate of false negatives. This means that the test *reports* that 2% of the people who don't have the disease do have it, and the test *reports* that 1% of the people who do have the disease don't have it. This problem is about the following question:

Question: If a person tests positive for the disease, what is the probability that he or she actually has it?

Before you start answering the questions below, go back and read the beginning of the problem again. Notice that there is a difference between *actually having* the disease and *testing positive* for the disease, and there is a difference between *not having* the disease and *testing negative* for it.

(a) Make a guess: what do you think the answer to the question above is?

(b) For parts (b) – (f), suppose there is a random group of 10,000 people and all of them are tested for the disease.

Of the 10,000 people, in the ideal, how many would you expect to actually have the disease and how many would you expect not to have the disease?

(c) Continuing part (b), of the people from the group of 10,000 who do not have the disease, how many would you expect to test positive, in the ideal? (Give a number, not a percentage.)

(d) Continuing part (b), of the people from the group of 10,000 who do have the disease, how many would you expect to test positive? (Give a number, not a percentage.)

(e) Using your work in parts (c) and (d), of the people from the group of 10,000, how many in total would you expect to test positive, in the ideal?

(f) Using your work above, what percent of the people who test positive for the disease should actually have the disease, in the ideal? How does this compare to your guess in part (a)? Are you surprised at the actual answer?

6. Suppose you have two boxes, 50 black pearls, and 50 white pearls. You can mix the pearls up any way you like and put all them back into the two boxes. You don't have to put the same number of pearls in each box, but each box must have at least one pearl in it. You will then be blindfolded and you will get to open a random box and pick a random pearl out of it.

(a) Suppose you put 25 black pearls in box one and 25 black pearls and 50 white pearls in box two. Calculate the probability of picking a black pearl by imagining that you were going to pick a random box and a random pearl in the box a large number of times.

In the ideal, in what fraction of the time would you pick box 1? In what fraction of those times that you picked box 1, would you pick a black pearl from box 1, in the ideal?

On the other hand, in the ideal, in what fraction of the time would you pick box 2? In what fraction of those times that you picked box 2, would you pick a black pearl from box 2, in the ideal?

Therefore, overall, in what fraction of the time would you pick a black pearl, in the ideal?

(b) Describe at least two other ways to distribute the pearls than what is described in part (a). Find the probability of picking a black pearl in each case.

(c) Try to find a way to arrange the pearls so that the probability of picking a black pearl is as large as possible. In particular, can you find a way to make the probability of picking a black pearl greater than $\frac{1}{2}$?

Appendix A

Cutouts for Exercises and Problems

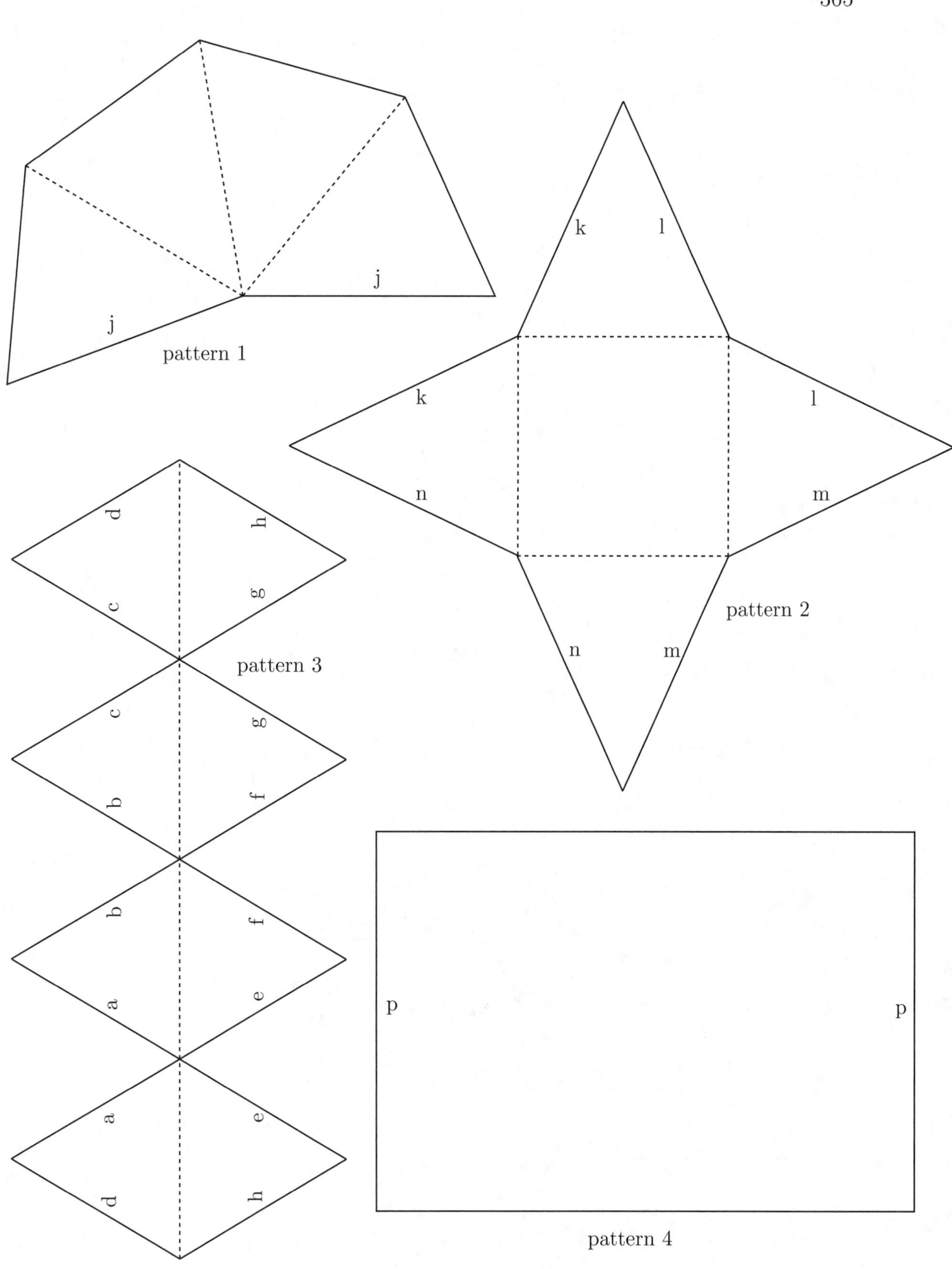

Figure A.1: Patterns for Shapes for Exercise 1 on Page 7

Figure A.2: Patterns for Shapes for Exercise 2 on Page 7

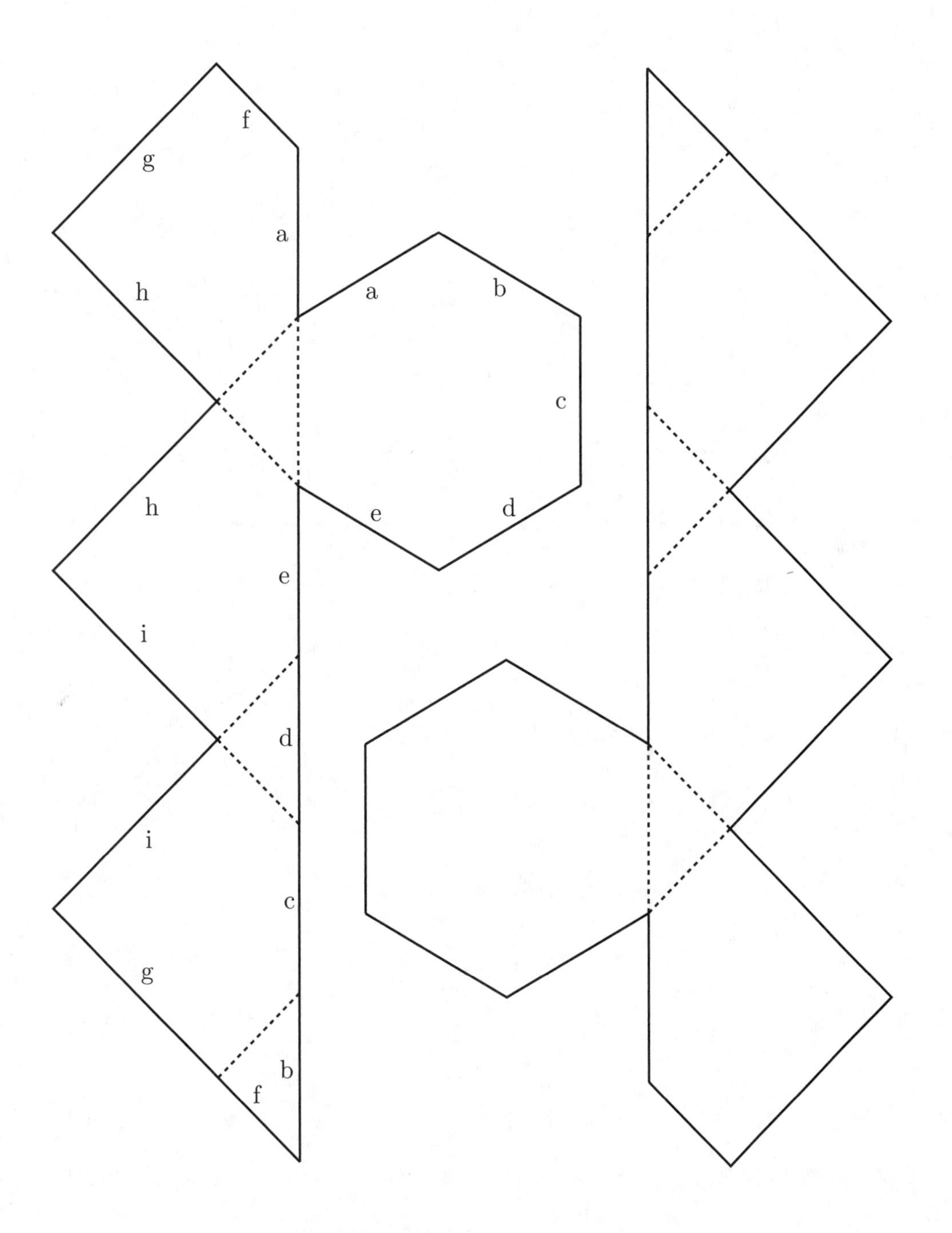

Figure A.3: Patterns for a Sliced Cube for Exercise 5 on Page 11

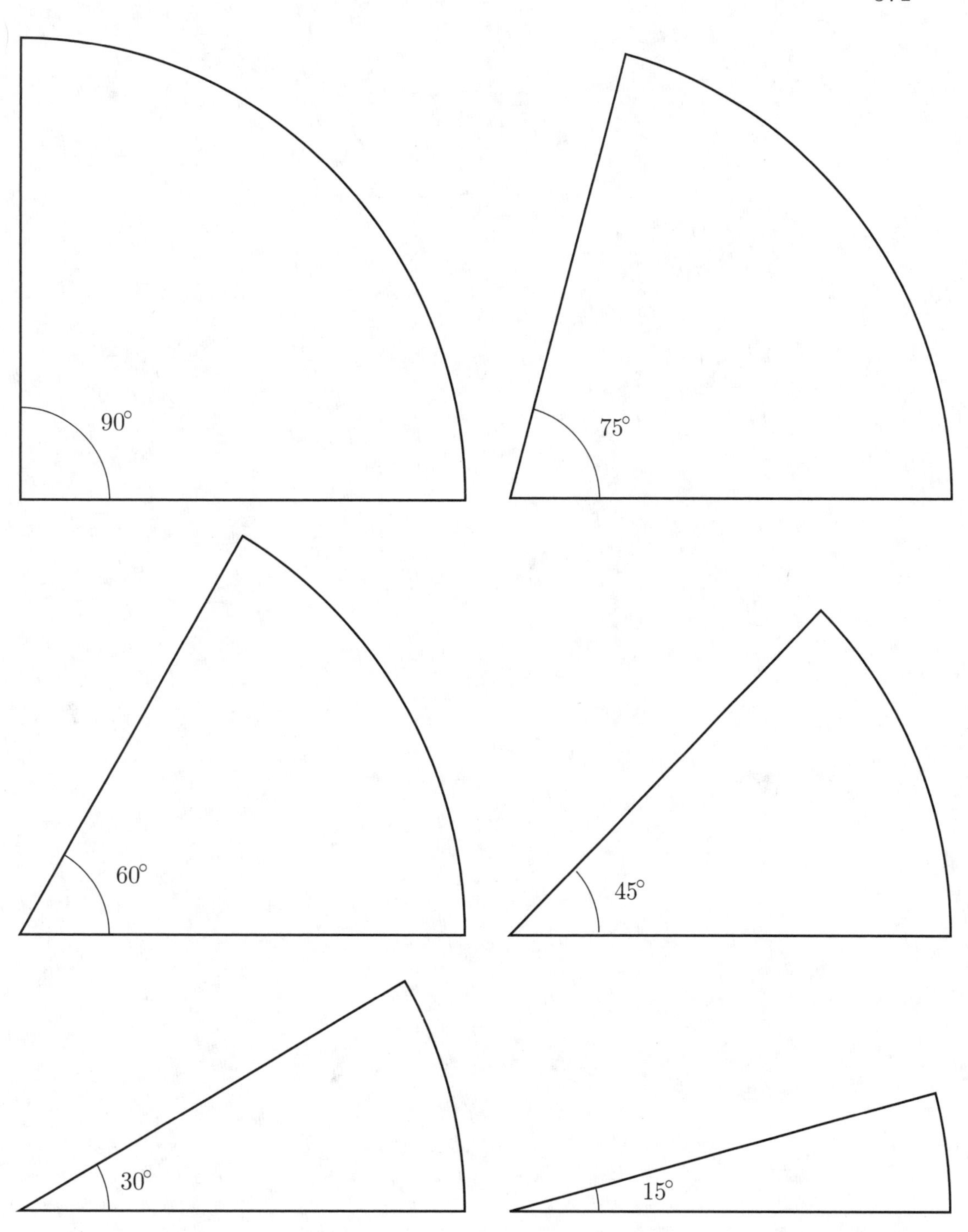

Figure A.4: Pie Wedges for Measuring Angles for Problem 1 on Page 34

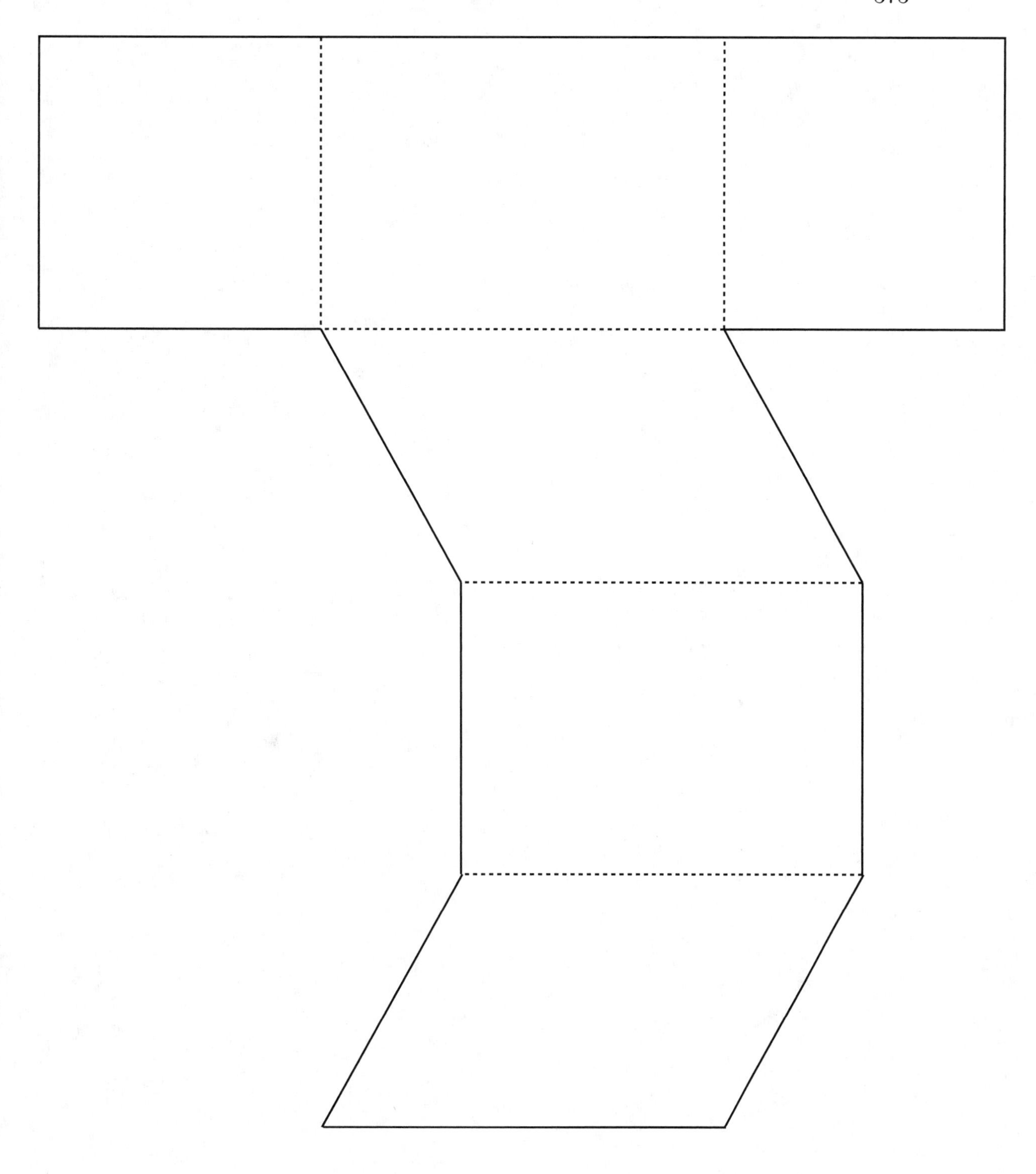

Figure A.5: For Exercise 2 on Page 89

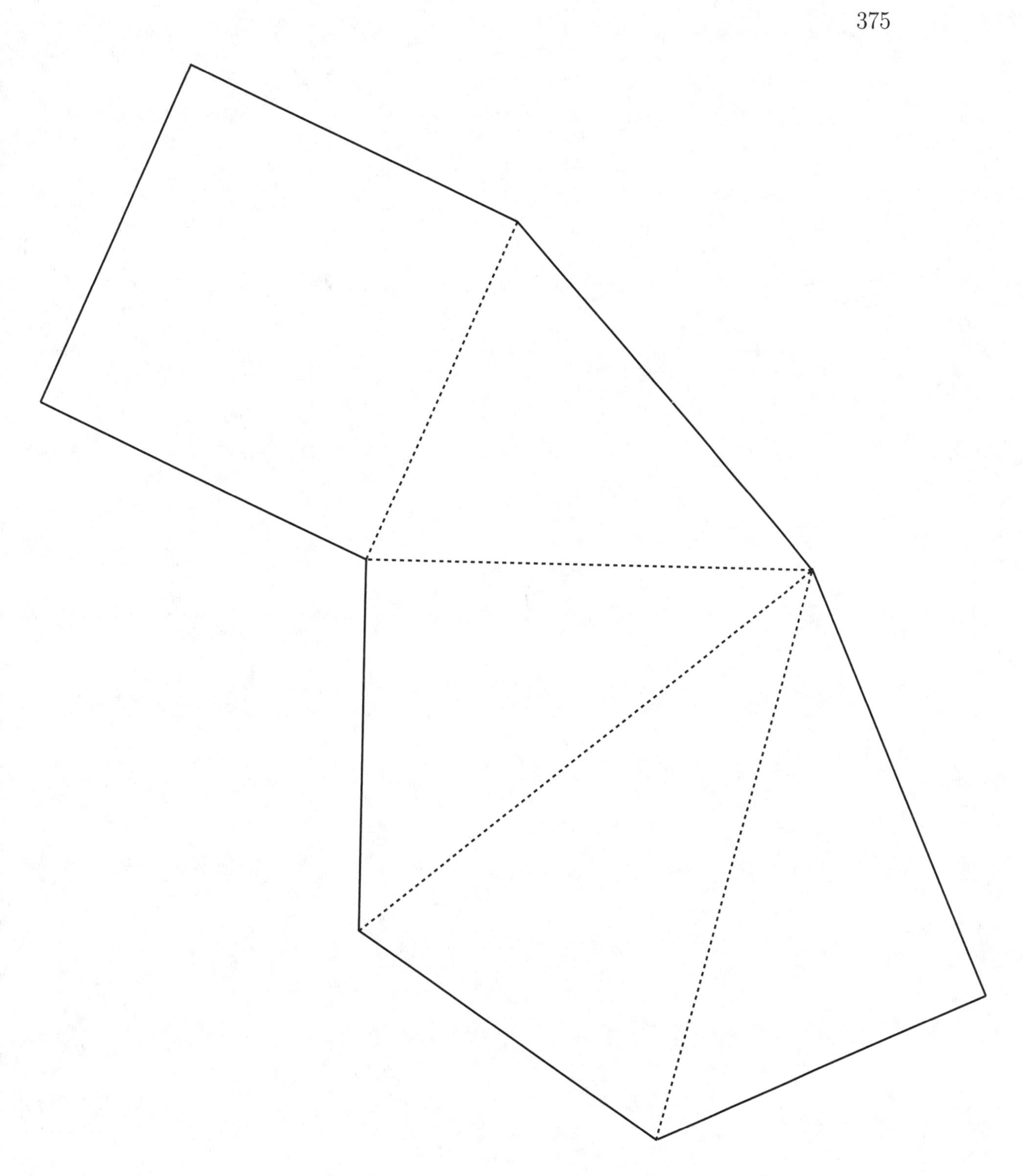

Figure A.6: For Exercise 2 on Page 89

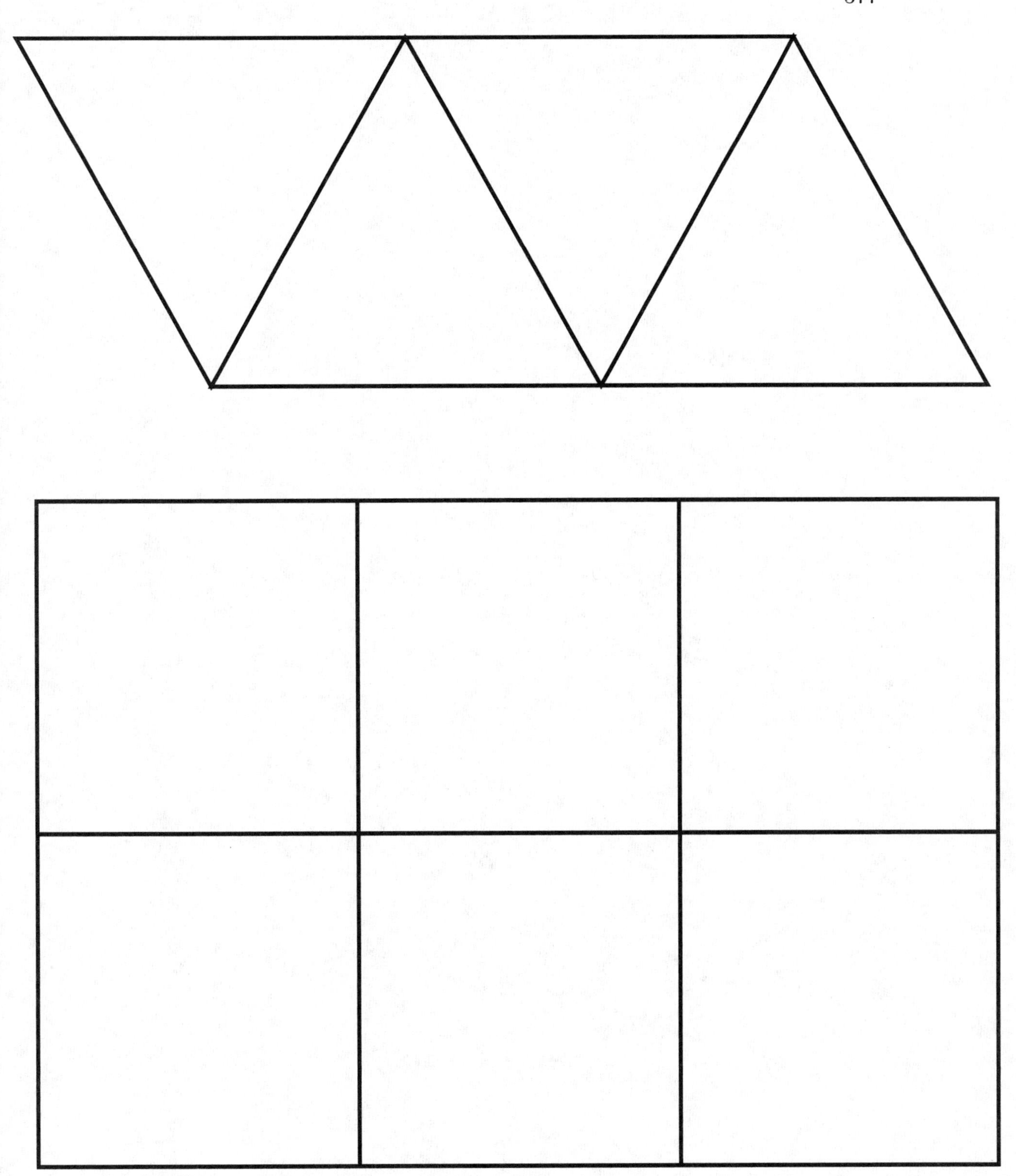

Figure A.7: For Exercise 5 on Page 90

Figure A.8: For Exercise 5 on page 90

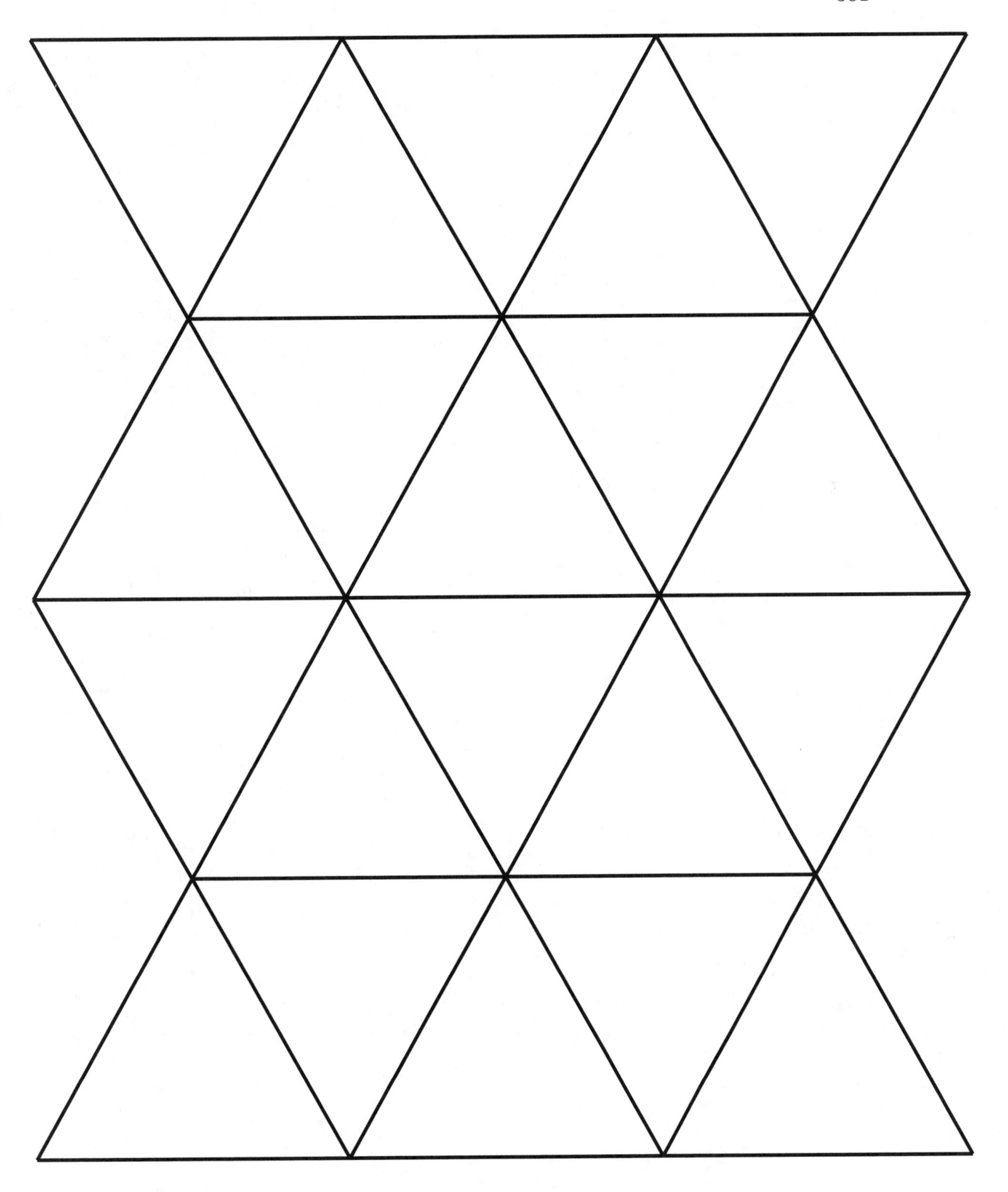

Figure A.9: For Exercise 5 on page 90

Figure A.10: Triangles for Problem 7 on Page 95

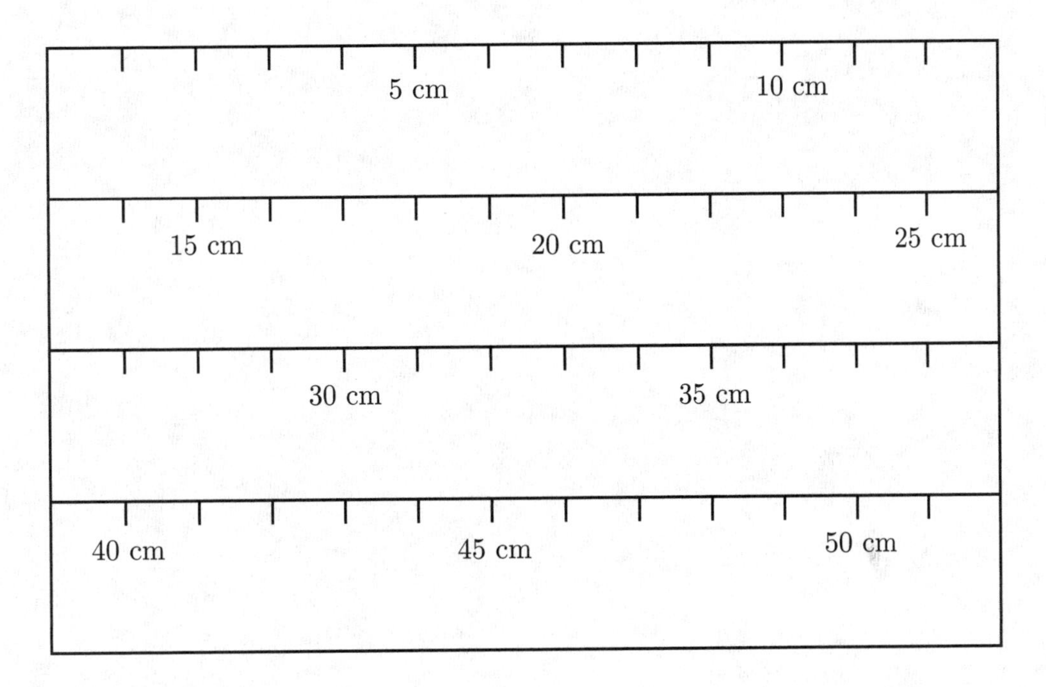

Figure A.11: A Centimeter Tape Measure for Problem 11 on Page 275

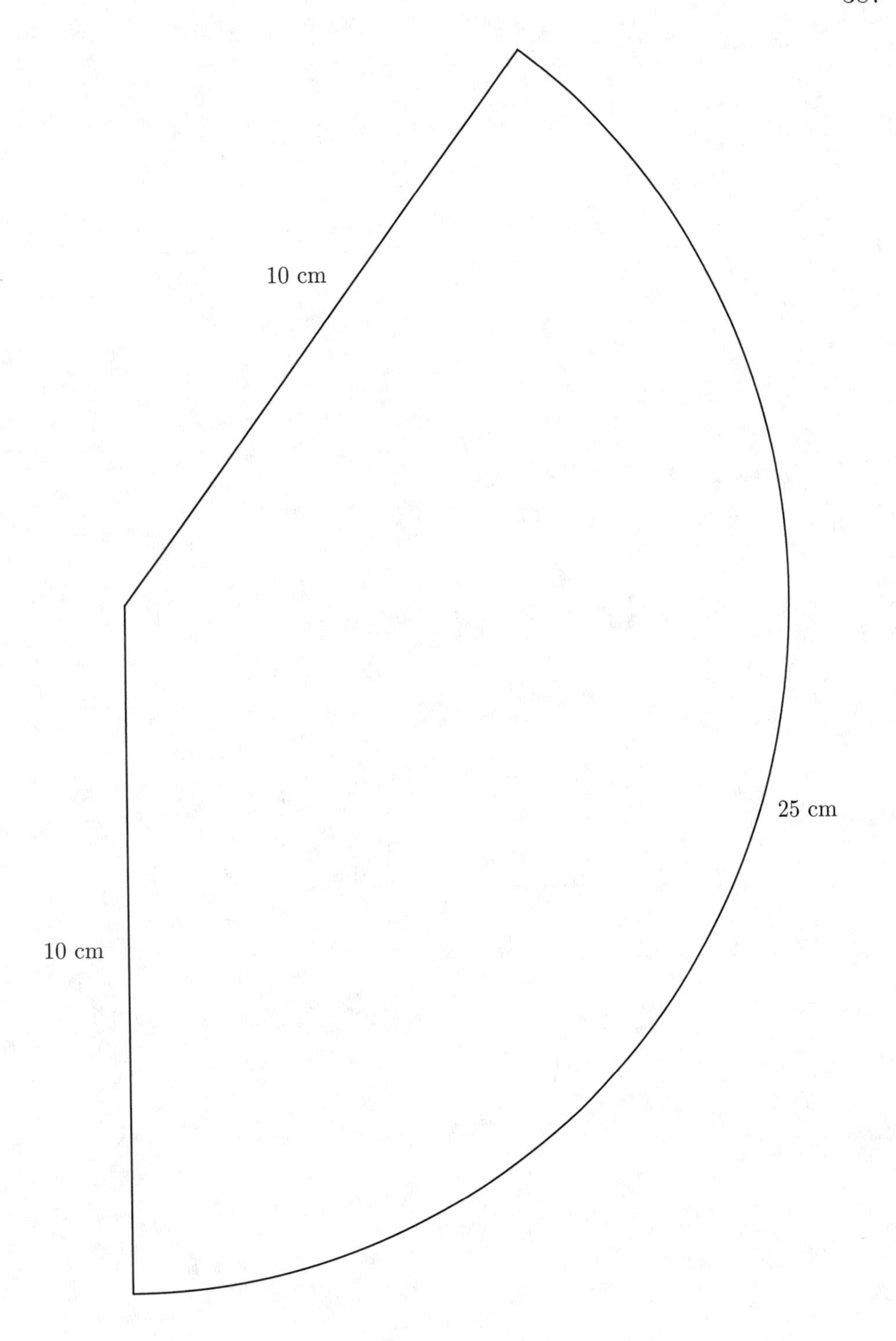

Figure A.12: A Pattern for a Cup for Problem 12 on Page 275

Bibliography

[1] David Freedman, Robert Pisani, and Roger Purves. *Statistics.* W. W. Norton and Company, 1978.

[2] H. G. Jerrard and D. B. McNeill. *Dictionary of Scientific Units.* Chapman and Hall, 1963.

[3] National Council of Teachers of Mathematics. *Principles and Standards for School Mathematics.* Author, 2000. See the website `www.nctm.org`.

[4] The Secretary's Commission on Achieving Necessary Skills. *What Work Requires of Schools, A SCANS Report for America 2000.* U. S. Department of Labor, 1991.

Index